Drinking Water Treatment 5

# Drinking Water Treatment 5

*Calco-carbonic Equilibrium and Disinfection*

Kader Gaid

iSTE

WILEY

First published 2023 in Great Britain and the United States by ISTE Ltd and John Wiley & Sons, Inc.



ISTE Ltd
27-37 St George's Road
London SW19 4EU
UK

www.iste.co.uk

John Wiley & Sons, Inc.
111 River Street
Hoboken, NJ 07030
USA

www.wiley.com



Cover illustration:
© imageBROKER.com/Matton Images

Library of Congress Control Number: 2022947894

British Library Cataloguing-in-Publication Data
A CIP record for this book is available from the British Library
ISBN 978-1-78630-787-3

# Contents

**Chapter 20. Calco-carbonic Equilibrium, Correction of Aggressivity and Remineralization** . . . . . 1

20.1. Characteristics of water leading to calco-carbonic equilibrium . . . . . 2
20.1.1. Chemical equilibria . . . . . 2
20.1.2. Aggressive water . . . . . 12
20.1.3. Scaling water . . . . . 12
20.1.4. Corrosive water . . . . . 12
20.2. The equilibrium reactions of water's constituents . . . . . 15
20.2.1. Equilibrium pH . . . . . 16
20.2.2. Langelier equation . . . . . 20
20.3. Hallopeau–Dubin diagram . . . . . 25
20.4. Indicative criteria to determine the aggressivity or corrosivity of water . . . . . 29
20.4.1. Indicators of aggressivity: concrete pipelines . . . . . 29
20.4.2. Corrosivity indicators . . . . . 32
20.5. The calco-carbonic equilibrium of water . . . . . 36
20.5.1. Water quality and regulations . . . . . 38
20.5.2. The correction of aggressivity . . . . . 38
20.5.3. Aggressivity correction treatments . . . . . 39
20.6. Remineralization treatments . . . . . 60
20.6.1. Graphic method . . . . . 61
20.6.2. Processes for implementing remineralization: chemical reactions in tanks . . . . . 62
20.7. Characteristics of the various reagents used . . . . . 85
20.7.1. Lime . . . . . 85
20.7.2. "Micronized" lime . . . . . 86
20.7.3. Caustic soda . . . . . 86

20.7.4. Caustic soda at 50% . . . . . 87
20.7.5. Caustic soda flakes or grains . . . . . 87
20.7.6. Sodium carbonate . . . . . 88
20.7.7. Sodium bicarbonate . . . . . 89
20.7.8. Calcium carbonate . . . . . 89
20.7.9. Acticalmag limestone . . . . . 90
20.7.10. Magno . . . . . 90
20.7.11. Calcium sulfate . . . . . 91
20.7.12. Calcium chloride . . . . . 92
20.7.13. Carbon dioxide . . . . . 92
20.7.14. Sulfuric acid (90–98%) . . . . . 93
20.7.15. Hydrochloric acid . . . . . 94
20.8. References . . . . . 96

**Chapter 21. Disinfection** . . . . . 99

21.1. Microorganisms present in the water . . . . . 99
21.1.1. Bacteria . . . . . 99
21.1.2. Indicator microorganisms or test germs . . . . . 102
21.1.3. Viruses . . . . . 104
21.1.4. Parasites . . . . . 105
21.1.5. Micro-algae . . . . . 107
21.2. Quality of potable water . . . . . 108
21.2.1. French regulations . . . . . 108
21.3. General rules of chemical disinfection . . . . . 110
21.3.1. Disinfection mechanisms . . . . . 110
21.3.2. The mode of action of chemical disinfectants . . . . . 111
21.3.3. Inactivation kinetics . . . . . 112
21.3.4. The notion of Ct . . . . . 113
21.4. Factors affecting the efficiency of chemical disinfection . . . . . 117
21.4.1. Contact time . . . . . 117
21.4.2. Turbidity . . . . . 118
21.4.3. Presence of oxidizable matter . . . . . 118
21.4.4. pH . . . . . 119
21.4.5. Injection mode and injection point . . . . . 119
21.4.6. Design of the contact tank . . . . . 119
21.5. Qualities of a good disinfectant . . . . . 120
21.6. Chlorine disinfection . . . . . 121
21.6.1. Gaseous chlorine . . . . . 122
21.6.2. Hypochlorite . . . . . 124
21.6.3. Chlorine application points . . . . . 128
21.6.4. Oxidant demand . . . . . 132

21.6.5. The implementation of chlorination . . . . . . . . . . . . . . . . . . 137
21.6.6. Disinfection performances with chlorine . . . . . . . . . . . . . . 146
21.6.7. Synthesis of chlorine reactions . . . . . . . . . . . . . . . . . . . . . 148
21.7. Calcium hypochlorite . . . . . . . . . . . . . . . . . . . . . . . . . . . . 149
21.8. Chlorine dioxide disinfection . . . . . . . . . . . . . . . . . . . . . . . 150
21.8.1. Chlorine dioxide preparation . . . . . . . . . . . . . . . . . . . . . 151
21.8.2. Chlorine dioxide performances . . . . . . . . . . . . . . . . . . . . 155
21.8.3. Dechlorination . . . . . . . . . . . . . . . . . . . . . . . . . . . . . . 157
21.8.4. The advantages of dioxide over chlorine . . . . . . . . . . . . . . 159
21.8.5. The special case of the use of chlorine dioxide at a station equipped with ozone . . . . . . . . . . . . . . . . . . . . . . . 160
21.8.6. Advantages and drawbacks of using chlorine dioxide . . . . . . . 160
21.9. Chloramination . . . . . . . . . . . . . . . . . . . . . . . . . . . . . . . 161
21.9.1. Principle . . . . . . . . . . . . . . . . . . . . . . . . . . . . . . . . . 161
21.9.2. Implementation . . . . . . . . . . . . . . . . . . . . . . . . . . . . . 163
21.9.3. Performances of monochloramine . . . . . . . . . . . . . . . . . . 164
21.9.4. Dechloramination . . . . . . . . . . . . . . . . . . . . . . . . . . . . 165
21.9.5. Advantages and drawbacks of chloramination . . . . . . . . . . . 166
21.10. Proportion of chlorine in chlorine disinfectants . . . . . . . . . . . . 167
21.11. Disinfection with ozone . . . . . . . . . . . . . . . . . . . . . . . . . . 168
21.11.1. General remarks on ozone . . . . . . . . . . . . . . . . . . . . . . 168
21.11.2. Ozone production . . . . . . . . . . . . . . . . . . . . . . . . . . . 170
21.11.3. Ozone demand . . . . . . . . . . . . . . . . . . . . . . . . . . . . . 172
21.11.4. The implementation of ozonation . . . . . . . . . . . . . . . . . . 174
21.11.5. Performances of ozone disinfection . . . . . . . . . . . . . . . . . 181
21.11.6. De-ozonation . . . . . . . . . . . . . . . . . . . . . . . . . . . . . . 183
21.12. Criteria for choosing a chemical disinfection technique . . . . . . . . 184
21.12.1. Practical implementation of chemical disinfection . . . . . . . . . 184
21.12.2. Comparative efficiency of the main techniques . . . . . . . . . . . 185
21.13. Another chemical disinfectant used: bromine ($Br_2$) . . . . . . . . . . 187
21.14. Disinfection by ultraviolet radiation . . . . . . . . . . . . . . . . . . . 187
21.14.1. General remarks on UV radiation . . . . . . . . . . . . . . . . . . 187
21.14.2. Inactivation mechanisms . . . . . . . . . . . . . . . . . . . . . . . 189
21.14.3. Lethal dose and inactivation kinetics . . . . . . . . . . . . . . . . 190
21.14.4. Implementation . . . . . . . . . . . . . . . . . . . . . . . . . . . . . 194
21.14.5. UV disinfection design parameters . . . . . . . . . . . . . . . . . 199
21.14.6. Factors affecting the efficiency of a UV treatment . . . . . . . . . 200
21.14.7. UV radiation performances . . . . . . . . . . . . . . . . . . . . . . 204
21.14.8. Photoreactivation . . . . . . . . . . . . . . . . . . . . . . . . . . . . 208
21.14.9. Advantages and drawbacks of UV disinfection . . . . . . . . . . . 209
21.14.10. Conclusions on UV disinfection . . . . . . . . . . . . . . . . . . . 209
21.15. Comparative criteria between the various chemical disinfectants . . 210
21.16. References . . . . . . . . . . . . . . . . . . . . . . . . . . . . . . . . . 212

**Chapter 22. Disinfection By-products** . . . . . . . . . . . . . . . . . . . . 217

22.1. General aspects . . . . . . . . . . . . . . . . . . . . . . . . . . 217
22.2. Reaction by-products . . . . . . . . . . . . . . . . . . . . . . . 218
22.3. Formation and evolution of chlorination by-products . . . . . . . . . . 222
22.4. Kinetics and formation mechanisms . . . . . . . . . . . . . . . . . 224
22.4.1. Formation kinetics . . . . . . . . . . . . . . . . . . . . . . 224
22.4.2. Mechanisms . . . . . . . . . . . . . . . . . . . . . . . . . 226
22.4.3. Chlorination of HS . . . . . . . . . . . . . . . . . . . . . . 228
22.4.4. Chlorination of carboxylic acids . . . . . . . . . . . . . . . . 230
22.4.5. Factors influencing the formation of DBPs . . . . . . . . . . . . 230
22.5. Regulations . . . . . . . . . . . . . . . . . . . . . . . . . . . 238
22.6. Predictive models of CBPs . . . . . . . . . . . . . . . . . . . . 239
22.7. Removal of THMs and HAAs . . . . . . . . . . . . . . . . . . . . 240
22.7.1. Aeration. . . . . . . . . . . . . . . . . . . . . . . . . . . 240
22.7.2. Activated carbon . . . . . . . . . . . . . . . . . . . . . . . 242
22.7.3. Biofiltration . . . . . . . . . . . . . . . . . . . . . . . . . 246
22.7.4. High-pressure membranes . . . . . . . . . . . . . . . . . . . 246
22.8. The case of nitrosamines and NDMA . . . . . . . . . . . . . . . . 247
22.8.1. Nitrosation mechanism with HOCl. . . . . . . . . . . . . . . . 247
22.9. Oxidation by-products related to chlorine dioxide . . . . . . . . . . 248
22.10. Ozonation by-products. . . . . . . . . . . . . . . . . . . . . . 251
22.11. Recommendations . . . . . . . . . . . . . . . . . . . . . . . . 254
22.12. References . . . . . . . . . . . . . . . . . . . . . . . . . . . 255

**Chapter 23. Sludge Treatment** . . . . . . . . . . . . . . . . . . . . . . 261

23.1. Choosing a treatment chain. . . . . . . . . . . . . . . . . . . . . 262
23.2. Characteristics of drinking water sludge . . . . . . . . . . . . . . . 263
23.2.1. The quantity of sludge produced . . . . . . . . . . . . . . . . 263
23.2.2. Sludge concentration estimate at different stages of the chain . . . . . . . . . . . . . . . . . . . . . . . . . 265
23.2.3. Sludge quality: physical and chemical properties . . . . . . . . . 266
23.3. Handling and storage: shovelable and stackable nature . . . . . . . . 268
23.4. Different classes of sludge . . . . . . . . . . . . . . . . . . . . 269
23.4.1. Hydroxide sludge . . . . . . . . . . . . . . . . . . . . . . . 269
23.4.2. Softening sludge. . . . . . . . . . . . . . . . . . . . . . . . 269
23.4.3. Metal species sludge treatment . . . . . . . . . . . . . . . . . 270
23.4.4. Biological sludge . . . . . . . . . . . . . . . . . . . . . . . 270
23.4.5. The case of mixed sludge . . . . . . . . . . . . . . . . . . . 270

23.5. Sludge composition depending on the characteristics of raw water . . . . . 271
23.5.1. Surface water sludge . . . . . 271
23.5.2. Treatment sludge with coagulants (Fe or Al) . . . . . 272
23.5.3. Borehole sludge . . . . . 272
23.6. Thickening of drinking water sludge . . . . . 273
23.6.1. Function and criteria for choosing a thickener . . . . . 273
23.6.2. Thickener design . . . . . 275
23.6.3. Implementation of thickeners . . . . . 281
23.6.4. Flotation . . . . . 286
23.7. Drinking water sludge dewatering . . . . . 288
23.7.1. Plate filter . . . . . 288
23.7.2. Centrifugation . . . . . 291
23.7.3. Belt filters . . . . . 294
23.7.4. Filter bags . . . . . 295
23.7.5. Drying beds . . . . . 296
23.7.6. Sludge lagoon treatment . . . . . 302
23.8. Advantages and drawbacks of the different sludge dewatering treatments . . . . . 304
23.9. References . . . . . 305

**Chapter 24. The Treatment Chain: Conception and Design** . . . . . 307

24.1. The treatment chain . . . . . 309
24.2. The definition of a treatment chain . . . . . 310
24.3. The stages of a treatment chain . . . . . 313
24.4. The renovation of water treatment plants . . . . . 315
24.4.1. Adaptation of new goals . . . . . 316
24.4.2. The choice of treatment technologies . . . . . 317
24.5. References . . . . . 324

**Chapter 25. The Future of Water** . . . . . 327

25.1. The major elements of the future of water . . . . . 327
25.2. Will there be enough water? . . . . . 330

**Index** . . . . . 333

**Summaries of other volumes** . . . . . 335

# 20

# Calco-carbonic Equilibrium, Correction of Aggressivity and Remineralization

Water follows a natural cycle in which the chemical elements it contains evolve. Rainwater naturally contains carbon dioxide ($CO_2$). When it crosses humus layers, rich in acids, it can become very rich in $CO_2$. When penetrating into calcareous soil, that is to say, soil rich in calcium carbonate ($CaCO_3$), it becomes loaded with calcium ($Ca^{2+}$) and bicarbonate ions ($HCO_3^-$). Calcium carbonate can in fact be dissolved by water loaded with $CO_2$. Such water is said to be scaling or encrusting. On the other hand, when rainwater crosses a rock poor in calcium (e.g. a region of granite), it remains highly loaded with dissolved $CO_2$ and is generally acidic. Such water is said to be aggressive.

The aggressivity of water can be defined as its propensity to dissolve the metals with which it is in contact.

In contrast, scaling water is water that tends to precipitate calcium carbonates that can leave a deposit on the metals with which it is in contact.

The species present in the water react with a tendency toward a chemical equilibrium. This is called "calco-carbonic equilibrium". It corresponds to a certain pH value known as the saturation pH (pHs) or equilibrium pH. This pH depends on the concentrations of the different chemical species at play. These waters also contain more or less dissociated dissolved salts, such as:

– salts from strong acids and bases;

– salts from weak acids and bases.

## 20.1. Characteristics of water leading to calco-carbonic equilibrium

### 20.1.1. *Chemical equilibria*

A chemical reaction whereby compounds A and B react and lead to compounds C and D is written out as:

$$Aa + bB \rightarrow cC + dD$$

Coefficients a, b, c and d correspond to the numbers of moles of compounds A, B, C and D. At equilibrium and according to the law of mass action, the equilibrium constant, related to the concentrations, is:

$$K = \frac{[C]^c [D]^d}{[A]^a \, [B]^b}$$

For a given reaction, the value of K varies with the solution's temperature and ionic strength. Concentrations are normally expressed in $mol \cdot L^{-1}$.

Due to ion–ion interactions, the activity of an ion is defined by:

$$ai = \Upsilon i \, [Ci]$$

where:

– $ai$: activity of ion i in $mol \cdot L^{-1}$;

– $\Upsilon i$: activity coefficient of ion i;

– [Ci]: concentration of ion i in solution in $mol \cdot L^{-1}$.

For example: $[CO_3^{2-}] = (a\ CO_3^{2-})/\Upsilon_{CO3}^{2-}$

By including the activities of ions, this leads to the equilibrium constant K′ related to the activities:

$$K' = \frac{[aC]^c [aD]^d}{[aA]^a \, [aB]^b} = \frac{(\Upsilon cC)^c (\Upsilon dD)^d}{(\Upsilon aA)^a \, (\Upsilon bB)^b}$$

The activity coefficient $\Upsilon$ is given by:

$$\log \Upsilon = -\frac{0.5(Zi)^2\sqrt{I}}{1+\sqrt{I}}$$

where:

– Zi: charge of ion i;

– I: ionic strength.

Ionic strength I is obtained by the equation:

$$I = \frac{1}{2} \Sigma Ci\, Zi^2 \text{ mol·L}^{-1}$$

The ionic strength can also be directly calculated from the total concentration of dissolved salts ($mg·L^{-1}$):

$$I = 2.5 \times 10^{-5} \text{ TDS}$$

This equation can give a quick estimation of I.

The equilibrium constant of a reaction including precipitation with its ions is called the solubility product (Ksp). For example, the reaction of calcium carbonate is given as:

$$CaCO_3 \rightarrow Ca^{2+} + CO_3^{2-}$$

where the concentrations are limited by the solubility product.

The solubility product is $Ksp = [Ca^{2+}][CO_3^{2-}]$, which can also be expressed as:

$$Ksp = \Upsilon Ca^{2+} [Ca^{2+}] . \Upsilon CO_3^{2-} [CO_3^{2-}]$$

For example, let us consider the ionic strength of water (25°C) whose chemical composition is shown in Table 20.1.

| **Ions** | **Concentrations Ci mg·L$^{-1}$** | **Concentrations mol·L$^{-1}$** | **Z$^2$** | **CZ$^2$** |
|---|---|---|---|---|
| **$Ca^{2+}$** | 80 | 80/40,000 = 2 × $10^{-3}$ | 4 | 8 × $10^{-3}$ |
| **$Mg^{2+}$** | 21 | 0.864 × $10^{-3}$ | 4 | 3.456 × $10^{-3}$ |
| **$Na^+$** | 42 | 1.826 × $10^{-3}$ | 1 | 1.826 × $10^{-3}$ |
| **$K^+$** | 12 | 0.307 × $10^{-3}$ | 1 | 0.307 × $10^{-3}$ |
| **$HCO_3^-$** | 221 | 3.622 × $10^{-3}$ | 1 | 3.622 × $10^{-3}$ |
| **$SO_4^{2-}$** | 98 | 1.020 × $10^{-3}$ | 4 | 4.080 × $10^{-3}$ |
| **$Cl^-$** | 79 | 2.225 × $10^{-3}$ | 1 | 2.225 × $10^{-3}$ |
| **$NO_3^-$** | 24 | 0.387 × $10^{-3}$ | 1 | 0.387 × $10^{-3}$ |
| **Total** | | | | 23.903 × $10^{-3}$ |

**Table 20.1.** *Estimation of water's ionic strength*

With Ksp ($CaCO_3$): $3.32 \times 10^{-9}$ at 25°C.

The ionic strength is:

$$I = \frac{1}{2}\, 23.903 \times 10^{-3} = 11.951 \times 10^{-3} \text{ mol·L}^{-1}$$

For monovalent ions (Z = 1):

$$\log \Upsilon = -\frac{0.5(1)^2\sqrt{11.951} \times 10^{-3}}{1 + \sqrt{11.951} \times 10^{-3}} = -0.0492$$

$$\Upsilon = 10^{-0.0492} = 0.892$$

For divalent ions (Z = 2):

$$\log \Upsilon = -\frac{0.5(2)^2\sqrt{11.951} \times 10^{-3}}{1 + \sqrt{11.951} \times 10^{-3}} = -0.1968$$

$$\Upsilon = 10^{-0.1968} = 0.635$$

The minimum solubility of calcium is obtained as follows. The solubility product is given as:

$$Ksp = \Upsilon Ca^{2+}[Ca^{2+}] . \Upsilon CO_3^{2-}\left[CO_3^{2-}\right] = \Upsilon^2 [C]^2$$

The concentration C is deduced as follows:

$$C = \sqrt{\frac{Ksp}{\Upsilon^2} = \sqrt{\frac{3.32 \times 10^{-9}}{0.636^2}} = 0.904 \times 10^{-4} \text{ mol} \cdot \text{L}^{-1}}$$

The calcium concentration is expressed in mg·L$^{-1}$:

$$Ca = 0.904 \times 10^{-4} \times 40{,}000 = 4.44 \text{ mg·L}^{-1}$$

#### 20.1.1.1. *Influence of temperature on reaction rates*

Solubility products, equilibrium constants and rate constants depend on temperature. This dependence is expressed by the Van't Hoff–Arrhenius equation:

$$\frac{d(Lnk)}{dt} = \frac{E}{RT^2}$$

where:

– k: rate constant;

– T: temperature in Kelvin (273.15 + T°C);

– E: activation energy ($J \cdot mol^{-1}$);

– R: universal gas constant (8.314 $J \cdot mol^{-1} \cdot K^{-1}$).

The integration between two temperatures T1 and T2 results in:

$$Ln\frac{k1}{k2} = \frac{E}{RT1T2}(T2 - T1)$$

We can deduce:

$$\frac{k1}{k2} = \theta^{(T2-T1)}$$

where θ is the temperature coefficient.

20.1.1.2. *Alkalinity*

Alkalinity is defined as the sum of all titratable bases until reaching a pH of approximately 4.5. It is found by experimentally estimating the quantity of acid required to lower water pH to 4.5. In most waters, the only significant contributions to alkalinity are the carbonate species and the free $H^+$ or $OH^-$ species:

$$\text{alkalinity} = [HCO_3-] + 2[CO_3^{2-}] + [OH^-]\text{-}[H^+]$$

Concentrations are normally expressed in $mol \cdot L^{-1}$.

The ions $OH^-$ are present only for pH > 8.3 (in practice, pH > 9).

Under natural conditions, $[H^+]$ and $[OH^-]$ are negligibly small compared to the carbonate species concentrations, leading to:

$$\text{Alkalinity} = [HCO_3^-] + 2[CO_3^{2-}]$$

$[CO_3^{2-}]$ is multiplied by two because it can accept two protons.

Alkalinity in $meq \cdot L^{-1}$ is: alkalinity = $[HCO_3^-] + [CO_3^{2-}]$.

In practical terms, alkalinity is expressed in $mg \cdot L^{-1}$ $CaCO_3$. To convert $meq \cdot L^{-1}$ into $mg \cdot L^{-1}$ $CaCO_3$:

$$\text{meq } CaCO_3 = \frac{100 \text{ mg} \cdot \text{mmole}^{-1}}{2 \text{ meq.mole}^{-1}} = 50 \text{ mg} \cdot \text{meq}^{-1}$$

Note that 1 $meq \cdot L^{-1}$ of alkalinity is equivalent to 50 $mg \cdot L^{-1}$ $CaCO_3$. It means also that 61 $mg \cdot L^{-1}$ $HCO_3$ corresponds to 40/2 = 20 $mg \cdot L^{-1}$ Ca.

Or

$Ca^{2+}$ ($mg \cdot L^{-1}$) = $HCO_3^-/3$ in $mg \cdot L^{-1}$

– Converting alkalinity carbonate from $mg \cdot L^{-1}$ as $CaCO_3$ to $mg \cdot L^{-1}$ as $CO_3^{2-}$

$CaCO_3$ has a molecular weight of 100 $g \cdot mol^{-1}$. The $CO_3^{2-}$ anion has a molecular weight of 60 $g \cdot mol^{-1}$. Therefore, each milligram of $CaCO_3$ contains 60/100 = 0.6 mg of $CO_3^{2-}$

The conversion is as follows:

Alkalinity carbonate as $CO_3^{2-}$ ($mg \cdot L^{-1}$) = 0.6·Alkalinity carbonate as $CaCO_3$ ($mg \cdot L^{-1}$)

– Converting bicarbonate alkalinity from $mg \cdot L^{-1}$as $CaCO_3$ to $mg \cdot L^{-1}$ as $HCO_3^-$

Consider the following reaction:

$$CaCO_3 + H_2O + CO_2 \longrightarrow Ca(HCO_3)_2$$

$CaCO_3$ has a molecular weight of 100 $g \cdot mol^{-1}$. The $HCO_3^-$ anion has a molecular weight of 61 $g \cdot mol^{-1}$.

Therefore, each mole of $Ca(HCO_3)_2$ corresponds to 1 mole of $CaCO_3$ (100 g) and contains 2 × 61 g = 122 g of $HCO_3^-$.

The conversion is as follows:

Alkalinity bicarbonate as $HCO_3^-$ ($mg \cdot L^{-1}$) = 1.22 × alkalinity bicarbonate as $CaCO_3$ ($mg \cdot L^{-1}$)

– Converting hydroxide alkalinity from mg/L as $CaCO_3$ to mg/L as OH the conversion is as follows:

Alkalinity hydroxide as $OH^-$ ($mg \cdot L^{-1}$) = 2×17/100 × Hydroxide alkalinity as $CaCO_3$ ($mg \cdot L^{-1}$)

or:

Alkalinity hydroxide as $OH^-$ ($mg \cdot L^{-1}$) = 0.34 × Alkalinity hydroxide as $CaCO_3$ ($mg \cdot L^{-1}$)

The relevant acid–base reactions are:

$H_2CO_3 \rightarrow H^+ + HCO_3^-$ $pK_1$: 6.35 at 25°C

$HCO_3^- \rightarrow CO_3^{2-} + H^+$ $pK_2$: 10.33 at 25°C

Based on the pK values, the species present are as follows:

– below a pH of 4.5, almost all of the carbonate species are present in the form of $H_2CO_3$ and alkalinity is negative (due to $H^+$);

– at a pH value of 8.3, most carbonate species are present as $HCO_3^-$ and alkalinity is equal to $HCO_3$;

– above pH 12.3, virtually all carbonate species are present in the [$CO_3^{2-}$] form and alkalinity is equal to [$CO_3^{2-}$] [$OH^-$].

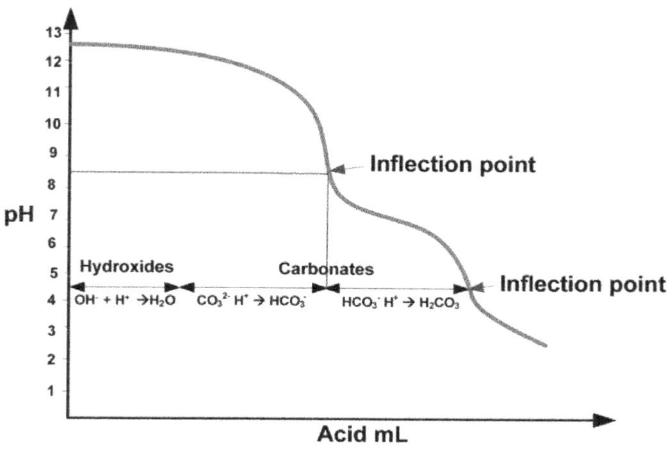


**Figure 20.1.** *Evolution of the different species in function of pH*

Figure 20.1 shows the change in species when pH decreases by adding acid to alkaline water. The pH starts over 12.3. After the addition of acid, the pH slowly decreases once the first acidities ($H^+$) are neutralized by hydroxides ($OH^-$). The acid is then consumed by carbonates ($CO_3^{2-}$) converted into bicarbonates ($HCO_3^-$).

At about pH 8.3, carbonates are completely converted into bicarbonates. This is the first inflection point. The supplementary addition of acid converts bicarbonates into carbonic acid. This is the second inflection point at pH 4.5.

Alkaline water has a pH above 7, whereas high alkalinity water has a high buffering capacity. Alkaline water may or may not have a high buffering capacity. Similarly, water with high alkalinity may or may not have a high pH.

By convention, alkalinity is expressed in $mg \cdot L^{-1}$ $CaCO_3$. To convert all $mg \cdot L^{-1}$ species into $CaCO_3$, multiply each species by the ratio of the weight of the $CaCO_3$ equivalent by the weight of the species equivalent:

$$mg \cdot L^{-1}\ CaCO_3 = mg \cdot L^{-1} \text{ of each species } (\text{equiv} \cdot CaCO_3/\text{equiv} \cdot \text{species})$$

$$\mathrm{C(CaCO3)} = \frac{50\ Ci}{M(eq)}$$

where:

– $C(CaCO_3)$: conversion of the ion in $mg \cdot L^{-1}$ $CaCO_3$;

– Ci: Concentration of the ion in $mg \cdot L^{-1}$;

– M(eq): Equivalent mass of the ion.

For example, consider water containing 100.0 $mg \cdot L^{-1}$ $CO_3^{2-}$ and 75.0 $mg \cdot L^{-1}$ $HCO_3^-$ at pH 8. The exact alkalinity at 25°C is obtained by converting the different $CO_3^{2-}$, $HCO_3^-$, $OH^-$ and $H^+$ species into $mg \cdot L^{-1}$ $CaCO_3$. The weight equivalents are as follows:

$CO_3^{2-}$: MW = 60 g, n = 2, $equiv \cdot g^{-1}$: 30

$Ca^{2+}$: MW = 40 g, n = 2 , $equiv \cdot g^{-1}$: 20

$HCO_3^-$: MW = 61 g, n = 1, $equiv \cdot g^{-1}$: 61

$Mg^{2+}$: MW = 24.4 g, n = 2, $equiv \cdot g^{-1}$: 12.2

$OH^-$: MW = 17 g, n = 1, $equiv \cdot g^{-1}$: 17

$Na^+$: MW = 23 g, n = 1, $equiv \cdot g^{-1}$: 23

$K^+$: MW = 39.1 g, n = 1 , $equiv \cdot g^{-1}$: 39.1

$SO_4^{2-}$: MW = 96 g, n = 2, $equiv \cdot g^{-1}$: 48

$H^+$: MW = 1 g, n = 1, $equiv \cdot g^{-1}$: 1

At pH 8, $H^+ = 10^{-8}\ mol \cdot L^{-1}$ and the species expressed in $mg \cdot L^{-1}$ is given by:

$$(mol \cdot L^{-1})\ (MW)\ (10^3\ mg \cdot g^{-1}) = 10^{-8} \times 1\ g \cdot mol^{-1} \times 10^3 = 10^{-5}\ mg \cdot L^{-1}\ H^+$$

The $OH^-$ species is deduced from water's ionization constant:

$$K_{H2O} = [OH^-][H^+] = 10^{-14}$$

That is:

$$OH^-: 10^{-14}/10^{-8} = 10^{-6}\ mol \cdot L^{-1}$$

and:

$$10^{-6} \times 17 \times 10^3 = 1.7\ mg \cdot L^{-1}$$

Then:

$$CO_3^{2-} = 100 \times 50/30 = 167\ mg \cdot L^{-1}\ CaCO_3$$

$$HCO_3^- = 75 \times 50/61 = 61\ mg \cdot L^{-1}\ CaCO_3$$

$$OH^- = 1.7 \times 50/17 = 5\ mg \cdot L^{-1}\ CaCO_3$$

$$H^+ = 10^{-8} \times 50/1 = 5 \times 10^{-7}\ mg \cdot L^{-1}\ CaCO_3$$

The exact alkalinity is then: $167 + 61 + 5 - 5 \times 10^{-7} = 228.05\ mg \cdot L^{-1}\ CaCO_3$, which can be simplified (2% error) as: $167 + 61 = 228\ mg \cdot L^{-1}\ CaCO_3$.

#### 20.1.1.3. *Relationships between free $CO_2$/aggressive $CO_2$/balancing $CO_2$*

During its cycle in nature, water dissolves a number of elements which are mainly:

– gases such as carbon dioxide, oxygen and nitrogen;

– there is a slight dissociation of $H_2S$ depending on pH;

– carbon dioxide dissolves partially in combination with water and forms carbonic acid $H_2CO_3$, which is the essential agent responsible for the aggressivity of water, where it is partially dissociated into ions.

The solubility of carbon dioxide in water obeys to Henry's law:

$$S = P_{CO2}\,[CO_2]$$

where:

– $[CO_2]$s: specific solubility of $CO_2$ equal to 0.033 mol·L$^{-1}$;

– $P_{CO2}$: partial pressure of $CO_2$ in the gas phase ($P_{CO2} = 3 \times 10^{-4}$ atm).

Therefore:

$$S_{CO2} = 0.033 \times 3 \times 10^{-4} = 9.9 \times 10^{-6}\ \text{mol}\cdot\text{L}^{-1}$$

As MW($CO_2$) = 44 g·mol$^{-1}$, then $S_{CO2}$ = 0.44 mg·L$^{-1}$.

This solubility corresponds to $CO_2$ in water when it is under atmospheric pressure, but if this pressure increases, Henry's law provokes an increase in solubility.

This is observed for groundwater, which is generally subject to pressure values higher than atmospheric pressure. As a result, groundwater presents higher $CO_2$ values than those observed for surface waters.

Total $CO_2$ is divided into free $CO_2$ and carbonate-related $CO_2$. Free $CO_2$ is the sum of balancing $CO_2$ and aggressive $CO_2$. Carbonate-related $CO_2$ is the sum of combined $CO_2$ ($CO_2$ present in the form of carbonates) and semi-combined $CO_2$ ($CO_2$ present in the form of hydrogen carbonates).

Water in equilibrium (pH = pHs) often contains a negligible – but not zero – quantity of $CO_2$. This is called the "balancing" $CO_2$. Aggressive water contains more $CO_2$ than water in equilibrium. Aggressive $CO_2$ is defined as the difference between free $CO_2$ and equilibrium $CO_2$.

Figure 20.2 shows the various forms in which carbon dioxide can be found in water.

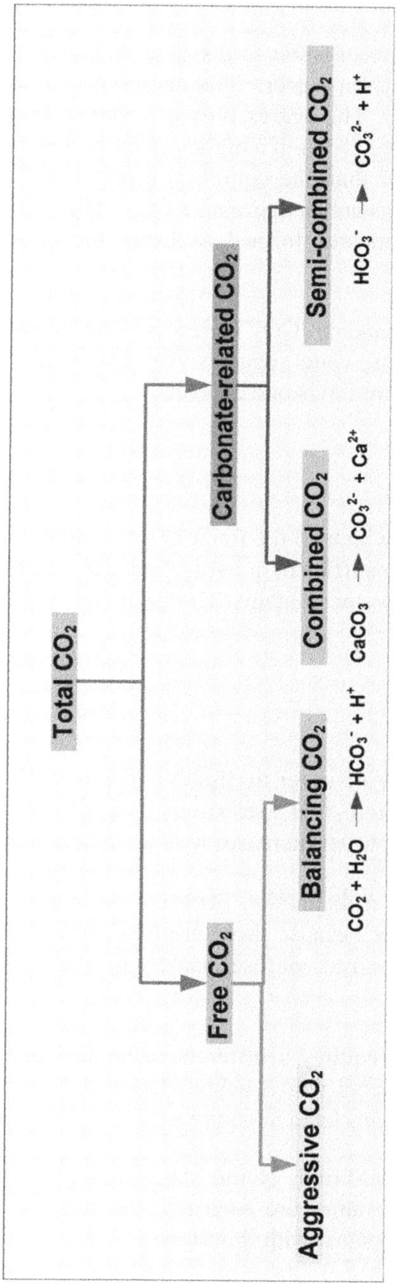


**Figure 20.2.** *The various forms of $CO_2$ in water*

### 20.1.2. *Aggressive water*

Aggressivity is the ability of water to dissolve limestone associated with a $CO_2$ excess (or aggressive $CO_2$) in water. Non-aggressive water is either water in equilibrium or scaling water. The aggressivity of water is related to the presence of dissolved $CO_2$ (known as free $CO_2$ ) in a higher quantity than balancing $CO_2$ . In this case, the water pH is lower than the equilibrium pH. The fraction of $CO_2$ excess likely to dissolve limestone is called aggressive $CO_2$ . The other fraction – serving to balance the hydrogen carbonates formed as limestone dissolves – is called the equilibrium potential of $CO_2$.

As the quantity of balancing $CO_2$ increases with temperature for the same hydrogen carbonate content, water aggressivity also changes with temperature. Aggressive water cannot form limestone deposits.

### 20.1.3. *Scaling water*

Scaling water is that in which all the free $CO_2$ is found only in its balancing form (no $CO_2$ excess), but in an insufficient quantity, because water tends to form calcium carbonate. However, this water contains a large quantity of semi-combined $CO_2$ (hydrogen carbonates).

### 20.1.4. *Corrosive water*

Corrosivity is the ability of water to dissolve metals, and which is related to its physicochemical composition (pH, resistivity, oxygen content, chlorides and sulfates). The corrosivity of water increases with its temperature.

It is necessary to draw a distinction between the aggressive character and the corrosive character of water because these are two physicochemical particularities that engineers sometimes tend to confuse, namely because they are close in certain aspects.

Each of these notions requires a different technical approach, associated with specifically adapted treatments which acknowledge such differences, and are not limited to what the two notions have in common.

The calco-carbonic equilibrium is the state of water whose $Ca^{2+}$, free $CO_2$, bicarbonate and carbonate values are such that the water's characteristics do not change when brought into contact with limestone.

The salts of carbonic acid and strong bases are calcium carbonate, magnesium, sodium, potassium, etc.

Hardness is defined as the sum of the calcium and magnesium ions. These are both expressed in $mg \cdot L^{-1}$ calcium carbonate ($CaCO_3$). Hardness can consist of both carbonate and non-carbonate hardness. If the hardness is equal to or less than the alkalinity (also expressed in $mg \cdot L^{-1}$ calcium carbonate), then all the hardness is referred to as "carbonate" hardness. If the hardness is greater than the alkalinity, the difference between the hardness value and alkalinity value is referred to as non-carbonate hardness.

Carbonate hardness is due to the presence of carbonate and bicarbonate salts of magnesium and calcium, while non-carbonate hardness arises from various alkaline salts of calcium and magnesium (such $CaCl_2$; $CaSO_4$). Since carbonates, bicarbonates and others compounds that contribute to non-carbonate hardness are alkaline, it means that hard water will present a pH > 7. Borehole waters are typically very hard because calcium and magnesium dissolve into the water from the soil.

The salts of weak acids and strong bases are almost completely dissociated into ions, among which the most important ones are the calcium salts:

– calcium bicarbonate $Ca(HCO_3)_2$ which exists only in an aqueous solution;

– slightly soluble calcium carbonate $CaCO_3$ (around 15 $mg \cdot L^{-1}$ at 25°C and 25 $mg \cdot L^{-1}$ at 100°C).

Other elements, including sodium and magnesium, give rise to the following salts: $MgCO_3$, $Mg(HCO_3)_2$, $Na_2CO_3$ and $NaHCO_3$.

Alkalinity is defined as the sum of carbonate ions. Alkalinity is the sum of all the titratable bases, and it is measured by titrating a water sample with an acid to a particular end-point (i.e. pH value). Therefore, reference is made to phenolphthalein alkalinity (phenolphthalein is an indicator that changes color at a pH of around 8.3) and total alkalinity (titration to a pH of 4.5). Generally, the alkalinity of surface water is a function of the carbonates, bicarbonates and hydroxides, although other compounds, such as borates, phosphates and silicates, can also contribute to the alkalinity. But alkalinity is predominantly due to carbonates, bicarbonates and hydroxides. The most predominant water soluble carbonates and bicarbonates are those of calcium and magnesium. Therefore, hard waters are generally also high in alkalinity and the pH of water, which is high in hardness and alkalinity, is therefore generally high.

It is expressed by two titles which describe the carbonate hardness of water:

– the simple alkalimetric title: measurement of $[OH^-]$ and ½ $[CO_3^{2-}]$;

– the total alkalinity (Alk): measurement of $[OH^-]$, $[CO_3^{2-}]$ and $[HCO_3^-]$.

The knowledge of the total alkalinity and the simple alkalimetric title makes it possible to estimate the distribution of alkalinity in bases, carbonates or bicarbonates. These titles can be expressed in various units:

– gram-equivalent, which is the quotient of the atomic or molar mass by the ion charge (polarity);

– milliequivalent or meq (per unit volume), which is the thousandth part of the equivalent;

– $mg \cdot L^{-1}$ $CaCO_3$.

The concentration of $HCO_3^-$ ions is evaluated from the total alkalinity (Alk), whereas the $Ca^{2+}$ and $Mg^{2+}$ concentrations are related to the hydrotimetric title (which expresses the calcium and magnesium hardness).

Salts of strong acids and strong bases are related to sulfates, chlorides, calcium nitrates, magnesium, sodium, etc. They dissociate into ions, among which the most important are sulfates and chlorides. For example, water having been in contact with gypsum soils (calcium sulfate, $CaSO_4$) becomes charged with $SO_4^{2-}$ anions and $Ca^{2+}$ cations, according to the following reaction:

$$CaSO_4 \rightarrow Ca^{2+} + SO_4^{2-}$$

The carbonate hardness or temporary hardness is the sum of calcium and magnesium carbonates and bicarbonates. Non-carbonate hardness or permanent hardness is the result from chlorides and sulfates associated with calcium and magnesium. Total hardness is the sum of temporary hardness and permanent hardness. It describes the calcium and magnesium content of the water. It is expressed in $mg \cdot L^{-1}$ $CaCO_3$ or $meq \cdot L^{-1}$:

$$\text{Total hardness} = \text{Permanent hardness} + \text{temporary hardness}$$

Calcium hardness corresponds to the overall content of calcium salts. Magnesium hardness corresponds to the overall content of magnesium salts:

$$\text{Total hardness} = \text{Ca hardness} + \text{Mg hardness}$$

pH is the acidity scale with $pH = -\log [H^+]$.

The pH of natural waters can be higher, equal or lower than the equilibrium pH. It depends on the course of water, climatic conditions and the soil's hydrogeology. There are many calculation methods and graphical methods to estimate the saturation pH of water with a specific chemical composition.

## 20.2. The equilibrium reactions of water's constituents

In natural waterways, calcium hydrogen carbonate is most common and the most prominent. It derives from a hypothetical acid, the carbonic acid $H_2CO_3$. This is particularly unstable and decomposes as soon as it is formed, releasing carbon dioxide and water:

$$H_2CO_3 \rightarrow CO_2 + H_2O$$

In an aqueous solution, calcium hydrogen carbonate and carbonic acid participate in an equilibrium traditionally called "calco-carbonic equilibrium".

For example, if the water is hard, this reaction is favored by the release of carbonate ions in the solid phase (precipitation) in the form of calcium carbonate, following the reaction:

$$CO_3^{2-} + Ca^{2+} \rightarrow CaCO_3$$

These two reactions (decomposition of hydrogen carbonates into carbonates, and the subsequent formation of calcium carbonate) are the opposite of the dissolution reaction:

$$\underset{\text{solid limestone}}{CaCO_3 + H_2O + nCO_2} \rightarrow \underset{\text{dissolved limestone}}{Ca(HCO_3)_2} + \underset{\text{balancing carbon dioxide}}{(n-1)CO_2}$$

It is a heterogeneous equilibrium, governed by the laws which are specific to this type of equilibrium. It is an equilibrium between calcium and the compounds from carbon dioxide $CO_2$ (carbonates and hydrogen carbonates). Any modification affecting the concentration of one of the elements involved in this equilibrium shifts such equilibrium. This equilibrium reaction can move in one direction (limestone dissolution) or in the other (limestone deposit), depending on the temperature and concentration of one or the other element involved in the reaction. For the same temperature, a decrease in the $CO_2$ concentration tends to shift the reaction from right to left (solid limestone formation). Conversely, an increase in the concentration of $CO_2$ tends to move the reaction from left to right (limestone dissolution).

Schematically, it is possible to represent the calco-carbonic equilibrium reactions as shown in Figure 20.3.

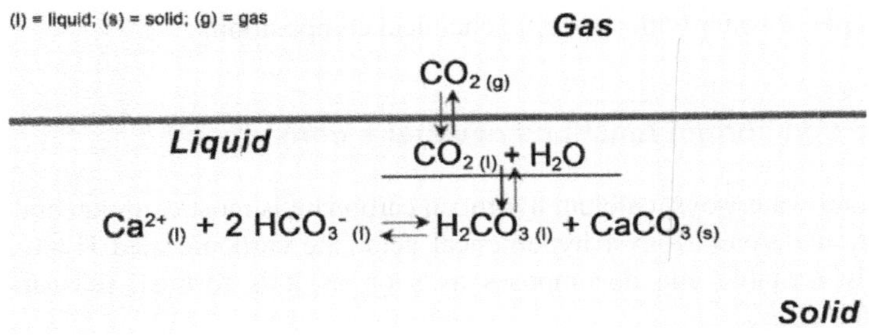


**Figure 20.3.** *Calco-carbonic equilibrium reactions*

Calcium bicarbonate tends to precipitate the calcium carbonate (limestone) contained in water. To keep the bicarbonate in solution, the introduction of a certain quantity of free $CO_2$ or $CO_2$ equivalent inhibits the precipitation reaction.

### 20.2.1. *Equilibrium pH*

The presence of $CO_2$ and hydrogen carbonates in water constitutes a so-called buffered medium, characterizing water with a well-defined pH. When, at a given temperature, water contains the exact quantity of $CO_2$ required for maintaining a situation of equilibrium such as the stability of hydrogen carbonates, it has a certain pH, which is known as equilibrium pH or saturation pH (pHs).

The basic equations are and the equilibrium conditions are quantified by the dissociation or acidity constants. The constants involved in developing a saturation equation represent the law of mass action applied to the following equilibria:

$$H_2CO_3 \Leftrightarrow H^+ + HCO_3^-$$

$$K'1 = \frac{[H3O+][HCO3-])}{[H2CO3]} \text{ related to concentrations [C]}$$

$$K1 = \frac{(H3O+)(HCO3-)}{(H2CO3)} = \frac{\gamma H+[H+]\gamma HCO3^-[HCO3^-}{\gamma H2CO3[H2CO3]} \text{ related to activities}$$

and

$$K'1 = \frac{[H3O+])[HCO3-]}{[H2CO3]} = K1\ \frac{\gamma H2CO3}{\gamma H3O+\ \gamma HCO3-} = 10^{-6.35}$$

$$HCO_3^{\ -} \Leftrightarrow H^+ + CO_3^{\ 2-}$$

$$K'2 = \frac{[H3O^+][CO3^{2-}]}{[HCO3^-]}$$

$$K2 = \frac{(H3O+)(CO3\ 2-)}{(HCO3^-)} = \frac{\Upsilon H+[H+]\Upsilon CO3^-[CO3^{2-}]}{\Upsilon HCO3-\ [HCO3^-]}$$

$$K'2 = \frac{[H3O^+][CO3^{2-}]}{[HCO3^-]} = K2\ \frac{\Upsilon HCO3^-}{\Upsilon CO3^{2-}\Upsilon H3O^+} = 10^{-10.33}$$

If the solution is saturated with calcium carbonate, the application of the law of mass action to the heterogeneous equilibrium is given as:

$$Ca^{2+} + CO_3^{\ 2-} \Leftrightarrow CaCO3$$

$$Ks = (Ca^{\ 2+})\ (CO_3^{\ 2-}) = \Upsilon Ca^{2+}\ [Ca^{2+}\ ]\ \Upsilon CO3^{2-}[CO3^{2-}]$$

$$K's = [Ca^{2+}][CO3^{2-}] = \frac{Ks}{\Upsilon Ca^{2+}.\ \Upsilon CO3^{2-}} = 3.32 \times 10^{-9}\ (\text{at } 25°C)$$

*Ks* is the solubility product of $CaCO_3$:

$$H^+ + OH^- \longrightarrow H_2O$$

$$Kw = (H^+)(OH^-) = \Upsilon_{H+}\ [H^+]\ \Upsilon_{OH-}\ [OH^-]$$

$$K'w = [H^+][OH^-] = \frac{Kw\ (H2O)}{\Upsilon OH^-} = 10^{-14}$$

where:

– $K'w$, $K'_1$ and $K'_2$: constants for a given temperature and for a given concentration [C];

– $Kw$, $K_1$ and $K_2$: constants for a given temperature and for a given ionic strength (the ionic strength is a function of the concentrations and valences of the ions present in the solution).

In natural waterways:

$$K' = f\,(K(T), \varepsilon, I\,([Ci], Zi))$$

where:

– I: ionic strength;

– ε: Debye and Huckel correction coefficient;

– Zi: valence of ion i;

$$I = \frac{1}{2} \Sigma CiZ^2i$$

Simplified calculation for ionic strength (I): dry residual (mg·L$^{-1}$) × 25 × 10$^{-6}$:

$$\mathcal{E} = \frac{\sqrt{I}}{1 + 1.4\sqrt{I}}$$

Alkalinity = 0.5 [$HCO_3^-$] + [$CO_3^{2-}$] + 0.5 [$OH^-$] – 0.5 [$H_3O^+$]

pH = – log[$H_3O^+$]

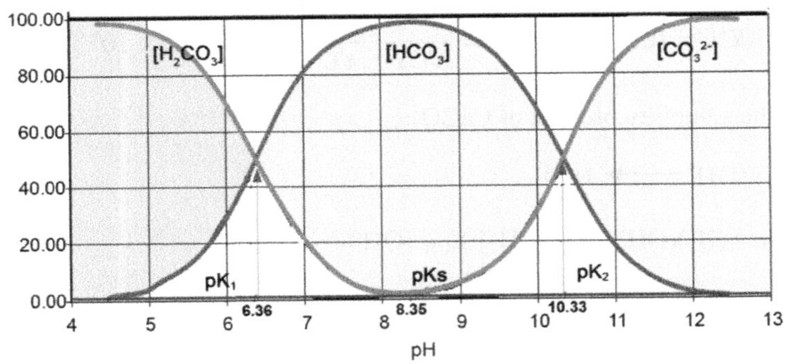


**Figure 20.4.** *Dissociation and form of carbonates as a function of pH. For a color version of this figure, see www.iste.co.uk/gaid/watertreatment5.zip*

Constants $K_1$, $K_2$, $Kw$ and $Ks$ are commonly expressed by their decimal cologarithms, respectively, denoted as p$K_1$, p$K_2$, p$Kw$ and p$Ks$. Their values for freshwater (ideal solution) and seawater are a function of the water temperature (0–40°C) and the water salinity. The $Kw$, $K_1$, $K_2$ and $Ks$ values can be deduced from:

$$pK_1 = 6.583 - 646 \times 10^{-9}\,T^3 + 163.5 \times 10^{-6}\,T^2 - 12.3 \times 10^{-3}\,T$$

$$pK_2 = 10.627 - 328 \times 10^{-9}\,T^3 + 135.3 \times 10^{-6}\,T^2 - 15.04 \times 10^{-3}\,T$$

$$pKw = 14.940 - 1.49 \times 10^{-9}\,T^3 + 270 \times 10^{-6}\,T^2 - 43.44 \times 10^{-3}\,T$$

$$pKs = 8.022 + 444 \times 10^{-9}\,T^3 - 61 \times 10^{-6}\,T^2 + 14 \times 10^{-3}\,T$$

$$pKs = 7.8156 + 0.03111T + 1502/T - 5.518 \log T$$

In these formulas, the temperature T is expressed in °C.

At a given mineralization (defined total hardness and Alk), Tillmans showed that there is a pH called saturation pH (pHs or equilibrium pH), beyond which a precipitation of calcium and bicarbonate ions takes place in the form of calcium carbonate. This precipitate (Tillmans layer) plays a protective role for the metallic parts in relation to water. Water can therefore be classified depending on the pH value in relation to the pHs:

– if the pH > equilibrium pH, it means that the $CO_2$ concentration is lower than it is at equilibrium. The shift in the equilibrium will tend to the right, that is to say, toward the precipitation of $CaCO_3$ from $Ca(HCO_3)_2$. Water tends to deposit $CaCO_3$, so is said to be "scaling";

– if the pH < equilibrium pH, this means that the $CO_2$ concentration is higher than it is at equilibrium. The shift in the equilibrium will tend to the left, that is to say, to an attack of $CaCO_3$ by the $CO_2$ excess: $CaCO_3 + CO_2 + H_2O \rightarrow Ca(HCO_3)_2$. Water with the ability to dissolve limestone in this way is called "aggressive water".

Calculation of free $CO_2$:

$$\text{Log10}\,[CO_2] = \text{Log10}\,[Alk_{CaO}] + pK_1 - pH + 0.2$$

| **Temperature °C** | **Equilibrium constants & Solubility product** | | |
|---|---|---|---|
| | $K_1 \times 10^7$ | $K_2 \times 10^{11}$ | $Ksp \times 10^9$ |
| **5** | 3.020 | 2.754 | 4.119 |
| **10** | 3.467 | 3.236 | 3.957 |
| **15** | 3.802 | 3.715 | 3.767 |
| **20** | 4.169 | 4.169 | 3.554 |
| **25** | 4.467 | 4.477 | 3.324 |
| **40** | 5.012 | 6.026 | 2.596 |

**Table 20.2.** *Equilibrium constants for calcium carbonate*

Dissolved carbon dioxide concentrations can be also estimated from water's alkalinity and pH for a given temperature, using this following equation:

$$CO_2\ (\mathrm{mg}\cdot\mathrm{L}-1) = 44{,}000\left\{\frac{(\mathrm{Alk\ as\ CaCO_3})}{50{,}000} - (10^{\mathrm{pH}-\mathrm{p}Kw}) - 10^{-\mathrm{pH}}\right)\{\frac{1}{(10^{\mathrm{pH}-\mathrm{p}KsK1}) + 2((10^{2\mathrm{pH}-\mathrm{p}KsK1-\mathrm{p}K2}))}\}$$

#### 20.2.2. Langelier equation

Instead of using the $CaCO_3$ concentration as a measure of saturation, Langelier calculated the theoretical pH of water (using alkalinity and calcium concentration) at which water would be just saturated in relation to $CaCO_3$. He called this pH the saturation pH (pHs).

The saturation state of water is given by the saturation index (SI) or Langelier index (LSI), which is defined as the difference between the water pH and the saturation pH, that is to say:

$$\mathrm{LSI} = \mathrm{pH} - \mathrm{pHs}$$

$$H_2CO_3 \rightarrow H^+ + HCO_3^-$$

$$K'1 = \frac{[H^+][HCO_3^-]}{[H2CO3]}$$

$$HCO_3^-\ H^+ + CO_3^{2-}$$

$$K'2 = \frac{[H_3O^+][CO_3^{2-}]}{[HCO_3^-]}$$

$$[H_3O^+] = \frac{K'2[HCO_3-]}{[CO_3^{2-}]}$$

$$\mathrm{pH} = -\log\,[H^+]$$

$$\mathrm{p}K'_2 = -\log K'_2$$

$$pH = pK'_2 - \log [HCO_3^-] + \log [CO_3^{2-}] = pHs$$

$$CaCO_3 \rightarrow Ca^{2+} + CO_3^{2-}$$

$$K's = [CO_3^{2-}][Ca^{2+}]$$

$$[CO_3^{2-}] = \frac{K's}{[Ca^{2+}]},$$

Then:

$$pK's = -\log K's$$

$$pHs = pK'_2 - \log [HCO_3^-] - \log [Ca^{2+}] - pK's$$

$$\text{and } pHs = pK'_2 - pK's - \log [HCO_3^-] - \log [Ca^{2+}]$$

$$pHs = pK_2 - pKs - \log [HCO_3^-] - \log [Ca^{2+}] + 2\epsilon$$

$$\mathcal{E} = \frac{\sqrt{I}}{1 + 1.4\sqrt{I}}$$

$$pK'_1 = pK_1 - \mathcal{E} \qquad pK'_2 = pK_2 - 2\mathcal{E} \qquad pK's = pKs - 4\mathcal{E}$$

$$pK'_1 = 6.35 \qquad pK'_2 = 10.33 \qquad pK's = 8.30 \quad \text{at } 25°C$$

$[Ca^{2+}]$ is expressed in $mol \cdot L^{-1}$ and $[HCO_3^-]$ = alkalinity is expressed in $equiv \cdot L^{-1}$.

The calculation of LSI needs only $Ca^{2+}$ and $HCO_3^-$ ions. The simplified formula is:

$$LSI = pH - pHs$$

$$LSI = pH - (pK'_2 - pK's + 9.16 - \log (0.4608 [HCO_3^-] - (\log 1.4 [Ca^{2+}])$$

Where the $Ca^{2+}$ and alkalinity are expressed in $mg \cdot L^{-1}$ CaO. The conversion is included in the equation.

For a given temperature, LSI can be calculated with the following equation:

$LSI = pH - (10.627 - 15.04(10^{-3}T) + 135.3(10^{-3}T)^2 - 328(10^{-3}T)^3 - 8.022\text{-}14 (10^{-3}T) + 61(10^{-3}T)^2 - 444(10^{-3}T)^3 + 9.356 - \log ([Ca^{2+}]) - \log ([HCO_3^-]))$ where the $Ca^{2+}$ and $HCO_3^-$ concentrations are expressed in $mg \cdot L^{-1}$ of Ca and Alk, respectively.

The LSI is an index of whether a water will precipitate or dissolve calcium carbonate.

If the pHs is equal to the measured pH, the water is considered “balanced”. This means that calcium carbonate will not be dissolved or precipitated.

If the pHs is less than the measured pH (the LSI is a positive number), the water will tend to deposit calcium carbonate and is scale-forming (non-aggressive).

If the pHs is greater than the measured l pH (the LSI is a negative number), the water is not saturated and will dissolve calcium carbonate (aggressive).

In summary:

If LSI = 0: water in equilibrium (pH = pHs).

If LSI < 0: aggressive water (pH < pHs).

If LSI > 0: encrusting water (pH > pHs).

Aggressive waters will tend to dissolve calcium carbonate due to their undersaturation of this element. The pH is higher than the pH acidity is largely due to the presence of dissolved carbon dioxide, which corresponds to the aggressive $CO_2$.

Balanced waters will behave indifferently to calcium carbonate. The pHs is equal to the pH.

Encrusting waters will cause precipitation of calcium carbonate due to oversaturation of this element. The free $CO_2$ content is lower than the balancing $CO_2$.

Table 20.3 shows an example calculating the LSI for a given water chemical composition and a temperature of 25°C.

The LSI can be quickly estimated from pH, T°C, $Ca^{2+}$ and $HCO_3^-$ ($mg{\cdot}L^{-1}$).

| Ions | mg·L$^{-1}$ | mol·L$^{-1}$ | Parameters | Ions | mg·L$^{-1}$ | Parameters |
|---|---|---|---|---|---|---|
| **Calculations according to the basic equation** | | | | **Calculations according to the simplified equation** | | |
| $Ca^{2+}$ | 104.8 | 0.00262 | Ionic strength 0.0094 | $Ca^{2+}$ | 104.8 | – |
| $Mg^{2+}$ | 7.0 | 0.00028 | - | $Mg^{2+}$ | – | – |
| $Na^{+}$ | 11.9 | 0.00052 | Correction factor ε 0.0853 | $Na^{+}$ | – | – |
| $K^{+}$ | 2.70 | 0.00007 | - | $K^{+}$ | – | – |
| $Cl^{-}$ | 22.1 | 0.00062 | p$K'_1$: 6.284 | $Cl^{-}$ | – | p$K_2$: 10.33 |
| $NO_3^{-}$ | 22.9 | 0.00037 | p$K'_2$: 10.33 | $NO_3^{-}$ | – | p$Ks$: 8.34 |
| $SO_4^{2-}$ | 24.9 | 0.00026 | p$K's$: 8.34 | $SO_4^{2-}$ | – | pHs: 7.00 |
| $HCO_3^{-}$ | 265 | 0.00434 | pHs: 6.90 | $HCO_3^{-}$ | 265 | – |
| Measured pH | 8 | – | LSI = 1.10 | Measured pH | 8 | LSI = 1.00 |

**Table 20.3.** *Estimation of the Langelier index at 25°C*

Another quick formulation is obtained from:

$$pHs = (9.3 + A + B) - (C + D)$$

where:

$A = (\log10(TDS) - 1)/10$;

$B = -13.12 \times \log10(T°C + 273) + 34.55$;

$C = \log10$ (hardness $Ca^{2+}$ as $CaCO_3$) – 0.4;

$D = \log10$ (alkalinity as $CaCO_3$).

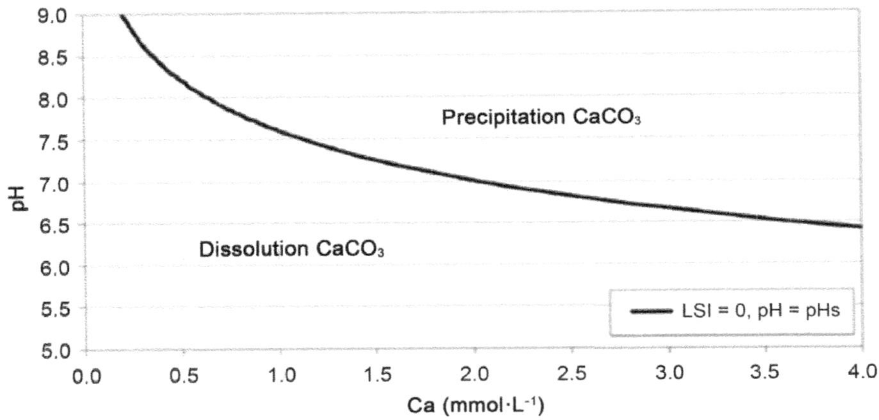


**Figure 20.5.** *$CaCO_3$ equilibrium curve (at 25°C)*

Figure 20.5 shows the $CaCO_3$ equilibrium curve at 25°C with $HCO_3^- = 2Ca^{2+}$.

EXAMPLE.– The precipitation potential of $CaCO_3$ at 20°C can be calculated with the LSI. For water whose composition is: $Ca^{2+} = 90\ mg \cdot L^{-1}$, $HCO_3^- = 65\ mg \cdot L^{-1}$, TDS = 350 $mg \cdot L^{-1}$ and pH = 7.5.

The ionic strength is estimated by the TDS concentration:

$$I = 2.05 \times 10^{-5}\ TDS = 2.05 \times 10^{-5} \times 350 = 7.17 \times 10^{-3}$$

The activity coefficients for calcium and bicarbonates are:

$$Ca^{2+}: 90\ mg \cdot L^{-1} = 2.25 \times 10^{-3}\ mol \cdot L^{-1}$$

$$\log \Upsilon\, Ca^{2+} = -\frac{0.5(Zi)^2\sqrt{I}}{1+\sqrt{I}} = -\frac{0.5(2)^2\sqrt{7.17 \times 10^{-3}}}{1+\sqrt{7.17 \times 10^{-3}}} = -0.156$$

$\Upsilon Ca^{2+} = 0.699$

$HCO_3^-$: 65 mg·L$^{-1}$ = 1.06 × 10$^{-3}$ mol·L$^{-1}$

$$\log \Upsilon HCO_3^- = -\frac{0.5(Zi)^2\sqrt{I}}{1+\sqrt{I}} = -\frac{0.5(1)^2\sqrt{7.17 \times 10^{-3}}}{1+\sqrt{7.17 \times 10^{-3}}} = -0.039$$

$\Upsilon HCO_3^- = 0.914$

The saturation pH is:

$$pHs = -\log \frac{Ka2\Upsilon Ca^{2+}\left[Ca^{2+}\right]\Upsilon HCO_3^-\left[HCO_3^-\right]}{Ksp}$$

$$pHs = -\log \frac{(4.17\ 10^{-11})(0.699)(2.25\ 10^{-3})(0.914)(1.06\ 10^{-3})}{3.34\ 10^{-9}} = 7.72$$

The LSI is:

pH – pHs = 7.5 – 7.72 = – 0.22

meaning that the LSI < 0, so it is an aggressive water.

The calculation of I based on the molar concentration of $Ca^{2+}$ and $HCO_3^-$ can also be given by:

$$I = [Ca^{2+}]\,[HCO_3^-]$$

## 20.3. Hallopeau–Dubin diagram

For water with a pH between 5.5 and 8.7, the concentrations of $[H^+]$ and $[OH^-]$ are negligible in comparison with the other terms, and $[CO_3^{2-}]$ is negligible in relation to $[HCO_3^-]$. Under these conditions, alkalinity is represented by the $[HCO_3^-]$ concentration:

$$\log [H_2CO_3] = \log [HCO_3^-] - pH + pK'_1$$

The carbonic acid (or $CO_2$) and bicarbonate concentrations are expressed in mol·L$^{-1}$, whereas the Hallopeau and Dubin representation defines the alkalinity in mg·L$^{-1}$ of CaO:

$$1\ \text{mol·L}^{-1}\ CO_2 = 44{,}000\ CO_2\ \text{in mg·L}^{-1}\ 1\ \text{mg·L}^{-1}\ CO_2 = 1.27\ \text{mg·L}^{-1}\ \text{CaO}$$

In calco-carbonic equilibrium, the solubility product of calcium carbonate is reached. As a result:

$$pH = pK'_2 - pK's - \log [HCO_3^-] - \log [Ca^{2+}]$$

$[HCO_3^-]$ in mol·L$^{-1}$ is transformed into 2 $[HCO_3^-]$/56,000 in mg·L$^{-1}$ of CaO

$[Ca^{2+}]$ in mol·L$^{-1}$ becomes $[Ca^{2+}]$/56,000 in mg·L$^{-1}$ of CaO

Then:

$$pH = pK'_2 - pK's + 2 \log 56{,}000 - \log 2 - \log [HCO_3^-] - \log [Ca^{2+}]$$

or:

$$pH = pK'_2 - pK's + 9.2 - \log [HCO_3^-] - \log [Ca^{2+}]$$

By posing:

$$m = \log \frac{\text{Alk}}{\text{lime}} = \log \frac{[HCO3^-]}{[Ca^{2+}]}$$

$[HCO_3^-]$ and $[Ca^{2+}]$ expressed in mg·L$^{-1}$ of CaO. It is a straight line having the form pH = a (log Alk) + b and whose slope a = –2 and b = $pK'_2 - pK's + 9.2 + m$, which depends on temperature and water mineralization.

The Hallopeau–Dubin carbon graph (Figure 20.6) includes:

– a network of straight lines of dissolved $CO_2$;

– an additional diagram making it possible to draw the saturation line, for a given water.

One mole of CaO corresponds to 2 mole of bicarbonates. Following the reaction:

$$CaO + H_2O + 2CO_2 \rightarrow Ca(HCO_3)_2$$

$[HCO_3^-]$ in mol·L$^{-1}$ becomes 2 $[HCO_3]$/56,000, that is to say:

$$\log \frac{[CO2]}{44{,}000} = \log \frac{2[HCO3-]}{56{,}000} - pH + pK'1$$

For waters of 15–20°C, p$K'1$ = 6.34. The Hallopeau–Dubin diagram represents the lines:

$$\log [CO_2] = \log [HCO_3^-] - pH + 6.54$$

The concentration of $[HCO_3^-]$ = Alk is expressed in mg·L$^{-1}$ of CaO and in mg·L$^{-1}$ $CaCO_3$.

The graphic method essentially comprises two axes of coordinates, which place the representative point of water depending on:

– the pH value plotted on a linear ordinate scale;

– alkalinity reported on a logarithmic scale on the abscissa.

Free $CO_2$ (mg·L$^{-1}$) is directly read in ordinate on oblique axes from a logarithmic scale. Depending on the temperature and alkalinity/CaO ratio, the equilibrium line is drawn between the dissolution of free $CO_2$ and the $CaCO_3$ saturation equilibrium line.

The diagram is thus divided into two areas (Figure 20.6):

– A zone, located below the saturation line, which represents water whose carbon dioxide content is higher than the saturation equilibrium. This designates aggressive water. In contact with calcium carbonate, this product will dissolve following the reaction:

$$CaCO_3 + CO_2 + H_2O \rightarrow Ca(HCO_3)_2$$

– A zone above the saturation line. The water located in this zone has a $CO_2$ content below the saturation equilibrium content. Bicarbonates are in unstable solution, ready to precipitate in the form of calcium carbonate by releasing $CO_2$. This designates encrusting water.

The water whose representative points lie on the saturation line is at carbonic equilibrium.

Lime (or soda) and calcium carbonate (or sodium carbonate) neutralization curves are deduced by translating the curves indicated on the diagram.

When water's figurative point has reached N1 following lime neutralization, there is precipitation of $CaCO_3$. The figurative point rises from N1 along the saturation line until the point corresponding to the limit solubility of $CaCO_3$, which is in the range of 15 mg·L$^{-1}$ at 15°C.

When the figurative point is N2 on the $CaCO_3$ saturation line following limestone neutralization, it becomes stable (even if there still remains a little dissolved $CO_2$). At saturation equilibrium, water's figurative point is on the saturation line.

Water is in equilibrium when its figurative point is on the line of equilibrium of free $CO_2$ and on the $CaCO_3$ saturation line, because it is only on this line that water is devoid of any aggressivity towards limestone. The two equilibria are achieved simultaneously only at the intersection point between the two equilibrium lines.

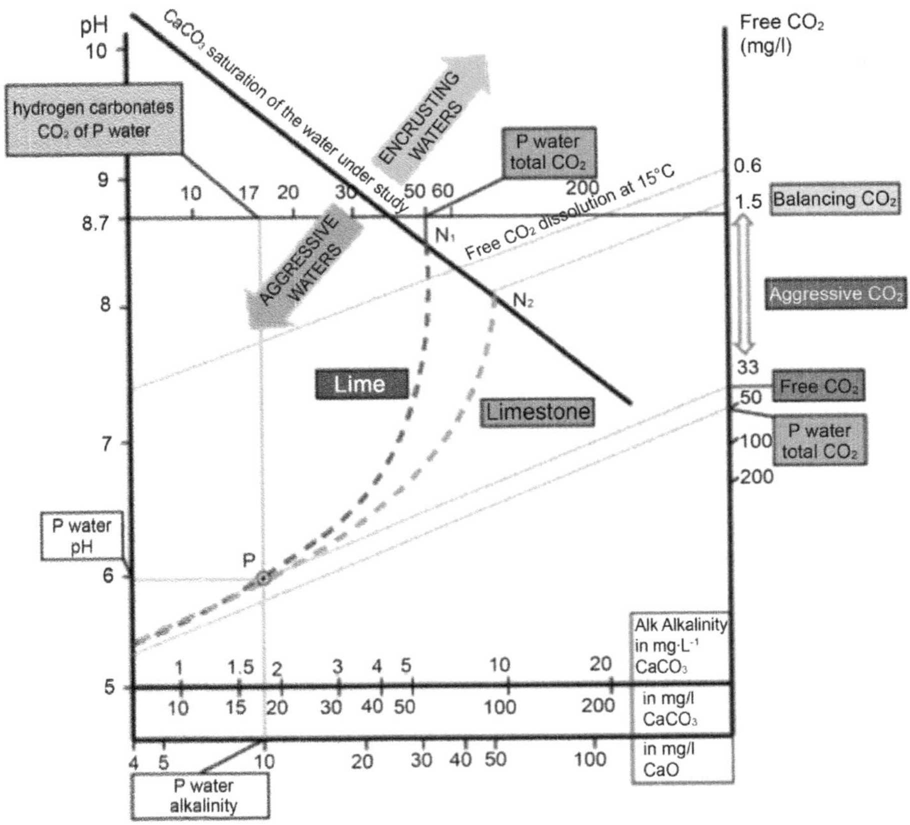


**Figure 20.6.** *Hallopeau–Dubin graph. For a color version of this figure, see www.iste.co.uk/gaid/watertreatment5.zip*

## 20.4. Indicative criteria to determine the aggressivity or corrosivity of water

The criteria the most often used to indicate the aggressivity or corrosivity of water are the following:

– the LSI, related to aggressivity;

– the SI, related to aggressivity;

– the Ryznar index (RSI), related to corrosivity;

– the Larson index (LRS), related to corrosivity;

– the Leroy index (LyR), related to corrosivity.

### 20.4.1. *Indicators of aggressivity: concrete pipelines*

Soft waters (hardness < 60 $mg \cdot L^{-1}$ $CaCO_3$, pH # 7) have a tendency toward acidity. The action of these waters is dissolving and hydrolyzing. When such water circulates in concrete pipelines, these gradually lose their most soluble constituents: lime, in particular. This leads to phenomena such as the hydrolysis of silicates and aluminates. The loss of substances opens concrete's pores and paves the way for deeper attacks.

#### 20.4.1.1. *The Langelier index*

This index is defined in the following manner:

$$LSI = pH - pHs$$

It helps to characterize the behavior of water in relation to limestone. It offers a qualitative assessment of the aggressive (LSI < 0) or encrusting (LSI > 0) nature of water.

This index is not a quantitative marker intended to provide a numerical measure of the amount of calcium carbonate that will be precipitated or dissolved. It simply denotes the tendency to precipitate or to dissolve calcium carbonate. If the water is extremely soft (deficient in calcium ions) and with low alkalinity, the protective capacity of water is minimal, even if a high pH and a positive LSI value are constantly maintained. Water must contain at least 80 $mg \cdot L^{-1}$ alkalinity and at least 80 $mg \cdot L^{-1}$ of calcium hardness in order to benefit from the protection of calcium carbonate.

However, maintaining a positive LSI does not exclude the possibility of corrosion. The conditions may be such that only a partial coating of the calcium carbonate is achieved, resulting in a type of corrosion in the uncoated areas known as "pitting". Pitting corrosion results in the loss of metal from the pipeline over relatively small areas. This pipeline can degrade fairly quickly by spreading the corrosive effect to the wall.

It is important to remember that the LSI value is not a quantitative measure of calcium carbonate saturation or corrosion. Because the protective scale formation is dependent on pH, bicarbonate ion, calcium carbonate, dissolved solids, and temperature, each may affect the water's corrosive tendencies independently. Soft, low-alkalinity waters with either low or excessively high pH are corrosive, even though this may not be predicted by the LSI. This is because insufficient amounts of calcium carbonate and alkalinity are available to form a protective scale. Waters with high pH values and sufficient hardness and alkalinity may also be corrosive, even if the LI predicts the opposite. This is the result of calcium and magnesium complexes that cannot actively participate in the scale-forming process. Analytical procedures do not distinguish between these complexes and available calcium and magnesium; therefore, the LSI value is not accurate in such situations. Corrosive tendencies may also be exhibited by water containing high concentrations of sulfate, chloride, and other ions which interfere with uniform carbonate film formation.

#### 20.4.1.2. *The SI related to IAP*

It is possible to know the level of the saturation of a water in opposite with a dissolved salt.

The following equation can be translated in term of SI which can be expressed by:

$$SI = \frac{IAP}{Ks} = \frac{(\mathrm{Ca}^{2+})(\mathrm{CO3}^{2-})}{Ks}$$

where IAP is the ionic activation product, and *Ks* is the solubility product.

The comparison between *Ks* and IAP makes it possible to test the saturation state of the solution vis-à-vis a salt. This comparison can be translated into terms of SI, which can be expressed as follows:

– water is said to be in equilibrium with a mineral when SI = 0;

– water is said to be undersaturated when SI < 0;

– water is said to be supersaturated when SI > 0.

$$HCO_3^- \longrightarrow H^+ + CO_3^{2-}$$

The associated equilibrium constant is:

$$K2\ (T,P) = \frac{\Upsilon H^{+}\ \Upsilon CO3^{2-}}{\Upsilon HCO3}$$

with $\Upsilon H^{+}$, $\Upsilon CO_3^{2-}$ and $\Upsilon HCO_3^{-}$ as the chemical activities of the corresponding species.

$$K'2 = \frac{[H3O^{+}][CO3^{2-}]}{[HCO3^{-}]} = K2\ \frac{\Upsilon HCO3^{-}}{\Upsilon CO3^{2-}\Upsilon H3O^{+}} = 4.67\ \times\ 10^{-11}$$

pK'2: $10^{-10.33}$

$Ca^{2+} + CO_3^{2-} \Leftrightarrow CaCO_3$

$Ks = (Ca^{2+})\ (CO_3^{2-}) = \Upsilon Ca^{2+}\ [Ca^{2+}\ ]\ \Upsilon CO3^{2-}[CO3^{2-}]$

$$K's = [Ca^{2+}][CO3^{2-}] = \frac{Ks}{\Upsilon Ca^{2+}.\ \Upsilon CO3^{2-}} = 3.32\ \times\ 10^{-9}\ (\text{at } 25°C)$$

*Ks* is the solubility product of $CaCO_3$.

By using log (SI) which includes the activities of ions, concentrations and *Ks*, it is possible to recover the LSI:

$$\text{LSI} = \log \text{SI} = \log \frac{\text{IAP}}{Ks}$$

For pH: 7.15; the activity is given by:

$$\frac{\Upsilon CO3^{2-}\Upsilon H^{+}}{\Upsilon HCO3\ -} = 10^{-10.33}$$

$$\Upsilon CO3^{2-} = 10^{-10.33}\,\frac{\Upsilon HCO3-}{\Upsilon H+} = \frac{10^{-10.33}\ \times\ 10^{-2.43}}{10^{-7.15}} = 10^{-5.58}$$

The ionic activity product is:

$$\text{IAP} = (\Upsilon CO3^{2-}\ )(\Upsilon Ca^{2+})$$

$$\text{IAP} = 10^{-2.85}\ \times\ 10^{-5.58} = 10^{-8.43}$$

*Ks*, the equilibrium constant of $CaCO_3$ at 25°C, is $10^{-8.47}$.

By comparison between IAP and *Ks*, we can deduce:

$$\frac{\text{IAP}}{Ks} = \frac{10^{-8.43}}{10^{-8.47}} = 10^{0.04} = 1.09$$

The water is supersaturated opposite with $CaCO_3$.

We can also write in other words as follows:

– IAP < *Ks:* water is said to be undersaturated;

– IAP = *Ks:* water is said to be in equilibrium;

– IAP > *Ks:* water is said to be supersaturated.

### 20.4.2. *Corrosivity indicators*

The corrosion of a metallic pipeline results in metal deterioration, which is caused by its chemical or electrochemical reaction with the water flowing through it.

The behavior of the various metals metallic pipelines (made of iron, cast iron, steel, copper and lead) essentially results from the reactions arising during their interactions with a certain number of physical, chemical and biological parameters.

Corrosion and the potential for precipitation are closely related. They must be taken into account in the design and operation of treatment facilities.

Reducing the potential for corrosion relies on adjusting the chemical quality of water as a practical means for controlling corrosion. The controllable factors are mainly calcium content, alkalinity and pH.

Oxygen and carbon dioxide in water have a significant effect on corrosion rates. Carbon dioxide lowers the pH and makes the water more aggressive. Carbon dioxide can be physically and chemically removed, but it is not possible to remove oxygen from the water produced. High concentrations of chloride and sulfate ions lead to high levels of corrosivity. However, their effects are slightly attenuated by alkalinity (carbonate, bicarbonate) and calcium ions. The appreciable effect of alkalinity is based on a concentration, which must be at least between 50 and 100 $mg \cdot L^{-1}$ $CaCO_3$. The higher the alkalinity and calcium concentrations, the greater the ability of water to retard corrosion. On the other hand, an excess of calcium and alkalinity leads to scaling water and it is necessary to find a compromise between corrosion and the formation of limescale.

| Pipeline material | Corrosive agent | Effects on the pipeline | Consequences on water quality |
|---|---|---|---|
| **Iron** | Dissolved oxygen, pH, dissolved salts, resistivity, low TH<br>Fe-oxidizing bacteria | General surface attack or pitting or cratering<br>Significant obstructions or pitting | Presence of iron |
| **Galvanized steel** | pH, dissolved oxygen, carbon dioxide, resistivity | Corrosion of the galvanized layer<br>Accelerated corrosion at high temperature | Presence of zinc and iron<br>Also cadmium and lead (impurities from the galvanizing process) |
| **Copper** | Soft and acidic water<br>High circulation velocity | General copper corrosion | Blue-green coloring<br>Copper and iron possibilities, zinc, arsenic, cadmium and lead associated with welding |
| **Lead** | Soft water and pH < 8<br>Hard water with high alkalinity | Generalized corrosion | Copper and iron possibilities, zinc, arsenic, cadmium and lead associated with welding<br>Lead |

**Table 20.4.** *Consequences on water quality after pipeline corrosion*

### 20.4.2.1. *The Larson index (LRS)*

This index is defined in the following manner:

$$LRS = \frac{0.028\left[Cl^-\right] + 0.0208\left[SO_4^{2-}\right]}{0.0164\,|HCO_3\text{-}]}$$

Concentrations are expressed in $mg \cdot L^{-1}$.

It is quantitatively representative of the corrosive nature of water mainly due to chlorides and/or sulfates. Table 20.5 shows the interpretation of the values associated with this index.

| LRS | Corrosion tendencies |
|---|---|
| <0.2 | No metal tendency |
| 0.2–0.4 | Light metal tendency |
| 0.4–0.5 | Low metal tendency |
| –1.0 | Medium metal tendency |
| >1 | High metal tendency |

**Table 20.5.** *Water corrosivity tendencies based on Larson's index*

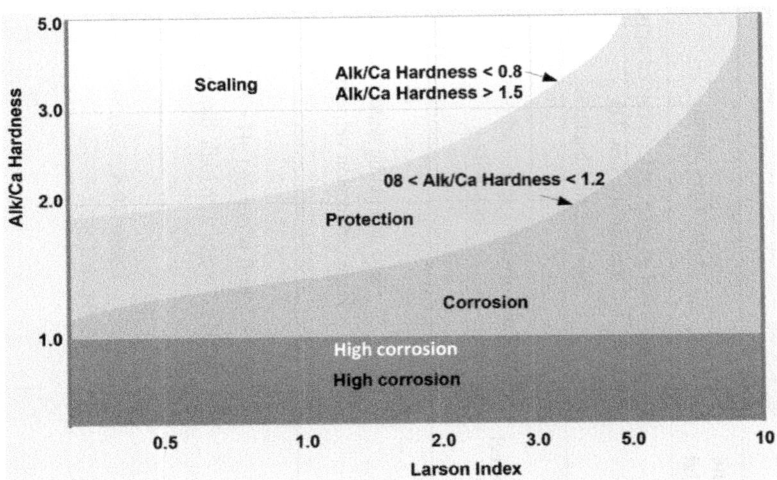


**Figure 20.7.** *Relationship between the Alk/TH ratio and the Larson index. For a color version of this figure, see www.iste.co.uk/gaid/watertreatment5.zip*

#### 20.4.2.2. *The Ryznar index (RSI)*

This index is defined in the following manner: RSI = 2pHs – pH.

It quantitatively describes the aggressive or scaling tendency of aerated water. Table 20.6 shows the interpretation of the values associated with this index. For water with a pH equal to 7, the water is scaling if the Ryznar index is lower than 7, and aggressive if it is higher than 7.

A scale has been determined from the Ryznar index, characterizing the various water tendencies (Table 20.6).

| Ryznar index | Water tendencies |
|---|---|
| 4–5 | Highly scaling water |
| 5–6 | Slightly scaling water |
| 6–7 | Water in equilibrium |
| 7–7.5 | Slightly corrosive water |
| 7.5–8.5 | Highly corrosive water |
| >8.5 | Rigorous corrosive tendency |

**Table 20.6.** *Ryznar indices and tendencies of water to be corrosive*

#### 20.4.2.3. *The Leroy index (LyR)*

This index is defined in the following manner:

$$\text{Leroy index} = \frac{[\text{Alk}]}{[\text{Total hardness}]}$$

The alkalinity and the total hardness are expressed in $mg \cdot L^{-1}$ $CaCO_3$.

A value of the Leroy index between 0.7 and 1.3 indicates a low tendency to corrosion.

#### 20.4.2.4. *The Puckorius scaling index (PSI)*

The Puckorius index quantifies the relationship between the saturation state and the formation of a protective layer ($CaCO_3$), including an estimate of the buffering capacity of the water. Water high in calcium, but low in alkalinity and buffering capacity, can have a high calcite saturation level. The high calcium level increases the ion activity product. The PSI index is calculated in a manner similar to the

Ryznar stability index, using an equilibrium pH rather than the actual system pH to account for the buffering effects:

PSI = 2 (pH)s – pHeq

pHs is the saturation pH.

pH eq = 1.465 × log10 [Alkalinity] + 4.54 with [Alkalinity] = $[HCO_3^-] + 2 [CO_3^{2-}] + [OH^-]$

These corrosivity indices result from theoretical calculations resulting from complex chemical reactions or obtained from experiments under well-defined operating conditions. These indices are only indicative tools in relation to the consequences of water's physicochemical parameters on metals. The choice of one index over another is adapted to the context of the treatment and the particular specificity of the drinking water station. It is chosen in accordance with the evolution of the parameters that contribute to corrosivity.

Nevertheless, they are often used with precaution in order to prevent metal degradation, though there are no real harmful consequences.

## 20.5. The calco-carbonic equilibrium of water

The study of the physicochemical characteristics of natural water as well as their action on the surrounding environment are often treated as special cases due to the complexity of the relationships between the ions at play. These equilibria are unstable due to the variations in temperature and dissolved gas content, which lead to new reactions in the medium in contact with the water.

Treatment, incrustation or corrosion problems are mainly caused by physicochemical equilibria variations and pose technical problems for the catchment, supply and the distribution of water.

For all these reasons, water must meet the following conditions:

– be in calcium saturation equilibrium, an essential condition for the spontaneous formation of a calcium carbonate deposit on the surfaces in contact, and for this deposit formed not to be attacked;

– have a suitable concentration of calcium ions so that the deposit in question is sufficient, without being excessive;

– not contain too high a proportion of $SO_4^{2-}$ or $Cl^-$ ions, which could render the deposit porous and ineffective;

– be at a the highest possible pH value so that corrosivity towards metals is minimal;

– contain 4–5 $mg \cdot L^{-1}$ of oxygen which condition the precipitation rate of insoluble deposits when metals are attacked.

To satisfy these conditions, water's figurative point on the carbonic graph must be located below the calcium saturation line, but close enough to the free $CO_2$ dissolution line. In this zone, Alk is around 60–100 $mg \cdot L^{-1}$ $CaCO_3$.

If scaling is not observed, the figurative point can be brought into the previously designated area above the saturation line, so that the SI is between 0 and 0.1.

Bringing water to equilibrium means providing it with stable characteristics by means of an appropriate treatment, throughout the time that elapses between its storage after treatment and its use by the end consumer.

Water equilibrium depends on two factors:

– its own characteristics;

– the nature of the materials likely to be in contact with it.

The search for this equilibrium requires the implementation of two means of action:

– making a judicious choice of these materials, which must be the most inert as possible in relation to the action of water so that the equilibrium is not compromised;

– endowing the water with stable intrinsic characteristics, acknowledging the presence of such materials.

The disadvantages resulting from a lack of water equilibrium are due to:

– water aggressivity toward limestone, concrete and cement;

– water corrosivity toward metals;

– water's encrusting character.

In the first two cases, the structures and equipment concerned are damaged and even degraded. Water acquires turbidity and coloring. In the last case, the pipelines are narrowed – sometimes even obstructed – and the expected flows can no longer pass through.

It is very important for water to be in calco-carbonic equilibrium during distribution. Water which is not in calco-carbonic equilibrium attacks the pipeline materials (in the case of aggressive water) or causes the formation of limestone deposits (in the case of scaling water). This results in the degradation of structures and water quality. Aggressive water attacks the network pipelines and releases toxic metals into the network (50% of connections present on the network are made of lead). On the other hand, scaling water clogs the pipelines. This results in an increase in turbidity, and therefore in the risk of bacterial proliferation. The consequences are high cleaning costs and mechanical problems on the valves. Network degradation translates into high renewal costs and major operating disruptions. In all cases, it is essential to control the characteristics of the water and to subject it to an appropriate treatment to achieve the calco-carbonic equilibrium.

Typically, two distinct problems exist: correcting aggressive water and correcting encrusting water. The correction of encrusting water can be done either by direct treatment or by reducing the potential for scaling by water softening. The direct treatment comes down to an addition of acid.

### 20.5.1. *Water quality and regulations*

In numerous countries, water hardness is not subject to a quality reference value.

The recommendation of the authorities is to deliver non-aggressive water and to provide indicator parameters for pH, conductivity, sulfates and chlorides. In all cases, the produced water must not be aggressive, corrosive or interfere with disinfection.

Water can become loaded during the transport of toxic substances (lead, nickel, cadmium), or loaded with undesirable (copper, zinc) substances.

In numerous drinking water plants, the operator often follows the recommendation to apply the "three eights rule" for water to be in equilibrium, even slightly scaling (Alk = 80 $mg \cdot L^{-1}$ $CaCO_3$, Ca hardness = 80 $mg \cdot L^{-1}$ $CaCO_3$, pH $\geq$ pHs + 0.2 (pH ~ 8)).

### 20.5.2. *The correction of aggressivity*

Natural water is often fraught with corrosion, aggressivity or encrustation problems. The resulting corrosion of the pipelines in the water distribution networks is a major concern to those responsible for the operation and maintenance of these networks.

Since aggressivity and corrosion problems and the specifically applicable treatments are in close relation to chemical composition, a distinction is drawn between the corrosive action of water, which is its power to attack pipeline metals, and the aggressivity of water, which is its power to attack limestone.

Indeed, corrosivity and aggressivity are two totally different notions, and not necessarily related. Thus, water can be aggressive but non-corrosive, corrosive but non-aggressive, or both aggressive and corrosive.

The aggressive power of water is explained by the existence of a double electric layer on the solid wall. The aggressivity of water leads to the dissociation of calcium carbonate into $Ca^{2+}$ ions and $CO_3^{2-}$ ions. As $Ca^{2+}$ ions are quickly carried away with water, the solid wall presents an excess of negative $CO_3^{2-}$ ions. As a result, the solid wall is composed of a negative inner layer comprising an excess of $CO_3^{2-}$ ions and a positive outer layer comprising an excess of cations attracted to the negative inner layer, such as $Ca^{2+}$ and $H_3O^+$ ions.

### 20.5.3. *Aggressivity correction treatments*

The normal process for treating aggressive water is to reduce its aggressive $CO_2$:

– by raising its pH by physically removing carbon dioxide. Due to the removal of $CO_2$ by aeration or stripping, the pH increases and approaches the equilibrium pH;

– by adding a suitable base (soda, lime, sodium carbonate). The neutralization of $CO_2$ aggressive is obtained, and the equilibrium pH is reached or approached;

– by filtration on alkaline-earth products (marble, neutralite) in order to obtain a slightly positive SI.

The choice of reagent depends on:

– the final Total Hardness for a given Alk value;

– the water flow to be treated and the equipment required for the incorporation of reagents;

– the treatment cost.

The reagents incorporation means and quantities to be injected can be calculated from the Hallopeau–Dubin diagram, focusing on the neutralization chemical reactions.

In all cases, the goal is to remove or neutralize the $CO_2$.

#### 20.5.3.1. *Aeration*

The physical removal of carbon dioxide is obtained by water aeration. $CO_2$ is entrained by air, provided that its partial pressure in the air is not limiting.

On the Hallopeau–Dubin diagram, $CO_2$ degassing takes place in the "aeration" zone (Figure 20.8). Water aeration changes its figurative point as follows:

– Zone A1: aggressive water with an Alk above approximately 250 $mg \cdot L^{-1}$ $CaCO_3$: supersaturated $CO_2$ degasses and the figurative point moves in parallel to the pH axis until the calcium saturation line. The calcium saturation equilibrium is then reached, but Alk is higher than the desired maximum. If aeration continues, degassing also persists and is accompanied by the deposition of calcium carbonate (water softening). The figurative point then rises along the calcium saturation line within the desired zone.

– Zone A2: aggressive water whose Alk is lower than approximately 250 $mg \cdot L^{-1}$ $CaCO_3$: the figurative point moves as before. In principle, it should become stable either on the $CO_2$ dissolution line in the open air, or on the calcium saturation line at their intersection point. In practice, aeration always leaves a residual of about 4-6 $mg \cdot L^{-1}$ of free $CO_2$. Below 150 $mg \cdot L^{-1}$ $CaCO_3$, equilibrium can only be reached by pursuing the operation with a base or a carbonate when the water's figurative point is located within the zones relating to these reagents.

– Zone A3: supersaturated water whose Alk is greater than approximately 70 $mg \cdot L^{-1}$ $CaCO_3$ and whose pH is lower than around 8.3: degassing is accompanied by the deposit of calcium carbonate and the figurative point joins the calcium saturation line following a slightly ascending trajectory. When this line is reached, the figurative point goes up and attains the desired location.

In all cases, an increase in water pH is observed.

Consider the aeration example for water with the following composition: pH: 6, T°C: 15°C, Alk: 100 mg $CaCO_3 \cdot L^{-1}$, Ca Hardness: 100 mg $CaCO_3 \cdot L^{-1}$ corresponds to a free $CO_2$ concentration of 210 $mg \cdot L^{-1}$.

It can be calculated using the ratio:

$$CO_2 = 10\text{^}(\text{Log (alkalinity}/1.78) + 0.2 - pH + 6.523 - 12.3*(10^{-3}\ T) + 163.5*(10^{-3}\ T)^2 - 646*(10^{-3}\ T)^3)$$

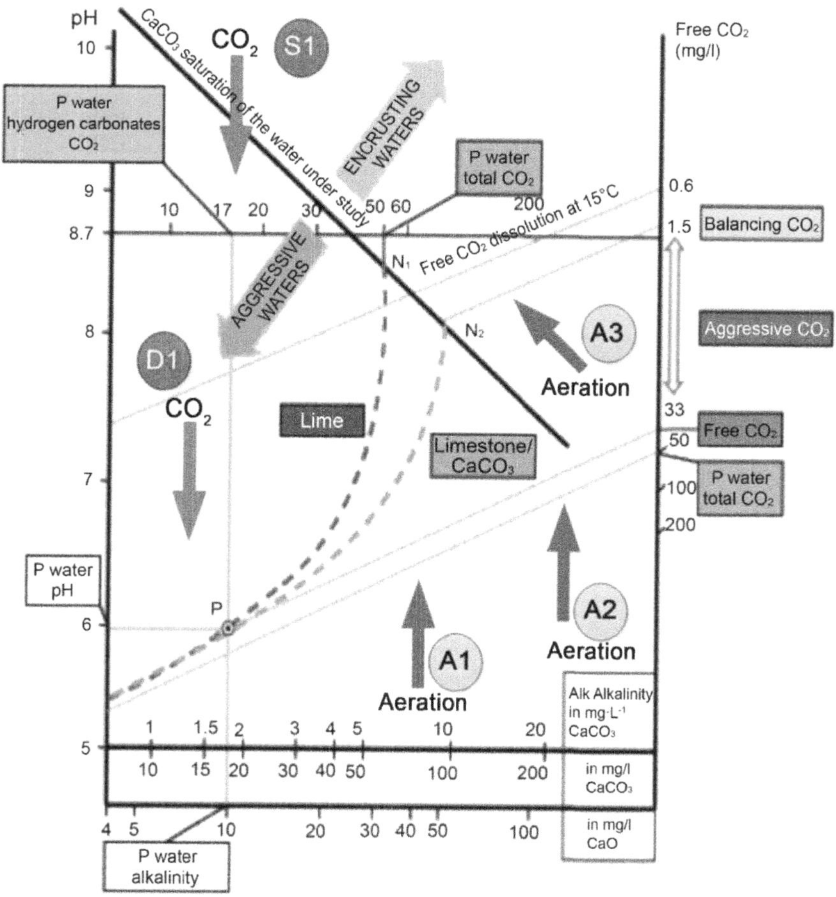


**Figure 20.8.** *Bringing water to equilibrium. For a color version of this figure, see www.iste.co.uk/gaid/watertreatment5.zip*

Alkalinity is expressed in mg $CaCO_3 \cdot L^{-1}$ and temperature in °C.

By placing the figurative point on the graph shown in Figure 20.9, and moving it vertically toward the $CaCO_3$ saturation curve, the final pH is equal to 8 and free $CO_2$ is about 2 $mg \cdot L^{-1}$. The Alk and Ca hardness concentrations remain the same.

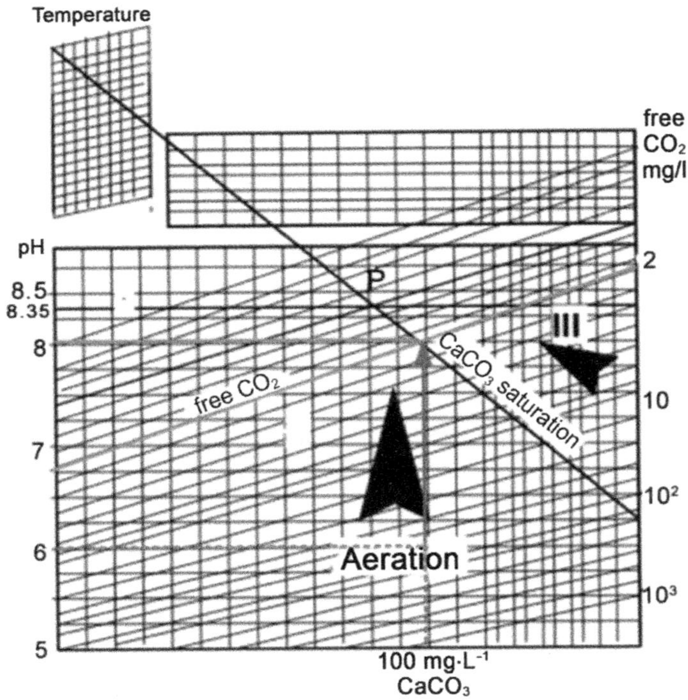


**Figure 20.9.** *$CO_2$ removal by aeration. For a color version of this figure, see www.iste.co.uk/gaid/watertreatment5.zip*

#### 20.5.3.2. *$CO_2$ degassing technologies*

Fresh water rich in $CO_2$ must be enriched with hydrogen carbonates in order to improve their properties in terms of aggressivity. The simplest method is to remove excess $CO_2$: the traditional methods "project" the water (using a pump), making it run off, or creating turbulent swirls. The solubility of gases decreases sharply when temperature increases and "degassing" is more efficient when temperature increases.

This technique is used when the aggressive carbon dioxide content is high, or when water mineralization should not increase.

##### 20.5.3.2.1. Aeration method: cascade

The design of the cascade is obtained using a ratio of the type:

$$(CCO_2)\text{outlet} = (CCO_2)\text{in}\ (1 - (1 - k)^n)$$

where:

– $(C_{CO2})$outlet and $(C_{CO2})$in: cascade inlet and outlet concentrations ($mg \cdot L^{-1}$);

– k: cascade performance coefficient;

– n: number of transfer steps.

Coefficient k is related to the cascade's total height and liquid–gas transfer in this process.

For an inlet concentration of 210 $mg \cdot L^{-1}$, a $CO_2$ outlet concentration lower than 7 $mg \cdot L^{-1}$ is obtained with a 4.8 m height, a number of transfer steps equal to 12 and a k value equal to 0.95 (see also Chapter 3).

#### 20.5.3.2.2. By injecting air into the liquid mass

*By water gushing*

This highly efficient process removes approximately 70% of $CO_2$. The devices used can be:

– nozzles fed with higher or lower pressure (0.5–1 bar);

– umbrella-like reversed cups onto which water is projected by a strained racket. Rackets are pierced with 8 mm diameter holes for a unit flow of 1–1.5 $m^3 \cdot h^{-1}$ and a head loss of approximately 2.50 m.

*Degassing tower method*

In a tower packed with contact materials (Levapac, Pall-type, Rashig rings, corrugated plates, etc.), the water to be treated runs down along the column against the air current. Degassing towers are cylindrical towers containing a packing that facilitates the transfer of oxygen into the water and accelerates the dissolved gases removal phenomena.

By injecting air through a porous surface at the base of a contact tower, one can theoretically achieve an almost complete removal of $CO_2$. This solution requires very high airflows.

The optimum design of the packing height and type of material favors the uniform distribution of the contact surface and the bed's porosity. The degassing tower includes two channel distributors in order to obtain a uniform distribution of the water to be treated. Depending on the total packing height, the towers are equipped with chimney trays installed between different bed heights in order to collect the water before redistributing it to the lower bed.

The design of degassing towers acknowledges various parameters (Henry's constant, mass transfer coefficient, exchange surface, liquid volume, gas concentration, etc.).

| Parameters | design |
|---|---|
| Water flow ($m^3 \cdot h^{-1}$); | 125 |
| Air flow ($Nm^3 \cdot h^{-1}$) | 2,500 |
| Air/water ratio | 20 |
| Air temperature (°C) | 2–20 |
| Water temperature (°C) | 6–20 |
| $CO_2$ inlet concentration ($mg \cdot L^{-1}$) | 85 |
| Pall rings (plastic) diameter (cm) | 3.81 |
| Packing height (m) | 2.00 |
| Tower diameter (m) | 1.60 |
| Packing volume ($m^3$) | 4.02 |
| Wetting rate ($m^2 \cdot s^{-1}$) | $130 \times 10^{-6}$ |
| Flow velocity in the tower ($m \cdot h^{-1}$) | 62.5 |
| $CO_2$ removal (%) | 94.4 |

**Table 20.7.** *Degassing tower design example*

Veolia has its own technique for designing and implementing degassing towers. By way of example, Table 20.7 presents the design of a carbon dioxide degassing tower installed on an industrial unit.

Veolia also has other technologies for removing $CO_2$ such as the Coplator, a reactor operating at high speed due to corrugated lacquers that accelerate the degassing mechanisms.

The conventional or forced aeration treatment results in a significant reduction of $CO_2$.

The safest method is the degassing tower. When there is residual $CO_2$, balancing is achieved by the addition of an alkaline reagent.

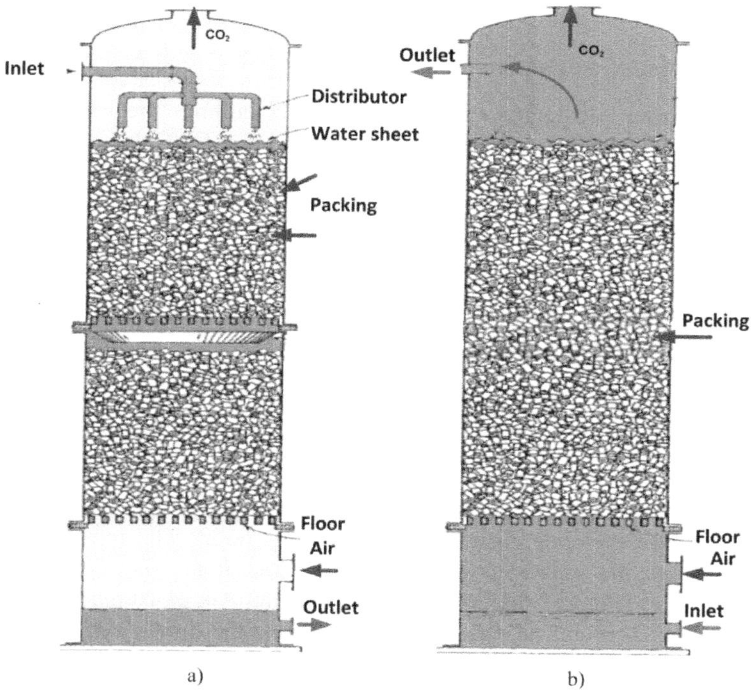


**Figure 20.10.** *Countercurrent (a) and co-current (b) degassing towers. For a color version of this figure, see www.iste.co.uk/gaid/watertreatment5.zip*

**Figure 20.11.** *Cascade (a) and Coplator (b). For a color version of this figure, see www.iste.co.uk/gaid/watertreatment5.zip*

### 20.5.3.3. *Chemical treatments for correcting water aggressivity*

#### 20.5.3.3.1. Dosage estimation using the Hallopeau-Dubin graph

The Hallopeau and Dubin carbon graph easily outlines the various balancing treatments, which can be implemented by adding a base ($Ca(OH)_2$, $Na_2CO_3$, $CaCO_3$, NaOH). As these treatments are carried out in tanks, the addition of $CaCO_3$ takes place within the filters instead.

On the graph, these treatments are represented by:

– *a cluster of "base" curves*: the points in this zone are brought to the desired location by the action of a base, in particular, lime or soda;

– *a cluster of "carbonate" curves*: the points in this zone which partially overlap the previous one are brought to the desired location by the action of a carbonate, in particular calcium or sodium carbonate.

From a figurative point representing the composition of the water to be treated, simply transpose the lime curve to this figurative point. The intersection of this curve gives the pH values at equilibrium and the final Alk. For example, for aggressive water (Figure 20.12) with a pH of 6.5, an Alk of 70 $mg \cdot L^{-1}$ $CaCO_3$ and 38 $mg \cdot L^{-1}$ of $CO_2$ (point A), the displacement along the lime curve produces water in equilibrium with a pH of 7.8, an Alk of 110 $mg \cdot L^{-1}$ $CaCO_3$ and a free $CO_2$ residual of about 3 $mg \cdot L^{-1}$ (point B).

Once the calcium saturation line has been reached, any additional reagent injection leads to water softening, in which case the figurative point rises along this straight line. It is thus possible to bring water's figurative point to the desired location below the indicated zones.

#### 20.5.3.3.2. Chemical reagents for implementing aggressivity correction

Technologies for correcting aggressivity by means of chemical reagents are relatively simple. They come down to mixing aggressive water with an alkaline reagent within a contact tank.

*Neutralization with lime milk, lime water or micronized lime*

The neutralization reaction is as follows:

$$2CO_2 + Ca(OH)_2 \rightarrow Ca(HCO_3)_2$$

In other words, 1 mg of $CO_2$ requires the injection of 0.84 mg of lime (100% pure) to produce 1 mg of alkalinity (as $CaCO_3$). The dosages are then adjusted depending on the reagents' purity.

Lime neutralization leads to the formation of calcium bicarbonate ($Ca(HCO_3)_2$). Neutralization with lime water leads to the same reactions, but the technology used is different (see section 20.6).

The lime injection point at the treatment station and the injection modality (lime milk, lime water, micronized lime) are directly related to the treatment. Whitewash is a concentrated product that involves a large increase in pH and demands a rigorous control of chemical reactions. It also leads to a high increase in turbidity and this solution is not recommended as a final remineralization stage. On the other hand, its application is appropriate for pre-remineralization or intermediate remineralization.

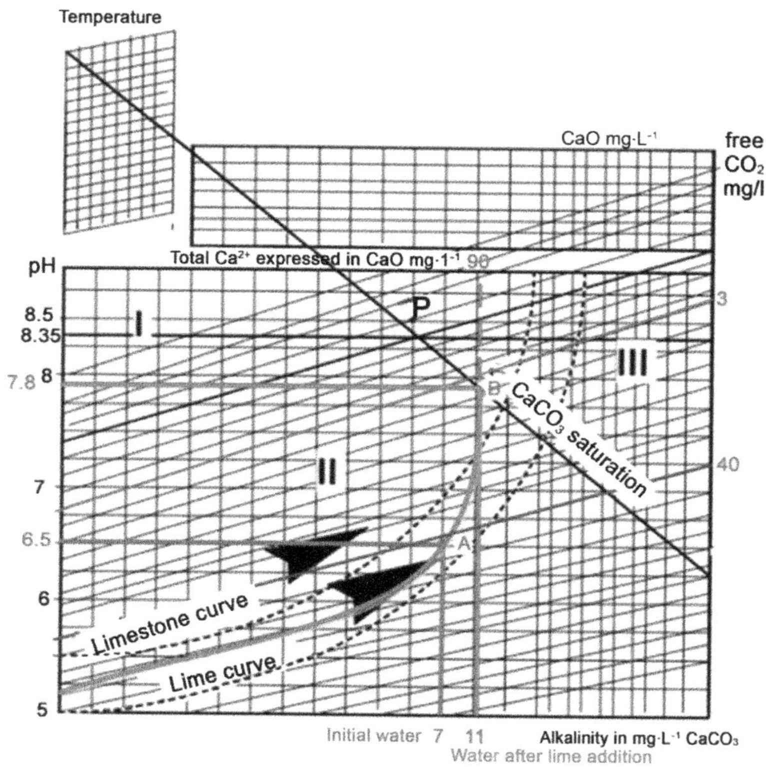


**Figure 20.12.** *$CO_2$ neutralization by lime addition. For a color version of this figure, see www.iste.co.uk/gaid/watertreatment5.zip*

Lime water is a less concentrated product and does not involve large pH variations. It contributes little to no turbidity. This solution is recommended as a final remineralization stage or inter-remineralization before sand filters.

Slaked lime is used in the form of lime milk or lime water. On average, commercial lime contains 70–80% CaO. The molar masses of quicklime and slaked lime are equal to 56 (CaO) and 74 ($Ca(OH)_2$), respectively. The theoretical ratio for passing from quicklime to slaked lime is 1.32.

Lime milk is prepared at a concentration of around 50–110 $g \cdot L^{-1}$, which is then injected into the contact tank. For dosing, volumetric or centrifugal pumps are used. The injection flow rate is calculated so as to provide the dose of $Ca(OH)_2$ exactly needed for aggressivity correction. The contact time varies between 3 and 5 min.

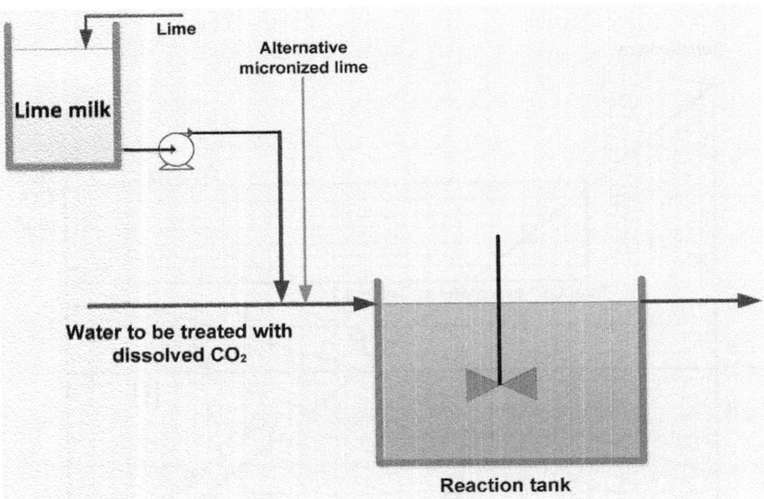


**Figure 20.13.** *Neutralization with lime milk or micronized lime. For a color version of this figure, see www.iste.co.uk/gaid/watertreatment5.zip*

| **CaO ($g \cdot L^{-1}$)** | **10** | **20** | **30** | **40** | **50** | **60** | **70** | **80** | **90** | **100** | **110** |
|---|---|---|---|---|---|---|---|---|---|---|---|
| **$Ca(OH)_2$ ($g \cdot L^{-1}$)** | 13.2 | 26.4 | 39.6 | 52.8 | 66 | 89.2 | 92.4 | 105.6 | 118.8 | 132 | 145.2 |
| **Density ($g \cdot cm^{-3}$)** | 1.008 | 1.017 | 1.025 | 1.032 | 1.039 | 1.043 | 1.054 | 1.061 | 1.068 | 1.075 | 1.083 |

**Table 20.8.** *Lime milk concentration at 15°C*

As an alternative to lime milk, micronized lime can be a solution. However, the latter is more expensive in terms of operating costs.

*Neutralization with caustic soda*

The neutralization reaction is as follows:

$$CO_2 + NaOH \rightarrow NaHCO_3$$

Caustic soda neutralization leads to the formation of sodium bicarbonate ($NaHCO_3$).

Therefore, 1 mg of $CO_2$ requires 0.91 mg of caustic soda (100% pure) to produce 1.13 mg of alkalinity (as $CaCO_3$). With caustic soda at 50% (soda lye), 1.82 mg of caustic soda must be added, whereas with 30% caustic soda, 2.73 mg of caustic soda must be added.

Caustic soda is used at 30% in the majority of cases (and sometimes at 50%). In winter, the premises have to be heated so as to prevent the reagent's crystallization.

The addition of caustic soda (strong base) neutralizes $CO_2$ (weak acid), entailing the production of soluble sodium bicarbonate. Caustic soda tends to significantly increase the pH, which is why the dosage has to be well controlled.

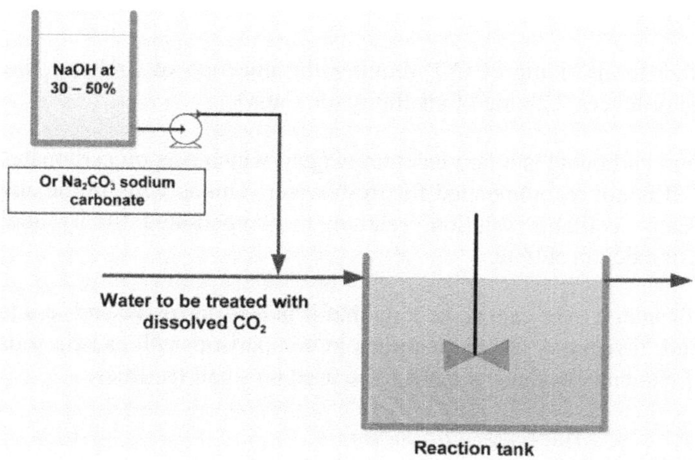


**Figure 20.14.** *Neutralization with caustic soda or sodium carbonate. For a color version of this figure, see www.iste.co.uk/gaid/watertreatment5.zip*

Caustic soda is a non-calcic reagent that only acts on carbonates. It is not indicated for very soft waters, which do not contain enough calcium for the formation of the protective layer in the networks' pipelines. This is the reason why, in the absence of any other locally available solution, calcium is compensated for by adding calcium chloride or calcium sulfate.

The use of caustic soda is recommended when the $Ca^{2+}$ concentration in water is greater than 30 mg·L$^{-1}$. In addition, the caustic soda injection leads to an increase in the sodium concentration in water. This is also the reason why this solution is only recommended to neutralize low concentrations of aggressive $CO_2$.

Caustic soda quickly raises the pH and can be injected directly. However, it is advisable to inject enough dilution water to the caustic soda in order to obtain a good dispersion in the water to be treated and to avoid calcifications at the injection points.

The commercial solution is usually 30%.

*With sodium carbonate and sodium bicarbonate*

The neutralization reaction is as follows:

$$CO_2 + Na_2CO_3 + H_2O \rightarrow 2NaHCO_3$$

Sodium carbonate neutralization leads to the formation of sodium bicarbonate ($NaHCO_3$).

In other terms, 1 mg of $CO_2$ requires the injection of 2.41 mg $Na_2CO_3$ (100% pure) and produces 2.28 mg of alkalinity (as $CaCO_3$).

Sodium carbonate is a non-calcium reagent which acts on carbonates, but not on calcium. It is not recommended for fresh water. This is why, in the absence of any other locally available solution, calcium is compensated for by adding calcium chloride or calcium sulfate.

Its advantage over caustic soda is that it is less corrosive and less toxic. On the other hand, it requires larger quantities in comparison with caustic soda. It is often prepared manually because it tends to be used on small facilities.

*Reminder of the necessary dosages*

| Neutralization reagents | Formulae | For neutralizing 1 mg $CO_2$ | Alkalinity produced (in mg $CaCO_3$) |
|---|---|---|---|
| **(Pure) Lime, mg·L$^{-1}$** | $Ca(OH)_2$ | 0.84 | 1 |
| **(Pure) Caustic soda, mg·L$^{-1}$**<br>**50%**<br>**30 %** | NaOH | 0.91<br>1.82<br>2.73 | 1.13 |
| **(Pure) Sodium carbonate, mg·L$^{-1}$** | $Na_2CO_3$ | 2.41 | 2.28 |

**Table 20.9.** *Dosages required to neutralize carbon dioxide*

### *With calcium sulfate and calcium chloride*

Calcium chloride ($CaCl_2$) is less soluble and easier to use. Solutions are prepared at a concentration of 200 g·L$^{-1}$. It takes 1.17–1.45 mg·L$^{-1}$ calcium chloride (depending on product purity) to raise hardness by 1 mg·L$^{-1}$.

Calcium sulfate ($CaSO_4$) is less soluble and easier to use. Solutions are prepared at a concentration of 20 g·L$^{-1}$. It takes 1.36 mg·L$^{-1}$ of (pure) calcium sulfate to raise hardness by 1 mg·L$^{-1}$. The water thus treated nonetheless becomes enriched with "aggressive" anions. It is also necessary to associate 1.68 mg of sodium bicarbonate. In each case, it should be checked that the water is in calco-carbonic equilibrium after the adjustment.

| Reagents | Chemical reactions | Consumption Reagents |
|---|---|---|
| **Sodium bicarbonate combined with calcium sulfate $NaHCO_3 + CaSO_4$** | $2NaHCO_3 + CaSO_4 \rightarrow CaCO_3 + CO_2 + H_2O + Na_2SO_4$ | 1.68 mg of $NaHCO_3$<br>1.36 mg $CaSO_4$ |
| **Sodium bicarbonate combined with calcium chloride $NaHCO_3 + CaCl_2$** | $2NaHCO_3 + CaCl_2 \rightarrow CaCO_3 + CO_2 + H_2O + 2NaCl$ | 1.68 mg of $NaHCO_3$<br>1.11–1.45 mg $CaCl_2$ |

**Table 20.10.** *Consumption of reagents with $CaCl_2$ and $CaSO_4$*

#### 20.5.3.3.3. Lime saturators

The correction of aggressivity with lime water requires the use of a lime saturator. The lime saturator is a device which prepares a lime-saturated solution, called lime water. This lime water is obtained by mixing water and lime milk. After settling, impurities (called unburnt lumps) are purged and lime-saturated water is extracted at the saturator's top.

Veolia has various types of saturators, characterized by differences related to geometry, operating speeds and the means used.

A distinction is made between:

– the classic saturator, with a cylindrical-conical shape for mirror velocities up to 2 m·h$^{-1}$;

– the conical lamella saturator with polymer use, for mirror velocities up to 6 m·h$^{-1}$;

– the Multiflo-type saturator for mirror velocities up to 8–10 m·h$^{-1}$.

The lime dosage in the form of a lime water saturated solution enables a rapid reagent reaction and prevents the introduction of impurities contained in the commercial product into the water to be treated. The lime water dosage is done by pH regulation or calculated by mass per unit area ($g \cdot m^{-3}$).

The saturator behaves like an updraft settler producing a clarified overflow (lime saturated at constant level), without any impurities. It also serves as a deposit containing various insoluble mineral substances such as calcium carbonate or calcium sulfate grains, or metal oxides (iron, magnesium) and silica. Some of these insolubles (e.g. $CaCO_3$ calcium carbonate) are generated by lime, due to the chemical reaction with the bicarbonates from the dilution water ($HCO_3^-$). The impurities are collected in a central cone and then evacuated toward the purging tank. This extraction of unburnt lumps is done discontinuously.

Supplying the saturator with lime milk is preferably continuous, but can be discontinuous for smaller facilities.

*The cylindrical-conical saturator*

It is equipped with the following elements:

– a lime milk supply at 50–100 $g \cdot L^{-1}$;

– an outlet for saturated lime water (submerged pierced crown, tulip or submerged collector);

– a carbonate collection cone with a periodic purge system;

– a sewage drain also enabling periodic purges of insoluble matter;

– a gushing water supply (for lime dilution);

– a make-up water supply (lime water flow adjustment).

Lime milk is stored in a constantly agitated tank, even when the station comes to a halt. Otherwise, it could become irreversibly clogged. The saturator's dimensions and the quantity of lime to be used in the mixing tank are estimated to ensure the desired operating autonomy, taking into account the $Ca(OH)_2$ content to be added. The saturator's lime milk feed can be either continuous or discontinuous, but preferably continuous.

Discontinuous feed saturators are calculated in such a way so as to respect a contact time of 40–60 min for an operating autonomy of 24 h.

**Figure 20.15.** *Cylindrical-conical lime saturator. For a color version of this figure, see www.iste.co.uk/gaid/watertreatment5.zip*

**Figure 20.16.** *Lime water quality with a cylindrical-conical saturator. For a color version of this figure, see www.iste.co.uk/gaid/watertreatment5.zip*

### *The lamella conical saturator*

This type of saturator is more compact than the cylindrical-conical saturator whose main improvements are the addition of lamellae and the possibility of injecting a polymer. The lime milk feed is continuous. It is necessary to add a polymer to operate at speeds over 3 $m \cdot h^{-1}$, bearing in mind that the maximum speed is 6 $m \cdot h^{-1}$. The lamellae are of the Greca type.

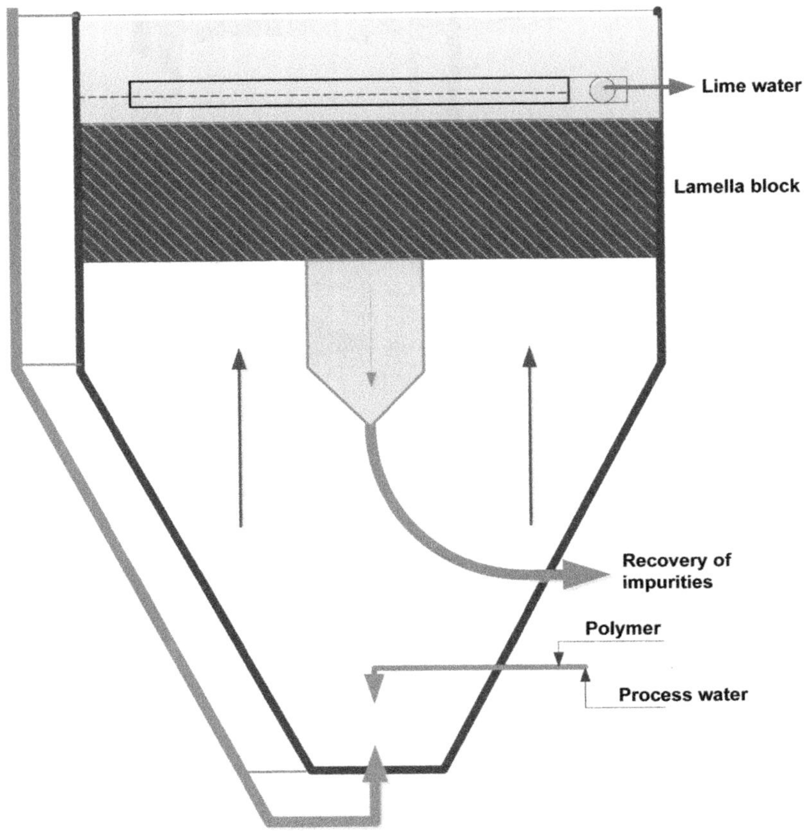


**Figure 20.17.** *Diagram of the lamella conical saturator. For a color version of this figure, see www.iste.co.uk/gaid/watertreatment5.zip*

### *The lamella saturator (Multiflo type)*

The lamella saturator (Multiflo type) is based on the two-stage decomposition of the functions of a lime saturator: to dilute and to dissolve lime milk as well as possible, something which requires strong agitation and settling the impurities in order to obtain lime water with low turbidity. Its design is similar to that of a Multiflo because in the first stage, in includes a rapid agitation in a contact tank, and then settling takes place in a lamella settler (Figure 20.18). It also features the recirculation of settled lime in order to maintain a high lime concentration. In addition, the concentration of impurities tends to accelerate their settling, promoting the production of water with low turbidity.

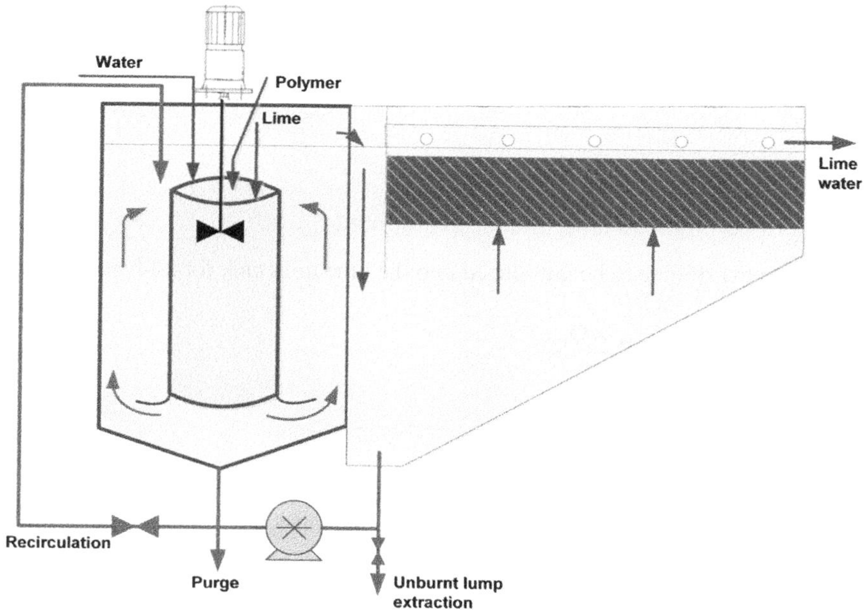


**Figure 20.18.** *Diagram of the lamella saturator (Multiflo type). For a color version of this figure, see www.iste.co.uk/gaid/watertreatment5.zip*

**Figure 20.19.** *The lamella saturator (Multiflo type). For a color version of this figure, see www.iste.co.uk/gaid/watertreatment5.zip*

Recirculation is done by extraction of the carbonates present at the bottom of the tremie, which is then pumped towards the rapid agitation tank, outside the draft tube and on the side opposite to the settler's feed, in order to prevent a settler short-circuit.

### *Design parameters of lime saturators*

#### Quantity of commercial lime to be implemented

The quantity of lime to be introduced into the lime milk tank for 24-h autonomy is:

$$q_{max} = \frac{24 \times T_{max} \times Q_{max}}{\eta \times 1{,}000}$$

where:

– qmax (kg): quantity of commercial lime to be implemented;

– Qmax ($m^3 \cdot h^{-1}$): maximum hourly flow of the facility;

– Tmax ($g^{-3}$): maximum $Ca(OH)_2$ treatment rate;

– η: lime water saturation efficiency which takes into account the purity of commercial lime and the efficiency of the device.

The following range of values is considered: 0.8 (lime with a purity of 80–90% $Ca(OH)_2$) to 0.9 (purity > 90% $Ca(OH)_2$).

#### Lime milk flow

The lime milk flow is given by:

$$Qmilk = \frac{Tmax \times Qmax}{\eta \times C\,milk}$$

where:

– $C_{milk}$ ($g \cdot L^{-1}$): $Ca(OH)_2$ concentration in lime milk;

– $Q_{milk}$ ($L \cdot h^{-1}$): flow rate of the lime milk to be injected.

The saturator's useful surface is:

$$S = \frac{Q}{V}$$

where V is the mirror velocity ($m \cdot h^{-1}$).

### Process water flow rate

The process water flow is between 50 and 100% of the saturator's nominal flow rate, with a gushing pressure on the orifice of around 0.7 bar. The gushing orifice is calculated according to the formula:

$$Q = KS\sqrt{2\,\mathrm{g\,h}}$$

where:

- Q: process water flow ($m^3$–$s^{-1}$);
- K: contraction coefficient (0.62);
- S: gush section ($m^2$);
- g: acceleration due to gravity (9.81 $ms^{-2}$);
- h: gush pressure (m).

| Parameters | Saturator Cylindrical-conical | Rapid saturator | Lamella saturator (Multiflo® type) |
|---|---|---|---|
| Flow ($m^3 \cdot h^{-1}$) | 40 | 15 without polymer<br>30 with polymer | 20–500 |
| Contact time (min) | 40–60 | 60 without polymer<br>30 with polymer | 25 |
| Preparation water | Turbidity < 1 NTU | Turbidity < 1 NTU | Turbidity < 1 NTU |
| Mirror velocity ($m \cdot h^{-1}$) | 2 | 3 without polymer<br>6 with polymer | 4 without polymer<br>8 with polymer |
| Lime milk concentration ($g \cdot L^{-1}$) | 50–110 | 50-110 | 50–110 |
| Operation mode | Continuous-discontinuous | Continuous-discontinuous | Continuous-discontinuous |
| Recirculation | – | – | 10% of flow |
| Lamellae | – | Yes | Yes |
| Polymer ($mg \cdot L^{-1}$) | – | 0.10–0.25 | 0.25 |

**Table 20.11.** *Operating characteristics of the three types of Veolia saturators*

### Purges

The purging flow rate will represent 3–5% of the saturator's nominal flow rate. It must guarantee the total removal of carbonates and insolubles. Carbonate purges are done discontinuously so that removals are more efficient. The diameter of the purge conduits is calculated to grant a minimum flow velocity of 2 $m \cdot s^{-1}$ in order to avoid deposits.

### Operating characteristics of Veolia saturators

The operating characteristics of Veolia saturators are summarized in Table 20.11.

### Purge flow adjustment

The calculation is based on a water loss of 3–5%:

$$Q_{purges} = 3 - 5\%\ Q_{feed}$$

### Saturation efficiency

The main difficulty for the incorporation of lime water is due to lime solubility (Figure 20.20), which is inversely proportional to temperature.

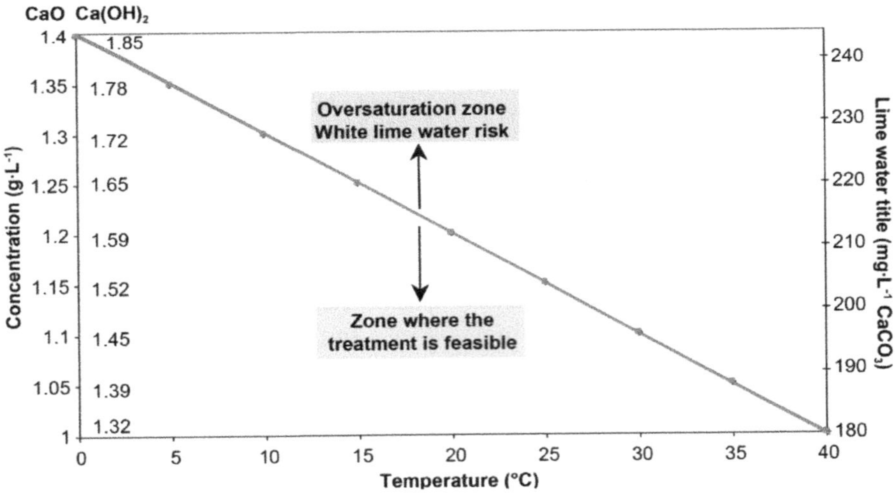


**Figure 20.20.** *Lime solubility as a function of temperature*

This low solubility makes it necessary to use a saturator producing lime water of a good quality (low turbidity, maximum concentration of solubilized lime). The title of the lime water produced must be close to the saturation line as a function of temperature. The quality of lime water is monitored by measuring Alk, pH and turbidity.

The differences in solubility between commercial lime and pure lime do not exceed 7%. These are due to the presence of sodium hydroxide and potassium hydroxide traces in commercial lime products. The presence of magnesia, silica and carbonates has no effect on the solubility of commercial lime products, although it affects their dissolution rate.

On the other hand, the dimensions of lime particles strongly affect solubility. Freshly slaked lime composed of fine particles is at least 10% more soluble than lime with large particles. In practice, the solubility of lime is between 1.2 and 1.33 $g \cdot L^{-1}$ of CaO or 1.59 and 1.76 $g \cdot L^{-1}$ of $Ca(OH)_2$.

**Figure 20.21.** *Lime water at the outlet of a cylindrical-conical saturator (without polymer addition). For a color version of this figure, see www.iste.co.uk/gaid/watertreatment5.zip*

To obtain lime water with a title at 1.6 $g \cdot L^{-1}$ of $Ca(OH)_2$, it is recommended to dose 1.8–2 $g \cdot L^{-1}$ of $Ca(OH)_2$, that is to say, a saturation efficiency of 80–90%. This efficiency can reach 90–95% when the dilution water used is weakly mineralized.

The suspension, whose concentration must not exceed 110 $g \cdot L^{-1}$ (to prevent the formation of deposits in the lime milk transfer conduit), is maintained homogeneous by a continuous impeller, including during the station's stop periods.

In summary, the appropriate treatment for aggressivity correction depends on the $CO_2$ concentration.

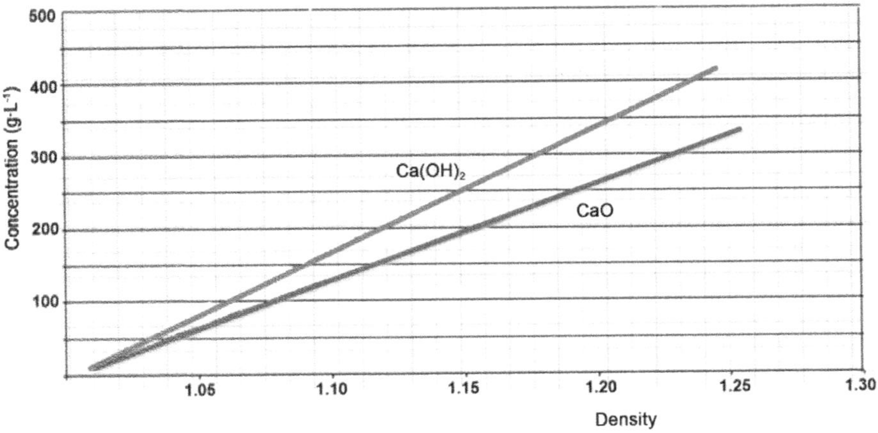


**Figure 20.22.** *Lime milk density depending on its concentration*

| | High $CO_2$ | Low $CO_2$ |
|---|---|---|
| **Main treatment** | Degassing tower | Direct neutralization with alkaline reagent |
| **Refining treatment** | Refining with alkaline reagent | – |
| **Complementary treatment** | Addition of $CaCl_2$ or $CaSO_4$ if caustic soda, or carbonate or bicarbonate are used | Addition of $CaCl_2$ or $CaSO_4$ if caustic soda, or carbonate or bicarbonate are used |

**Table 20.12.** *Neutralization treatment depending on the $CO_2$ concentration*

## 20.6. Remineralization treatments

The goal of remineralization is to endow water with stable chemical characteristics between the time it is manufactured, stored and then distributed and consumed. This stability is ensured by sufficient concentrations of alkalinity and calcium to maintain water's calco-carbonic equilibrium. This is needed because the above-mentioned elementary neutralization treatments cannot confer the desired optimal characteristics to all waters. This is the case of fresh water containing little or no carbon dioxide, as well as supersaturated water with a pH greater than 8.3.

To obtain the desired result, it is necessary to apply a remineralization treatment to these waters. Such treatment can give them a higher mineralization to the one obtained by an elementary carbonate neutralization treatment (or not). Unlike the correction of aggressivity, remineralization involves a supplementary injection of $CO_2$ apart from the one already present in water.

The most common solutions are:

– the combination of $CO_2$ and lime (or lime water) with a static mixer or mixing tank;

– the combination of $CO_2$ or acid with a limestone filter.

Other solutions such as the addition of calcium chloride ($CaCl_2$) and sodium bicarbonate ($Na_2CO_3$) are used in regions where carbon dioxide is not available. This is often the case in islands.

### 20.6.1. *Graphic method*

#### 20.6.1.1. *Remineralization by $CO_2$ injection, followed by neutralization with neutralizing reagents*

During the $CO_2$ injection, the figurative point moves downwards, in parallel to the pH axis.

The $CO_2$ injection stops when the point is inside the "carbon or base" zone.

Neutralization by one of these reagents then leads to the desired result. It is possible to act separately on Alk or Ca hardness by means of a judicious choice of one or more reagents.

#### 20.6.1.2. *Remineralization by the injection of sodium bicarbonate and a strong calcium salt*

This treatment concerns low content freshwater, but supersaturated with free $CO_2$.

The injection of sodium bicarbonate moves the figurative point along a free $CO_2$ line with a constant $CO_2$ concentration and orients it toward the "carbonate or base" zone where it is possible to pursue neutralization with one of these reagents. It is also possible for the sole action of bicarbonate to help reach the desired area.

The treatment must be completed or preceded by the addition of strong calcium salts such as, for example, calcium chloride or calcium sulfate. Their purpose is to contribute hardness to the desired value.

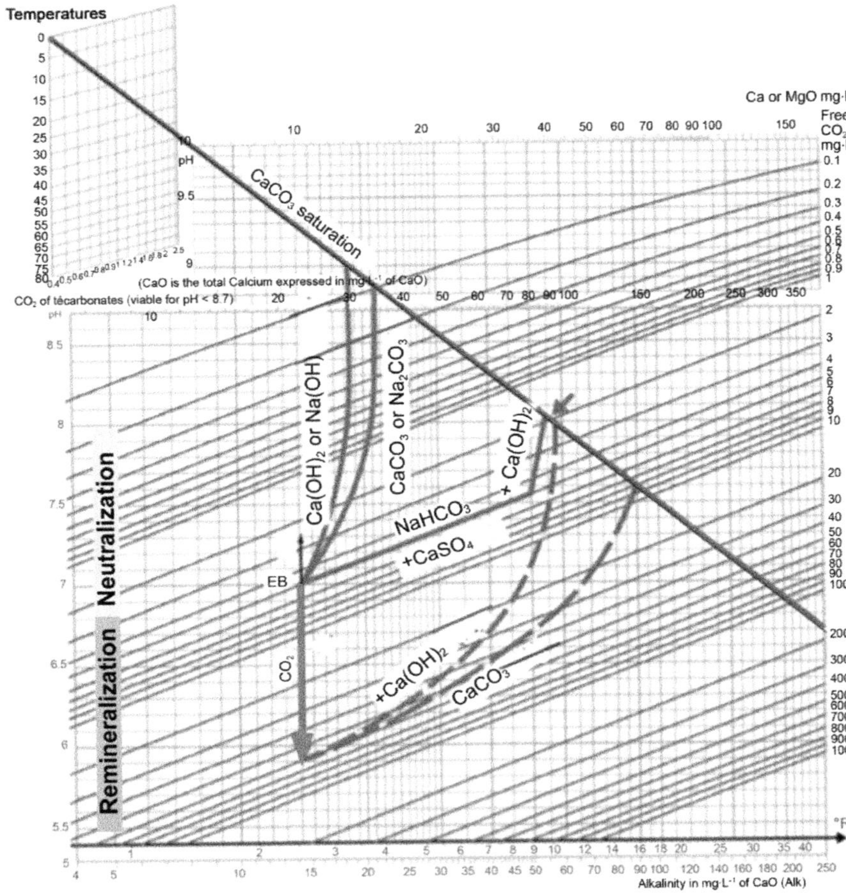


**Figure 20.23.** *Graphic method illustrating remineralization with $CO_2$ injection and neutralization with a base*

## *20.6.2. Processes for implementing remineralization: chemical reactions in tanks*

### 20.6.2.1. *Carbon dioxide/lime solution*

The most commonly used solution is the $CO_2$ injection, followed by an addition of lime or lime water.

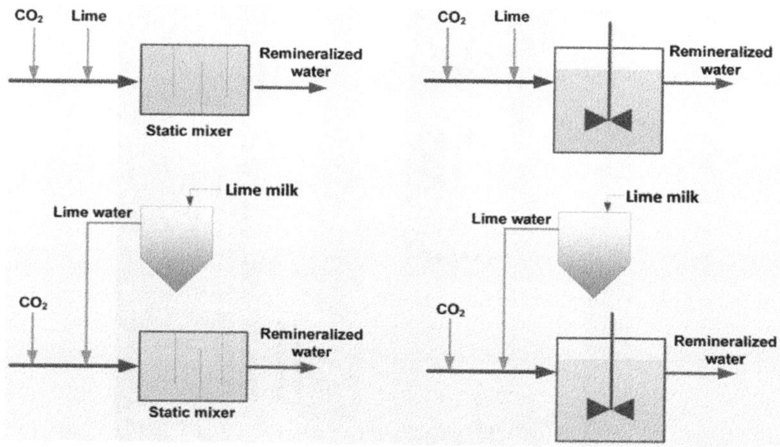


**Figure 20.24.** *Solutions with $CO_2$ and lime injection*

Remineralization with $CO_2$-lime can be performed in one or more stages, depending on the quality of the raw water and the treatment process. There are several possibilities (Figure 20.24):

– Partial remineralization at the beginning of the treatment, by dosing carbon dioxide and lime milk. The quantity of lime injected at this stage is deliberately limited to enable flocculation at pH 6.0, and optimize the removal of organic matter.

– A dosage of lime milk or lime water before the sand filters. The quantity of lime injected at this stage of the treatment is estimated to keep the water slightly aggressive on the sand filters and remove any risk of carbonate deposits on the sand grains.

– A final dosage of lime water downstream of filtration to achieve the desired pH and Alk.

The static mixer requires a constant flow and a rigorous control of dosages and pH. Due to the short contact time, this solution is not widely adopted for cold water if the desired alkalinity and hardness values are greater than 80–100 $mg \cdot L^{-1}$.

Lime water of a good quality is recommended at the end of the process in order to avoid injecting lime impurities and prevent the increase in turbidity in the water produced to values greater than 1 NTU. Note that 0.74 $mg \cdot L^{-1}$ of lime is required to raise alkalinity and calcium hardness by 1 $mg \cdot L^{-1}$ of $CaCO_3$.

**Figure 20.25.** *$CO_2$ storage and lime silo. For a color version of this figure, see www.iste.co.uk/gaid/watertreatment5.zip*

The dose of $CO_2$ applied is estimated by the operator depending on the treated water desired Alk. The quantity of lime is proportional to the quantity of $CO_2$ applied as a function of the desired optimal pH.

The lime dosage is automatically calculated, relying on a regulation chain that takes into account the flow to be treated and the pH adjustment.

The most common method uses the $CO_2$ stored in the liquid state in heat-insulated and refrigerated pressure tanks or, for lower flow rates, in bottles. The extracted liquid $CO_2$ is vaporized and then expanded to fit the conditions of use. It is finally injected in a gaseous state into the water to be treated via porous diffusers submerged under a few meters of water (large units) in contact tanks or is diffused by means of a hydro-ejector (the operating principle being identical to the chlorine injection).

In most cases, $CO_2$ is stored in canisters in liquid form at low pressure and injected by means of a hydro-ejector. The $CO_2$ storage capacity is calculated on the basis of the quantity of $CO_2$ to be injected for water remineralization, the desired storage autonomy, the supply period and the access conditions to the site.

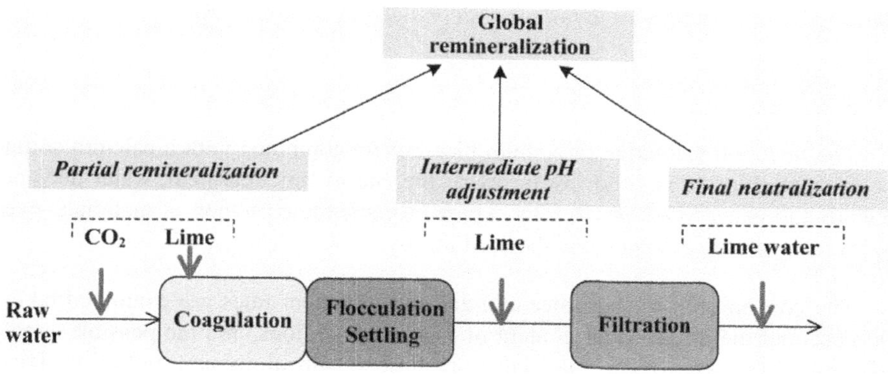


**Figure 20.26.** *Multiple stage remineralization. For a color version of this figure, see www.iste.co.uk/gaid/watertreatment5.zip*

A $CO_2$ station mainly includes:

– a refrigeration unit (storage of liquid $CO_2$ at –20°C and 20 bar);

– a tank containing the gas;

– a carbo-vapor granting the vaporization of the liquid $CO_2$ at the reservoir's outlet, by supplying it with the calories required for the change of state and gas superheating from –20 to +60°C;

– $CO_2$ dosing kit with control of the flow to be treated and including the various accessories (valves, regulator, flow meter, etc.).

Figure 20.26 shows an example of three-stage remineralization at a water treatment station. The evolution of Alk, Ca Hardness and pH parameters throughout the station is summarized in Table 20.13.

| | | Raw water | Water after partial remineralization and coagulation | Water before intermediate adjustment | Water after final neutralization |
|---|---|---|---|---|---|
| **Alk** | **$mg{\cdot}L^{-1}$ $CaCO_3$** | 1.25 | 2.3 | 7.5 | 8.0 |
| **Ca hardness** | **$mg{\cdot}L^{-1}$ $CaCO_3$** | 0.55 | 2.9 | 8.1 | 8.6 |
| **pH** | | 6.7 | 6.0 | 7.5 | 8.2 |

**Table 20.13.** *Example of results obtained with progressive remineralization at a treatment station*

### 20.6.2.2. *Solution using strong base salts and carbonates*

The injection technique of sodium bicarbonate combined with a calcium sulfate or calcium chloride is valid insofar as, at the end of this treatment, water does not contain too many $SO_4^{2-}$ or $Cl^-$ ions, which promote the corrosion of materials, even at levels much lower than the authorized limits.

The convenience of the procedure and the treatment rates are examined taking into account the initial water content of $SO_4^{2-}$ and $Cl^-$ ions, and the possible supply of ions of this kind by a coagulation treatment with aluminum sulfate or ferric chloride.

Sodium carbonate or soda ash is easy to use and safe. It is generally employed in the form of a 10% solution, injected by a dosing pump. Note that 1.06 $mg{\cdot}L^{-1}$ of sodium carbonate is required to raise alkalinity by 1 $mg{\cdot}L^{-1}$ of $CaCO_3$.

Bicarbonate is a weak base which barely raises the pH. It is relatively soluble in water and generally used in the form of a 5% solution, injected by a dosing pump. Note that 1.67 $mg{\cdot}L^{-1}$ of bicarbonate is required to raise alkalinity by 1 $mg{\cdot}L^{-1}$ of $CaCO_3$.

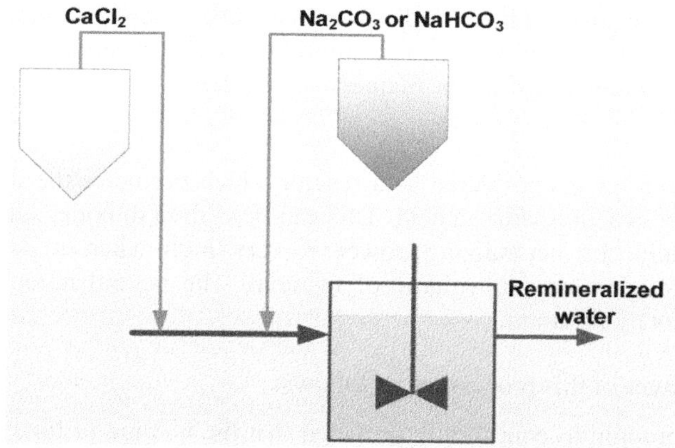


**Figure 20.27.** *Remineralization with strong base salts and carbonates. For a color version of this figure, see www.iste.co.uk/gaid/watertreatment5.zip*

### 20.6.2.3. *Chemical reactions when using a limestone filter*

#### 20.6.2.3.1. Solution combining $CO_2$ (or acid) and a limestone filter

Unlike conventional filtration on an inert material such as sand, which is intended to retain suspended solids, filtration is used as a remineralization treatment on a granular mass made up of limestone, and is part of a broader process involving the dissolution of limestone in water. In virtue of the reaction between the $CO_2$ present in water and limestone, remineralization is achieved by the formation of alkalinity and calcium hardness.

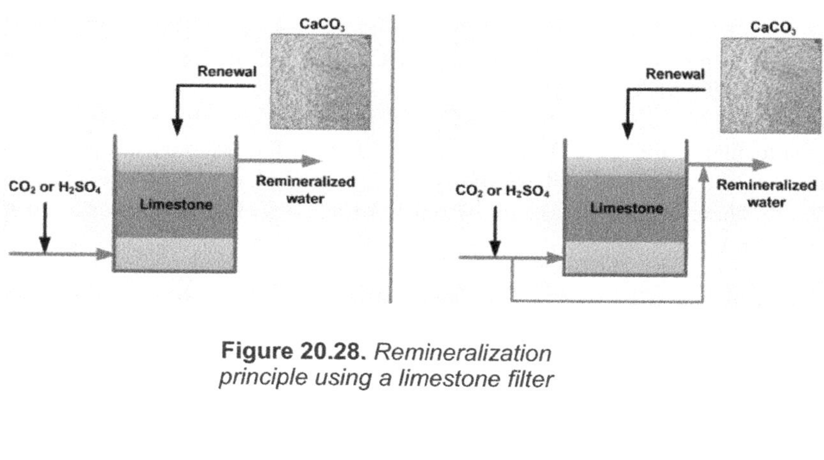


**Figure 20.28.** *Remineralization principle using a limestone filter*

The addition of an acid (sulfuric acid or hydrochloric acid) is an alternative solution to $CO_2$ that also leads to the dissolution of limestone and to the formation of sodium bicarbonate, as well as calcium chloride or calcium sulfate. However, the addition of hydrochloric acid is not frequently used due to its many implementation disadvantages.

A limestone filter is considered as a reactor which promotes the dissolution of limestone by the reaction with an acid. This can be carbon dioxide, sulfuric acid or hydrochloric acid. Its neutralizing power is very high when it acts by water percolation through a certain volume of material. The neutralization by calcium carbonate occurs by filtration.

The advantages of this process are as follows:

– water is brought to equilibrium provided that the volume of limestone used is sufficient;

– no potential risk of excessive limestone treatment;

– facility monitoring is greatly simplified.

As water passes through the filter, calcium carbonate acts as a buffer and pH remains between 7.5 and 8.5. It has the advantage of performing an automatic neutralization, with no further adjustment. There is no risk of overdosing, which could lead to the production of highly scaling water. On the other hand, it requires reloading with support material, as well as periodic washing and backwashing.

#### 20.6.2.3.2. Chemical reactions involving $CO_2$-limestone and acid-limestone with carbon dioxide ($CO_2$)

The chemical reaction is as follows:

$$CaCO_3 + CO_2 + H_2O \rightarrow Ca(HCO_3)_2$$

100 mg of $CaCO_3$/L + 44 mg of $CO_2$/L + 18 mg of $H_20$/L = 162 mg of $Ca(HCO_3)_2$/L

Note that 1 mg of $CO_2$ neutralizes 2.27 mg of limestone, and 1 mg of $CO_2$ increases the hardness and alkalinity of 3.68 mg of $Ca(HCO_3)_2$ or 2.28 $mg \cdot L^{-1}$ $CaCO_3$. There is an increase in 10 $mg \cdot L^{-1}$ $CaCO_3$ in Ca hardness per 10 $mg \cdot L^{-1}$ $CaCO_3$ in Alk.

An LSI of 0.0 to +0.1 can be obtained when a sufficient contact time (greater than 20 min) is considered.

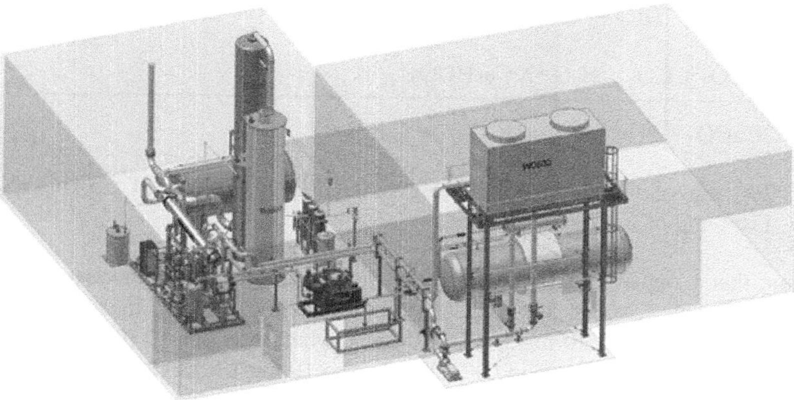

**Figure 20.29.** *3D view of a $CO_2$ facility*

*With sulfuric acid ($H_2SO_4$)*

The chemical reactions are as follows:

$$CaCO_3 + H_2SO_4 \rightarrow CO_2 + H_2O + CaSO_4$$

$$CaCO_3 + CO_2 + H_2O \rightarrow Ca(HCO_3)_2$$

---

$$2CaCO_3 + H_2SO_4 \rightarrow Ca(HCO_3)_2 + CaSO_4$$

200 mg of $CaCO_3$/L + 98 mg of $H_2SO_4$/L = 162 mg of $Ca(HCO_3)_2$/L + 136 mg of $CaSO_4$/L

Note that 1 mg of $H_2SO_4$ neutralizes 2.08 mg of limestone and increases hardness and alkalinity by 1.65 mg of $Ca(HCO_3)_2$, that is to say, 1.02 mg in $CaCO_3$. There is an increase in 20 $mg \cdot L^{-1}$ $CaCO_3$ in Ca hardness per 10 $mg \cdot L^{-1}$ $CaCO_3$ in Alk. There is also an increase in $SO_4^{2-}$ equivalent to 10 $mg \cdot L^{-1}$ $CaCO_3$. Calcium sulfate is formed during the remineralization reaction, whose presence will have to be taken into account in order to avoid deposits (gypsum) in the reactor, as well as when calculating the corrosivity of treated water. An LSI of –0.4 is often obtained when acid dosages are above 20 $mg \cdot L^{-1}$.

*With hydrochloric acid (HCl)*

The chemical reactions are as follows:

$$CaCO_3 + 2HCl \rightarrow CO_2 + H_2O + CaCl_2$$

$$CaCO_3 + CO_2 + H_2O \rightarrow Ca(HCO_3)_2$$

$$2CaCO_3 + 2HCl \rightarrow Ca(HCO_3)_2 + CaCl_2$$

200 mg of $CaCO_3$/L + 72.9 mg of HCl/L = 162 mg of $Ca(HCO_3)_2$/L + 110.9 mg of $CaCl_2$/L

Note that 1 mg of HCl neutralizes 2.81 mg of limestone and increases hardness and alkalinity by 2.28 mg of $Ca(HCO_3)_2$, that is to say, 1.41 mg in $mg{\cdot}L^{-1}$ $CaCO_3$. There is an increase in 20 $mg{\cdot}L^{-1}$ $CaCO_3$ in Ca hardness per 10 $mg{\cdot}L^{-1}$ $CaCO_3$ in Alk. There is also an increase in $Cl^-$ equivalent to 10 $mg{\cdot}L^{-1}$ $CaCO_3$.

Calcium chloride is formed during the remineralization reaction, whose presence will have to be acknowledged when calculating the corrosivity of treated water. An LI of –0.4 is often obtained when acid dosages are above 20 $mg{\cdot}L^{-1}$. In this way, the treatment with sulfuric acid requires far more than twice the calcareous material than with $CO_2$ in order to generate the same final alkalinity.

The treatment with hydrochloric acid requires one and a half additional quantity of calcareous material than with $CO_2$ in order to generate the same final alkalinity. The treatment with sulfuric or hydrochloric acid leads to a calcium hardness twice as high as with $CO_2$.

For these reasons, the choice will be made based on the data shown in Table 20.14.

| **Calco-carbonic parameters obtained** | **Reagents** | **Consequences depending on the choice of reagent** |
|---|---|---|
| Final alkalinity $\geq$ 100 $mg{\cdot}L^{-1}$ $CaCO_3$ | $CO_2$ | Water in calco-carbonic equilibrium where LSI: (0.0) to (+0.1) |
| Final alkalinity < 100 $mg{\cdot}L^{-1}$ $CaCO_3$<br>High calcium hardness | $H_2SO_4$<br>HCl | LSI: (–0.03) to (–0.04)<br>Addition of soda to obtain a LI $\geq$ 0<br>Beware of sulfates or chlorides at the outlet: impact on corrosion<br>Higher consumption of limestone materials |

**Table 20.14.** *Conditions for choosing between $CO_2$, sulfuric or hydrochloric acid*

#### 20.6.2.4. *Neutralization materials*

The materials used in limestone filters are mostly:

– dolomite ($CaCO_3$, $MgCO_3$); calcined dolomite ($CaCO_3 + MgO$); magno ($CaCO_3$, $MgO$);

– Akdolit CM G (Ca0, MgO); neutralite = *maërl* = lithotamnium = acticalmag ($CaCO_3$ + 5% $MgCO_3$);

– jurapearl ($CaCO_3$); marble ($CaCO_3$).

These materials are assimilated as consumable filtration materials whose appearance is granular (>1,000 μm), leading to particle size control by sifting (according to ISO 2591-1). They are characterized by properties that should be checked before implementation:

– particle size distribution curve represented by:

- the actual particle size or d10 effective size (breach enabling the passage of 10% (w/w) of the particles),

- the UC uniformity coefficient (d60/d10),

- minimum particle size d1 (breach enabling the passage of 1% (w/w) of the particles);

– solid density (absolute), dry density (apparent in air), wet density (saturated with water), porosity;

– estimated composition, case by case, depending on:

- purity level ($CaCO_3$ content, etc.),

- solubility in acid or in water,

- the content of metallic impurities,

- the shape: spherical, round, granular, cylindrical, oval, irregular, angular,

- hardness and friability.

Table 20.15 summarizes the characteristics of some materials used for re-mineralization with limestone filters.

The crushed terrestrial limestone is mechanically ground to reach the desired particle size. This is the case of Filtracarb SB and L-SB, Alga-filter Ca, Akdolit C, etc. The ultraporous modified terrestrial limestone is thermally treated to increase porosity, and then calibrated into micropellets depending on the desired particle size.

This is the case of Akdolit Hydro-Calcit CG. Marine limestone (such as NeutralG or Neutrimar) comes from the seabed.

Juraperle, Filtracarb L-SB and Algafilter Ca crushed limestone products have similar characteristics (same particle size, same density, similar composition, etc.). Akdolit C crushed limestone also has similar characteristics, but has a slightly smaller particle size and higher density than its competitors. Finally, despite having an equivalent composition, Akdolit Hydro-Calcit CG stands out from other products by its modified structure (pellets), which makes the product ultra-porous. These materials have very similar performances apart from marine limestone, which is far more efficient with faster kinetics. Calcined dolomites ($CaCO_3$ + MgO) contain a high proportion of alkaline earth oxides. Very rich in free bases, these products are extremely reactive.

| | Characteristics | | | | |
|---|---|---|---|---|---|
| | **Neutral** | **OMYA** | **ALGAVI** | **LHOIST** | **LHOIST** |
| **Products** | Neutralite | Filtracarb L-SB | Algafilter Ca | Akdolit C | Akdolit Hydro-Calcit CG |
| **Types** | Marine limestone | Crushed terrestrial limestone | Crushed terrestrial limestone | Crushed terrestrial limestone | Modified terrestrial limestone |
| **Compliances** | NF EN 1018 | NF EN 1018 | NF EN 1018 | NF EN 1018 | NF EN 1018 |
| **Particle size** | 2–4 mm | 1.5–2.5 mm | 1.5–2.5 mm | 0.5–2 mm | 0.5–3.15 mm |
| **Minimum $CaCO_3$ levels** | 85% | 84% | 98% | 98% | 97% |
| **Apparent density** | 1.1 | 1.3 | 1.3 | 1.5 | 1.3 |
| **Country of origin** | Brittany Iceland | Midi-Pyrenees | Lorraine | Rhone-Alpes | Germany |

**Table 20.15.** *Characteristics of some calcareous materials*

### 20.6.2.5. *Remineralization mechanism in the limestone filter*

The remineralization mechanism takes place in one or two stages, depending on the acid concentrations ($CO_2$ or mineral acid). In the presence of high acid concentrations, acid-limestone neutralization can be implemented upstream of the filter. The reaction rate is written as:

$$-\frac{dC}{dt} = K\, S\, Cr$$

where:

– dC/dt: reaction rate;

– K: rate constant per unit area;

– S: total $CaCO_3$ available surface;

– Cr: acid concentration.

Bearing in mind the proportionality between the acid consumed and bicarbonate formation in the water, then:

$$\frac{\mathrm{dC}\ [Ca(HCO_3)_2]}{dt} = K'(Coutlet - Cin)$$

where:

– dC $[Ca(HCO_3)_2]$/dt: bicarbonate formation rate;

– K′: reaction rate constant;

– Coutlet and Cin: final and initial calcium bicarbonate concentrations.

### 20.6.2.6. *Remineralization performances with a limestone filter*

In the majority of cases, filter performances are evaluated depending on the finally obtained remineralization, assessed in terms of calcium hardness and alkalinity. Treatment goals tend to achieve a Ca hardness of 80 $mg{\cdot}L^{-1}$ $CaCO_3$ and an Alk of around 80 $mg{\cdot}L^{-1}$ $CaCO_3$.

Limescale filters lead to turbidities below 0.5 NTU. When the concentration of injected aggressive $CO_2$ is too low, the sufficient mineralization of treated water is not reached. On the contrary, when the injected $CO_2$ concentration is too high, it can result in a significant $CO_2$ leak due to too short a contact time, too high a filtration speed, unsuitable material, etc. The quantity of $CO_2$ to be removed must be carefully calculated and adapted to the design of the limestone filter.

When limestone filter remineralization is well calibrated, an LSI between 0 and 0.1 is obtained with the $CO_2$ injection, but it only reaches –0.2 to –0.4 when using sulfuric or hydrochloric acid. On filtered water, a readjustment with soda is required to obtain a positive LSI.

The equilibrium pH of the neutralization reaction depends on many factors such as the initial pH, the initially present $CO_2$, the $CO_2$ dosage, raw water alkalinity, dry residuals and temperature.

Water losses from the filters are frequently low due to the fairly spaced washing frequency, thus letting filters be washed only once a week, or even once a fortnight.

In the case where only part of the flow is remineralized by limestone filtration, and later mixed with the other part to constitute the final mixture before distribution, care is taken for the mixture's turbidity to respect the guarantee required, which may differ depending on the specifications: 0.2 NTU in final for some, and up to 0.5–1 NTU for others.

### 20.6.2.7. *Limestone filters design parameters*

The design is identical to that of the open or pressurized filters used at the water treatment stations. "Classic" open concrete filters can follow an ascending or gravitational flow and have a rectangular reinforced concrete tank. When the filters are ascending, the floor is adapted to the acidity conditions of the supply water. Instead, Leopold-type floors are recommended. Pressure filters are metallic structures with water passing under pressure through a limestone bed.

**Figure 20.30.** *OTV pressure filters (Veolia DWT plant). For a color version of this figure, see www.iste.co.uk/gaid/watertreatment5.zip*

Both filtration directions are possible. When the flow type is descending, filtration enables not only the neutralization, but also the settlement of the material's insoluble particles. Backwashing is necessary to remove those particles.

When the flow type is ascending, the hydraulic flow tends to entrain the particles into the treated water, which may lead to a deterioration of turbidity in the treated water.

The quantities of material required to neutralize water depend on the following factors:

– temperature, whose rise increases the neutralization velocity and makes it possible to reduce the volume of the materials;

– the nature of the product as well as its particle size;

– initial and residual aggressive $CO_2$ levels before and after neutralization, or its acid equivalent (sulfuric or hydrochloric);

– the choice of reagent used to react with limestone depends on the final water quality goals in terms of Ca hardness and Alk.

#### 20.6.2.7.1. Empty bed contact time

The contact time refers to the time length during which water passes through the filtering material. It is closely related to the volume of material present in the filter, the temperature and the inlet $CO_2$ concentration. A short contact time associated with a high $CO_2$ concentration leads to a still aggressive filtered water (with an important $CO_2$ residual) and to a final mineralization, which does not meet the expected standard. The contact time varies between 15 and 50 min. An optimum contact time favors the kinetics leading to the calco-carbonic equilibrium, and therefore to an LSI equal to 0.

With marine limestone, it is generally accepted that in order to neutralize 40–50 $mg{\cdot}L^{-1}$ of aggressive $CO_2$, contact times range between 15 and 20 min.

When working with Filtracarb-type products, and considering that terrestrial limestone is less reactive than marine limestone, longer contact times are needed. When the $CO_2$ concentration is lower than 30 $mg{\cdot}L^{-1}$, a 20–30 min contact time suffices. When this concentration is higher than 30 $mg{\cdot}L^{-1}$, contact times are between 30 and 50 min.

With Akdolit CMG-type products whose reactivity is equal to or greater than that of marine limestone, contact times are around 20 min. However, their reactivity requires a close monitoring of the contact time in order to avoid times longer than required, in which case pH could rise over 9. In addition, this significant reactivity (with a rise in pH) is highly restrictive when recharging filters or after a prolonged interruption of filtration due to a discontinuous production of drinking water.

#### 20.6.2.7.2. Volume of limestone consumed

The volume of limestone consumed depends on the $CO_2$ or equivalent acid concentration present at the filter's inlet. This process consumes 2.3–2.9 mg of $CaCO_3$ (depending on product purity) per mg of aggressive $CO_2$. It also leads to an increase in hardness and alkalinity representing 0.91 mg of $CaCO_3$ per mg of limestone consumed.

Since $CO_2$ or acid are always in deficit in relation to the injectable required doses to reach the desired final mineralization, these reagents are added on demand.

A reduction in limestone activity is observed as the temperature decreases. For the same remineralization level, the volume of limestone is 1.3–2 times higher at 4°C compared to 15°C. The dissolution of limestone results in an increase in water's pH, Ca hardness and alkalinity. A regular decrease in the volume of material is observed. If this volume is not regularly monitored, remineralization performances are affected because the contact time, the material's height and kinetics are reduced and the final desired quality of water is not reached.

The consumption of materials such as magno- and magnedolite are summarized in Table 20.16.

In broader terms, the consumption of calcareous material can be deduced from the increase in the $Ca^{2+}$ concentration.

Consider flow × $\Delta Ca^{2+}$ × 2.493 mg $CaCO_3/Ca^{2+}$ = quantity of $CaCO_3$ (100%) consumed, which must be corrected depending on the material's purity.

| **Reagents** | **Chemical reactions** | **Increases in mineralization mg/mg material** | **Reagents consumed mg/mg $CO_2$** |
|---|---|---|---|
| **Calcium carbonate and magnesia $CaCO_3$, MgO (magno)** | $CaCO_3$, MgO + $3CO_2 + 2H_2O \rightarrow Ca(HCO_3)_2 + Mg(HCO_3)_2$ | 0.7–0.86 | 1.06–1.3 |
| **Lime and magnesia, CaO, MgO (magnedolite)** | CaO, MgO + $4CO_2 + 2H_2O \rightarrow Ca(HCO_3)_2 + Mg(HCO_3)_2$ | 0.48–0.53 | 0.55–0.60 |

**Table 20.16.** *Other materials used in neutralizing filters*

#### 20.6.2.7.3. Material height

The filtration layer of limestone filters can measure between 1 and 5 m. It is often considered as a basic parameter from which to deduce the filtration area as well as the filtration rate.

#### 20.6.2.7.4. Filtration velocity

As filter remineralization is carried out within a "reactor", the filtration velocity is not a limiting parameter. However, due to the hydraulic entrainment of fine particles into treated water, it is advisable to stay within a maximum velocity of 15 $m \cdot h^{-1}$ on N filters, and 20 $m \cdot h^{-1}$ in $N^{-1}$ filters.

The application of a high filtration velocity requires an important material height, resulting in an increase in the total filter height.

### 20.6.2.8. *Limescale filter technology (Filtraflo® F type)*

These open filters are the most widespread, thanks to their simplicity and low investment cost.

Veolia's downflow limestone filters are equipped with the following (Figure 20.31):

– a feeder granting an even distribution of the water along the entire filter's surface;

– a strained floor enabling a homogeneous distribution of water along the entire filter's surface during normal operation, as well as a homogeneous distribution of air and water along the entire filter's surface during backwashing operations;

– a gravel layer on the floor to protect the strainers;

– a device under the filtering floor to obtain a uniform distribution of the air mattress during washing operations;

– a device for collecting dirty washwater ensuring a homogeneous washwater recovery along the entire filter's surface. Washwater is evacuated by discharge over 1 or 2 sides, depending on the cell's width, as shown in the diagram in Figure 20.31.

Veolia's upflow limestone filters are suitably equipped for co-current washing. The material loading is always done from above.

Either with $CO_2$ or acid, the implementation configuration may include remineralization for the entire flow (Figure 20.32) or remineralization over part of the flow. The remineralized water is then mixed with the bypassed water.

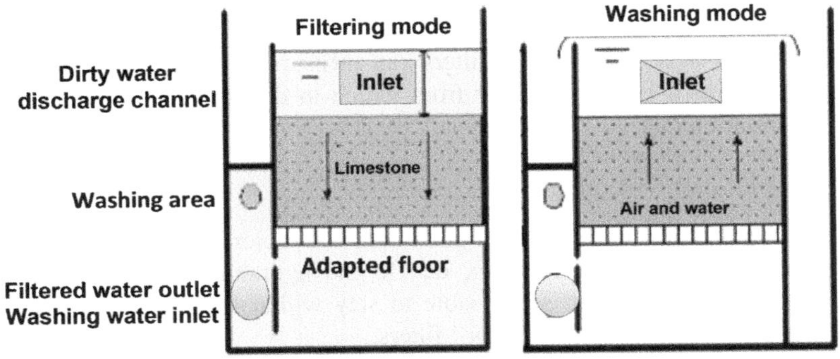


**Figure 20.31.** *Simplified diagram of Filtraflo® gravity filters. For a color version of this figure, see www.iste.co.uk/gaid/watertreatment5.zip*

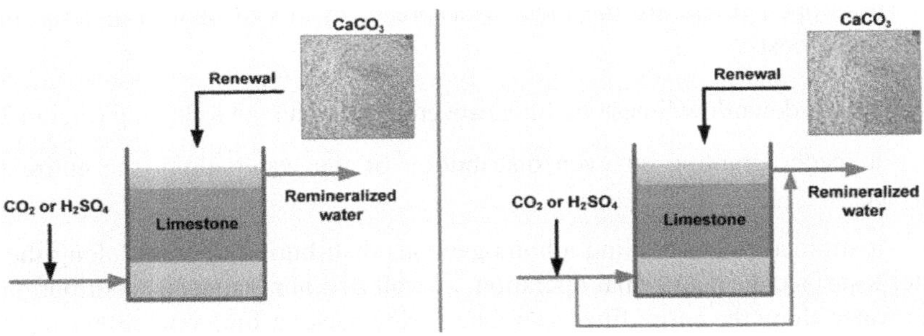


**Figure 20.32.** *$CO_2$-limestone filter remineralization diagram with bypass of part of the water to be remineralized*

The choice of $CO_2$ remineralization induces a contact tower of 2 min contact time so as to ensure the proper dissolution of the reagent in the water. This contact tower can be replaced by a static mixer. The advantage of a bypass in part of the water to be treated considerably reduces the filtration surfaces and the footprint, as well as the equipment associated with this stage. A bypass of 75–80% of the global flow is often implemented. As the total quantity of reagents must be identical with or without a bypass, this implies that a bypass configuration requires very high dosages of reagents and, all the more, a high limestone consumption.

Particular attention should be paid:

– to the injection and dissolution of reagents ($CO_2$, acid);

– ways in which to avoid $CO_2$ leaks, in particular during its mixing;

– the protection of structures against aggressivity and the right choice of equipment and pipelines against corrosion.

#### 20.6.2.8.1. Example of the benefits of bypassing part of the water to be treated

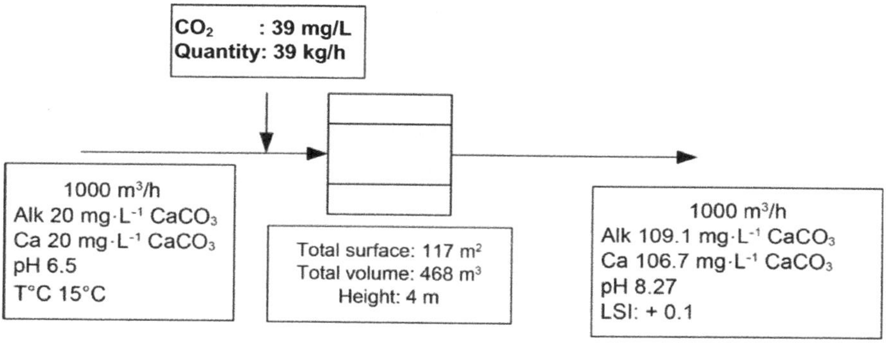


**Figure 20.33.** *Without a bypass*

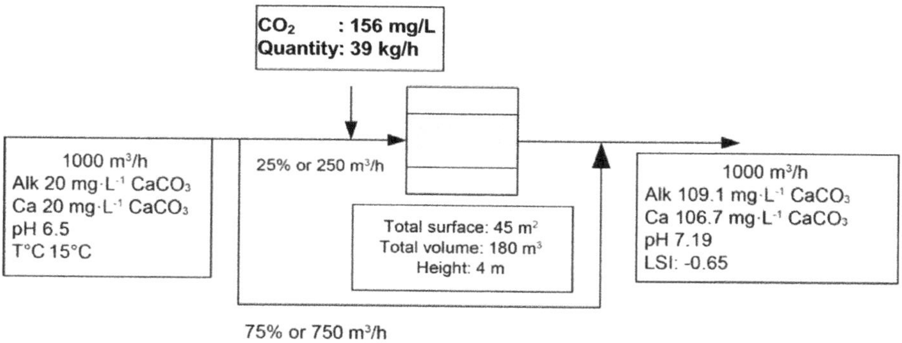


**Figure 20.34.** *With a bypass*

The gain in filtering surface is significant (see section 19.12.3.1) and can be a great advantage in terms of investment. The LSI of final water is negative and it is important to add caustic soda at the end of the treatment in order to neutralize the pH and distribute water in calco-carbonic equilibrium.

The implementation and operating conditions are summarized in Table 20.17.

Veolia's experience with these materials shows reaction times oscillating between 30 and 40 min to reach the alkalinity (100 mg·L$^{-1}$ $CaCO_3$) and calcium (100 mg·L$^{-1}$ $CaCO_3$) targets.

The operating conditions of neutralite and terrestrial limestone are similar.

The refill frequency and loading conditions are alike, and the density of the terrestrial limestones used is relatively similar.

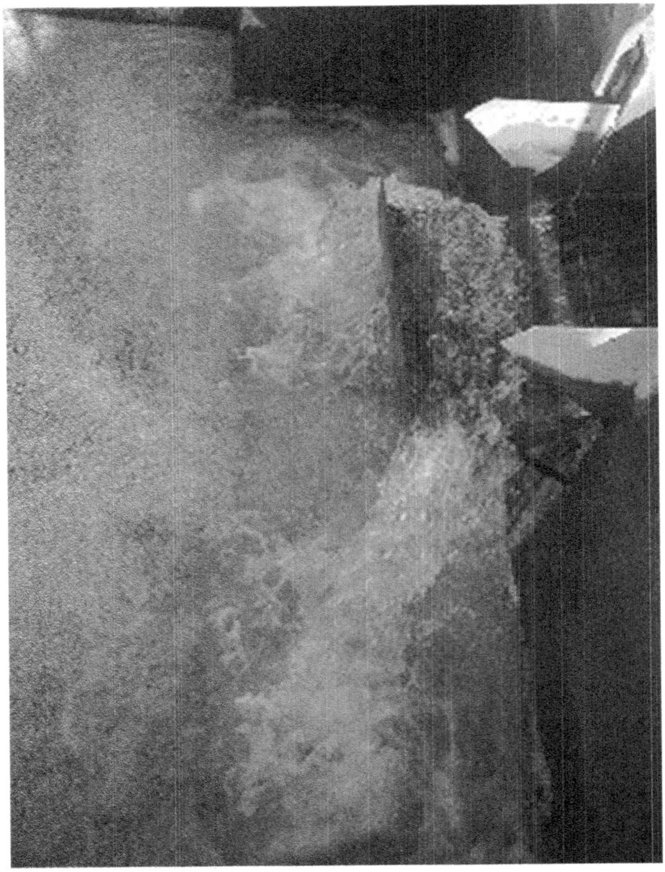

**Figure 20.35.** *Gravity limestone filter (Filtraflo® F-type). For a color version of this figure, see www.iste.co.uk/gaid/watertreatment5.zip*

| | Implementation (supplier data) | | | | |
|---|---|---|---|---|---|
| | **Neutral** | **OMYA** | **ALGAVI** | **LHOIST** | **LHOIST** |
| **Products** | Neutralite | Filtracarb L-SB | Algafilter Ca | Akdolit C | Akdolit C hydro-calcite |
| **Filtration direction** | Descending | Descending | Descending | Descending | Descending |
| **Contact time (min)** | 20 | 30 | 30 | 30-40 | 20 |
| **Washing** | Air + water | Air + water | Air + water | Air + water | Air + water |
| **Washing recommendations (supplier data)** | | | | | |
| **Declogging** | – | 1 min<br>Air at 75–90 $m \cdot h^{-1}$ | 8–10 min<br>Air at 30 $m \cdot h^{-1}$ | 5 min<br>Air at 60 $m \cdot h^{-1}$ | 5 min<br>Air at 60 $m \cdot h^{-1}$ |
| **Washing** | – | 3–10 min<br>Air at 75–90 $m \cdot h^{-1}$<br>Water at 12–20 $m \cdot h^{-1}$ | 5–10 min<br>Air at 50 $m \cdot h^{-1}$<br>Water at 20–25 $m \cdot h^{-1}$ | 10 min<br>Air at 60 $m \cdot h^{-1}$<br>Water at 8–12 $m \cdot h^{-1}$ | 10 min<br>Air at 60 $m \cdot h^{-1}$<br>Water at 8–12 $m \cdot h^{-1}$ |
| **Rinsing** | – | 5–10 min<br>Water at 15–20 $m \cdot h^{-1}$ | 5–10 min<br>Water at 30 $m \cdot h^{-1}$ | Water at 20–25 $m \cdot h^{-1}$ until clear water | Water at 20–25 $m \cdot h^{-1}$ until clear water |
| **Frequency of periodic washes** | Every 15 days | Every 15 days | Every 15 days | Every 15 days | Every 15 days |

**Table 20.17.** *Implementation of limestone materials (supplier data)*

The washing frequency remains the same (at least once every 15 days). Without any modification of the washing equipment, the wash duration and water volume required for rinsing are higher than for neutralite. The operating conditions of modified limestone require more attention than neutralite, at least while changing the material. For modified terrestrial limestones, the implementation conditions of the first load are more delicate because the material – which is more reactive – requires frequent pH monitoring.

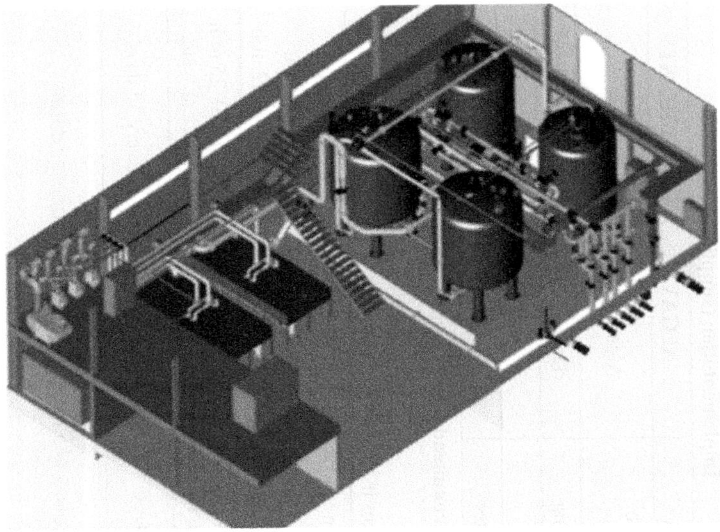

**Figure 20.36.** *Setting up limestone filters under pressure. For a color version of this figure, see www.iste.co.uk/gaid/watertreatment5.zip*

### 20.6.2.9. *Complementary treatment with caustic soda*

When the goal of remineralization is to reach calco-carbonic equilibrium with an LSI equal to 0 (±0.02), the $CO_2$-filter solution (with marine limestone) is a possibility. Veolia has numerous references operating with this solution. With terrestrial limestone, water can still be slightly aggressive (LSI between –0.1 and –0.25). This is explained by the limestone dissolution kinetics, which requires very long contact times.

When the goal of remineralization is to attain a calco-carbon equilibrium with an LSI equal to +0.1 to +0.3, regardless of the material used, these values are not reached. In that case, a complementary treatment with soda should be implemented downstream of filtration.

The calculation of supplementary caustic soda is based on the residual acid or $CO_2$ on the filtered water requiring neutralization.

As the dosage of pure caustic soda (mg·L$^{-1}$) = 0.91 × residual $CO_2$ (in mg·L$^{-1}$), the pHs can be estimated from the pH obtained after neutralization with caustic soda.

In general, caustic soda (100%) doses can be between 1 and 10 mg·L$^{-1}$, depending on the $CO_2$ or acid injected doses and contact times.

Since commercial caustic soda is likely to crystallize at low temperature, caustic soda storage, dosing and injection stations include:

– a temperature sensor for the control of a heating pin in the storage tank;

– storage tank protection;

– an insulated soda injection circuit, lined with a heating cable controlled by a temperature sensor (if necessary), depending on the site's specific conditions;

– an in-line caustic soda dilution device using previously softened water to avoid carbonation phenomena at the caustic soda injection point.

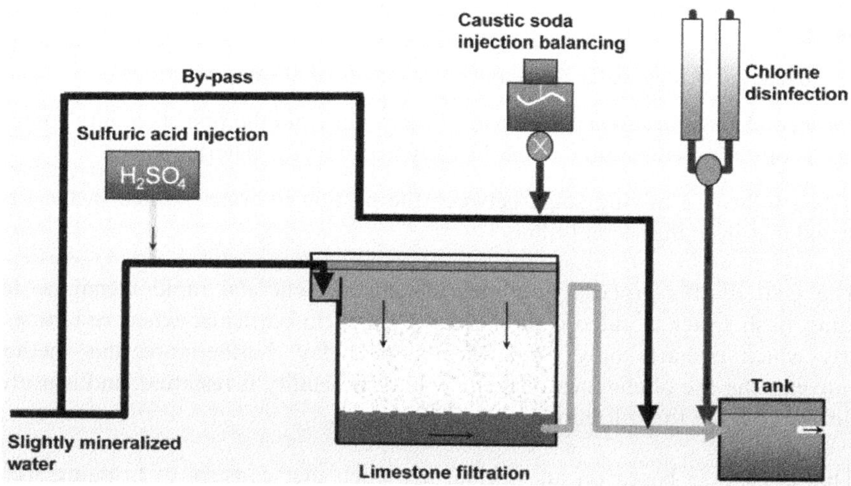


**Figure 20.37.** *pH adjustment with caustic soda on an acid-downflow filter solution, with a bypass. For a color version of this figure, see www.iste.co.uk/gaid/watertreatment5.zip*

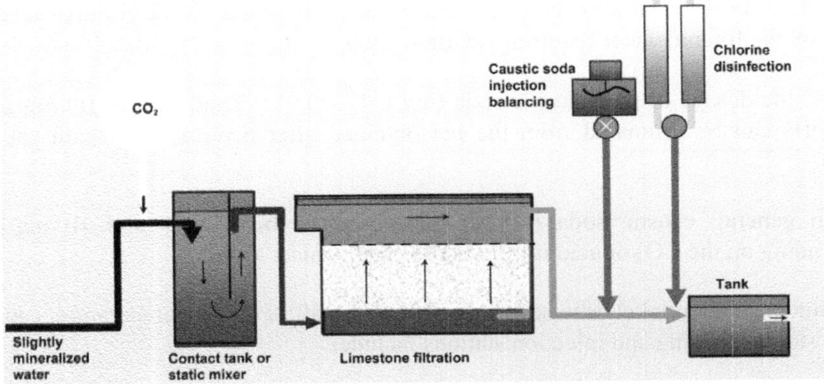


**Figure 20.38.** *pH adjustment with caustic soda on an acid-upflow filter solution, without a bypass. For a color version of this figure, see www.iste.co.uk/gaid/watertreatment5.zip*

#### 20.6.2.10. *Innovative Veolia remineralization process for small facilities*

The remineralization process involves treating water at a treatment facility, with the purpose of bringing it into calco-carbonic equilibrium. This process includes the following:

– a first stage brings water into contact by means of a limestone filter;

– a second stage in series brings part of the water from the first stage into contact by means of a filter containing a high-basicity neutralizing material;

– then follows a mixing stage between the portion of water treated in the first stage and that from the second filter.

The goal of this process is to provide an efficient and rapid technique for bringing fresh water to calco-carbonic equilibrium, in particular when restarting a facility which operates only for a few hours a day. Furthermore, this method circumvents the use of chemical reagents when the facility is restarted, and is useful for discharging the first filtration waters into sewage.

This process is based on an original approach that consists of bringing fresh water into calco-carbonic equilibrium by passing it successively through a lime contactor, then a neutralizing contactor with high-basicity, mixing part of the water from the limestone contactor stage with the water in contact with the high-basicity neutralizing material, and measuring the pH of the mixture obtained. The question comes down to adjusting the quantity of water coming from the first limestone filter, and redirecting it toward the second filter, in function of the mixture's pH.

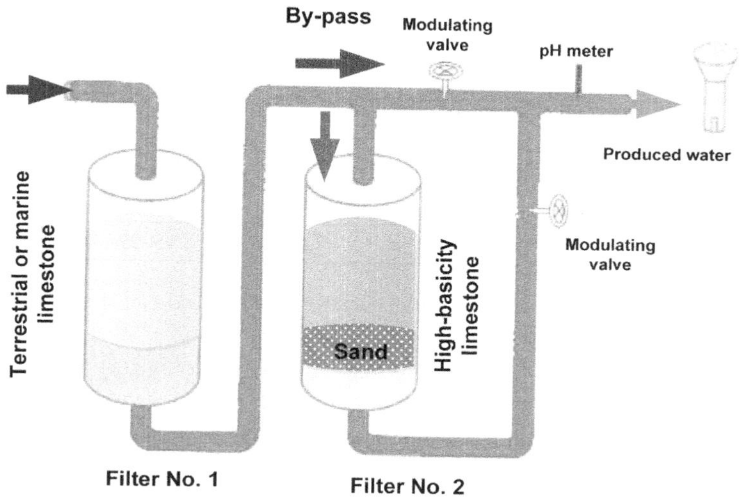


**Figure 20.39.** *Veolia process with limestone filtration in series. For a color version of this figure, see www.iste.co.uk/gaid/watertreatment5.zip*

This process quickly produces water in calco-carbonic equilibrium (or slightly scaling) under all circumstances, in particular when restarting the treatment facility.

The materials used in both filters are marine limestone and crushed terrestrial limestone. The high-basicity materials used in the second filter are chosen from the following group: semi-calcined dolomite, fully calcined dolomite and recarbonated lime.

## 20.7. Characteristics of the various reagents used

### 20.7.1. *Lime*

Lime is obtained by calcining natural calcium carbonate (temperature > 900°C). It exists in the form of quicklime or slaked lime.

Pure lime is relatively insoluble in water. Lime is delivered either as quicklime or as slaked lime.

Quicklime is rarely preferred over slaked lime, because it is only in this last form that it can be implemented. The direct injection of quicklime into the water leads to a derisory neutralization efficiency.

| Types | Granulometry | Purity by weight in CaO % | Purity by weight in $Ca(OH)_2$ % | Apparent density |
|---|---|---|---|---|
| **Quicklime:** | | | | |
| **– as rocks** | 20–50 mm | 92 | | 1.2 |
| **– in lumps** | 20–50 mm | 92 | – | 1–1.4 |
| **– as powder** | 0–2 mm | 80–95 | | 0.8 |
| **Slaked lime** | 20–80 µm | 72 | 95–97 | 0.5 |

**Table 20.18.** *General characteristics of commercial limes*

Slaked lime can be dosed:

– when dry, provided that it does not result from on-site quicklime slaking;

– as milk lime;

– as lime water.

In general, slaked lime is used in a highly pulverized form in order to compensate for the product's low solubility.

### 20.7.2. *"Micronized" lime*

It is a very finely divided slaked lime solution (lime milk). The particles of hydrated lime suspended in water, at 30%, have an average particle size of 6–7 µm.

– Advantage: this highly concentrated milk lime settles very slowly, thereby enabling a controlled distribution without any difficulty.

– Disadvantage: overall, it is more expensive than a conventional system for the preparation and distribution of milk lime.

### 20.7.3. *Caustic soda*

Caustic soda is mainly produced by the electrolysis of sodium chloride. Caustic soda solutions crystallize easily and their viscosity rises considerably as the concentration increases, or when temperature decreases. For this reason, it is essential to dilute to 30%. For higher concentrations (>30%), it is advisable to heat above 16°C.

The heat of dissolution for caustic soda in water is very important. At 25°C, and depending on the concentration, it varies from 777 to 1,045 kj·kg$^{-1}$ of NaOH.

Caustic soda is available in two forms:

– as white and translucent flakes or grains containing 99% NaOH, delivered in 50 kg bags or 200 kg drums, possibly in "commercial" containers or disposable cardboard containers with a polyethylene bag to be emptied by turning;

– as colorless caustic soda containing 47–50% in NaOH weight (density = 1.52), delivered by tank trucks of 25–26 tons or in containers of 2,500 L.

| **Absolute viscosity, in centipoise, of caustic soda, at 40–50% NaOH and 15–50°C** | | | | | | | | |
|---|---|---|---|---|---|---|---|---|
| **NaOH** | **Viscosity** | | | | | | | |
| g/100 g caustic soda | 15°C | 20°C | 25°C | 30°C | 35°C | 40°C | 45°C | 50°C |
| 40 | 48 | 31 | 23 | 18 | 13 | 11 | 9 | 8 |
| 42 | 65 | 41 | 28 | 22 | 16 | 13 | 10.5 | 9 |
| 44 | 82 | 50 | 34 | 25 | 19 | 15 | 11.5 | 10 |
| 46 | 100 | 60 | 40 | 29 | 22 | 17 | 13 | 11 |
| 48 | – | 70 | 46 | 33 | 26 | 19 | 15 | 13 |
| 50 | – | 80 | 54 | 40 | 30 | 22 | 18 | 15 |

**Table 20.19.** *Viscosity of soda solutions*

### 20.7.4. *Caustic soda at 50%*

The dosage of the undiluted solution is done with a dosing pump or dispenser. The discharge pipelines, located at a certain height outside a heated room, have thermal insulation and are heated by a heating cable controlled by a thermostat which keeps the internal temperature at 40°C.

### 20.7.5. *Caustic soda flakes or grains*

The product must be introduced very slowly into the water in order to prevent its vaporization from being accompanied by splashes, due to the high heat of dissolution. These splashes are strictly kept to a minimum. As well, dissolving tanks must be covered.

### 20.7.6. *Sodium carbonate*

Sodium carbonate is manufactured with the Solvay process at the same time as calcium chloride, resulting from sodium chloride and calcium carbonate. Its main characteristics are a saturation at 0°C for a concentration of around 70 $g \cdot L^{-1}$, and a pH at 25°C for solutions at 50 $g \cdot L^{-1}$ of 11.5.

The product is delivered in two forms, as detailed in Table 20.20.

| Forms | Apparent density | Particle size |
|---|---|---|
| **Light powder** | 0.51 | – |
| **Dense granules:** | | |
| **– M (metallurgical)** | 0.80 | 2–4 mm |
| **– Glass industry** | 1.05 | 1–2 mm |

**Table 20.20.** *Commercial forms of sodium carbonate*

In these two presentations, the product can be delivered in 50 kg bags or in bulk by truck containers. Its main features are:

– purity: 99% $Na_2CO_3$;

– higher hygroscopy than sodium bicarbonate;

– pneumatic handling under the same conditions as lime;

– neither toxic nor corrosive; it nonetheless emits a fine dust making the atmosphere unbreathable, in the same way as lime does. It is important not to store the two products in the same room, because their simultaneous presence in a humid atmosphere can lead to the formation of a very dangerous caustic soda;

– possible aging on the surface by the transformation of carbonate into bicarbonate, under the action of carbon dioxide and humidity in the air.

Sodium carbonate can be stored in 50 kg bags or in a silo (if supplied in bulk). It is treated as lime.

Dosing can be done in two ways:

– in the dry form. In that case, the product in dense granules is recommended;

– in solution. The light powder presentation is then recommended. It will be prepared as a 50 $g \cdot L^{-1}$ solution.

### 20.7.7. *Sodium bicarbonate*

The pH at 20°C for 50 g·L$^{-1}$ solutions is 8.5. The solubility of $NaHCO_3$ at 0°C is 69 g·L$^{-1}$ and 96 g·L$^{-1}$ at 25°C.

As carbonate, bicarbonate is delivered in two presentation forms: as anhydrous white powder (density 0.85) or in 13/50 granules (density 0.85–0.95).

The product is 99% pure $NaHCO_3$. It is sensitive to humidity, although less hygroscopic than sodium carbonate. It is often delivered in 50 kg bags or in bulk by 20 ton trucks. In this form it can be handled pneumatically, just like lime.

### 20.7.8. *Calcium carbonate*

Calcium carbonate is a natural product – usually limestone or marble – which is used after grinding and sifting. The main characteristics are a variable particle size between 0.5 and 6 mm and an apparent density of 1–1.4. In this form, it is not recommended to use it as a neutralizer because the reaction kinetics are very slow with efficiencies lower than 20%. On the other hand, when used in percolation within a filter, the kinetics are faster, with very high efficiencies.

The advantages of this process are:

– water is brought to equilibrium provided that the volume of limestone used is sufficient;

– no potential risk of excessive treatment;

– facility monitoring is greatly simplified.

The quantities of material required to neutralize water depend on the following factors:

– temperature, whose rise increases the neutralization velocity and makes it possible to reduce the volume of the materials;

– the nature of the product as well as its particle size;

– initial and residual aggressive $CO_2$ levels, before and after neutralization.

The volume of material to be used will also be reduced if calcium carbonate filtration is accompanied by a neutralizing reagent injection, in particular, by sodium carbonate. In that case, there are two possible ways to proceed:

– a sodium carbonate injection before filtration: for this type of application, the aforementioned advantages are retained;

– a sodium carbonate injection after filtration: although the above-mentioned advantages are lost, in this case the consumption of sodium carbonate is reduced, because the calcium carbonate in contact with a more aggressive water than in the previous case contributes to its neutralization.

In all its commercial forms, calcium carbonate is delivered in 50 kg bags or in bulk by container trucks. It can be stored in a silo and handled either pneumatically or hydraulically.

### 20.7.9. *Acticalmag limestone*

– Origin: algae fossils dredged in various places, notably off the coast of Glénan. These algae are dried, crushed and sieved. This product is commonly called *maërl*. It is also sold under the trade name "neutralite".

– Apparent density: 1;

– Composition:

- $CaCO_3$: 70–80%;
- $MgCO_3$: 8–10%;

– Particle size: No. 0: from 1.5 to 2 mm; No. 1: from 2 to 3 mm; No. 2: from 3 to 4.5 mm.

Its neutralizing power decreases with temperature.

### 20.7.10. *Magno*

Magno is an equimolecular complex of calcium carbonate $CaCO_3$ and magnesia MgO. It results from the moderate calcination (720°C) of dolomite, which is a natural carbonate of calcium and magnesium. Its main features are:

– chemical formula: $CaCO_3MgO$;

– molecular weight: 140;

– particle size and density:

100% < 5 mm; 85% < 4 mm; 13% < 2 mm; 3%

< 1 density from 1.05 to 1.2

The commercial product has an average purity in $CaCO_3MgO$ between 94 and 98%. New magno has a very strong neutralizing power due to the action of magnesia, which quickly leads to oversaturation. But, as the product gets washed by filtration, the magnesia is entrained and the neutralizing power of magno approaches asymptotically that of Acticalmag limestone.

Magno filtration does not seem to entail an increase in turbidity, except in the event of a halt, in which case a strong increase in turbidity and pH is observed.

### 20.7.11. *Calcium sulfate*

Calcium sulfate is prepared from natural gypsum by grinding, sifting and steaming at 50°C, or heating to a higher temperature. Its main characteristics are shown in Table 20.21.

| Varieties | Stable at temperature t | Chemical formulae | Molecular weights |
|---|---|---|---|
| **Anhydrous** | $t < 98°C$ | $CaSO_4$, 2 $H_2O$ | 172 |
| **Gypsum** | $98°C < t < 170°C$ | $CaSO_4$, 1/2 $H_2O$ | 145 |
| **Plaster** | $t > 170°C$ | $CaSO_4$ | 136 |

**Table 20.21.** *Chemical formula and molecular mass of calcium sulfate*

It has an apparent density equal to 0.6 and a solubility that depends on temperature.

| Temperatures | 0°C | 8°C | 20°C | 30°C |
|---|---|---|---|---|
| **Solubilities in $g \cdot L^{-1}$** | 1.8 | 1.9 | 2 | 2.1 |

**Table 20.22.** *Solubility of calcium sulfate*

The variety used is always natural gypsum, whose formula is $CaSO_4 \cdot 2H_2O$. The particle size is around 20–30 μm and has a variable purity between 97 and 98%. When delivered in bulk, calcium sulfate can be stored in silos and handled pneumatically like lime.

Calcium sulfate is more tempting than calcium chloride, because it is less expensive. However, it should be used with caution for treating water which naturally contains a certain quantity of sulfate ions, as well as for water requiring high levels of aluminum sulfate. The latter produces 60% of its weight in anhydrous calcium sulfate. As $SO_4$ ions are unfavorable for the compactness of the protective deposits in pipelines, it is advisable to keep the final content of $SO_4$ ions well below the tolerated limit of 250 mg·L$^{-1}$.

### 20.7.12. *Calcium chloride*

Calcium chloride is manufactured at the same time as sodium carbonate, via the Solvay process, from sodium chloride and calcium carbonate. Calcium chloride is a highly hygroscopic product. It dissolves very easily into water.

Calcium chloride is available in one of the following forms:

– ordinary production at 72–75% $CaCl_2$ (little used);

– ordinary flakes at 77–80% $CaCl_2$;

– high titer flakes (anhydrous) at 95–97% $CaCl_2$.

Like sulfate, calcium chloride should be used with caution, because the $Cl^-$ ion favors the attack of metals. One should pay careful attention to waters naturally containing (or being likely to receive) calcium chloride during coagulation treatments with ferric chloride.

The $Cl^-$ ion content of distributed water must be well below the tolerable 250 mg·L$^{-1}$ limit.

### 20.7.13. *Carbon dioxide*

Carbon dioxide is prepared from the synthesis of ammonia, from which it is a by-product extracted at 99.5% purity. Its prominent characteristics are the following:

– colorless gas, non-toxic, pungent odor;

– density in relation to water: 1.529;

– specific gravity in gaseous form under normal conditions: 1.977 g·L$^{-1}$;

– specific weight in liquid form at –20°C and 20 bar: approximately 1 kg·L$^{-1}$.

The conditions of use of carbon dioxide are close to those leading to its change of phase which can be solid, liquid or gaseous. To design or operate a carbon dioxide distribution facility, it is essential to know and interpret its enthalpy diagram, according to the following recommendations:

– the liquid phase only exists between 31.1 and –56.6°C and for an absolute pressure greater than 5.18 bar;

– the liquid and gaseous phases can only coexist under the above-mentioned conditions, but for an absolute pressure lower than 74 bar. The equilibrium pressure only depends on the temperature;

– above 31.1°C only the gas phase can exist, whatever the pressure. At 31.1°C, liquefaction begins at an absolute pressure of 74 bar;

– for an absolute pressure lower than 5.18 bar, the gaseous and solid phases can exist and even coexist at a pressure which depends only on temperature. At atmospheric pressure, the equilibrium temperature is –80°C.

For the reasons exposed, it follows that liquid carbon dioxide must be kept under pressure. Its value depends only on temperature.

The solubility of carbon dioxide in water depends on temperature and the partial pressure of the gas above the water. It is 2.2 $g \cdot L^{-1}$ at 10°C and 1.5 $g \cdot L^{-1}$ at 20°C.

Carbon dioxide is delivered with the liquid phase in and the gaseous phase in equilibrium:

– in bottles with no thermal insulation;

– in canisters with no thermal insulation;

– in low pressure storage at –20°C and 20 bar in a tank with thermal insulation.

### 20.7.14. *Sulfuric acid (90–98%)*

It has the following properties:

– Physical state: liquid.

– Odor: odorless.

– pH: 0.3 (1 N solution).

– Steam pressure: <0.001 mm Hg at 20°C.

– Vapor density: 3.38 (air = 1).

– Evaporation rate: slower than ether.

– Viscosity: 21 mPas at 25°C.

– Boiling point: 290–338°C.

– Fusion point: 10°C.

– Solubility: soluble with release of heat.

– Density: 1.84.

– Molecular weight: 98.07.

– Chemical stability: it reacts violently and explosively with certain organic compounds, minerals and even with water.

### 20.7.15. *Hydrochloric acid*

Hydrochloric acid is an aqueous solution of hydrogen chloride HCl. It is a strong acid (it ionizes completely in an aqueous solution). It is an acid commonly used as a chemical reagent in the industry. Since hydrochloric acid is a highly corrosive liquid, it must be handled with care. Concentrated hydrochloric acid can have a pH below 1. It is a mono-acid which dissociates (ionizes) and gives an $H^+$ ion (a proton). The other ion, formed during the dissociation, is the $Cl^-$ chloride ion. In an aqueous solution, the $H^+$ ion binds to a water molecule, forming an $H_3O^+$ ion:

$$HCl + H_2O \rightarrow H_3O^+ + Cl^-$$

Mono-acids have a single acid dissociation constant, Ka, which is related to the dissociation rate of the acid in water. For a strong acid like HCl, Ka is very high (>1).

Among the common strong acids in chemistry, all of which are inorganic, hydrochloric acid is the mono-acid least likely to undergo a parasitic oxidation-reduction reaction. It is one of the dangerous acids to handle. Medium-concentrated hydrochloric acid solutions are stable and their concentration does not vary over time. These properties, combined with the fact that it is available as a pure reagent, make hydrochloric acid a very good acidic reagent, or a very good acid for the titration of basic solutions of unknown concentrations.

The physical properties of hydrochloric acid, such as boiling or melting points, depend on the concentration or molarity of HCl in an aqueous solution. They vary from the physical properties of water for 0% HCl to those of fuming hydrochloric acid for fractions greater than 40% HCl.

Hydrochloric acid is produced in solution in concentrations of up to 38% HCl. Highly concentrated hydrochloric acid forms acid fumes. Both these vapors and the acid solution have a corrosive effect on human tissues and can damage the respiratory organs, eyes, skin and intestines.

| Conc. (w/w) kg HCl/kg | Conc. (w/v) kg HCl/m$^3$ | Densities ρ: kg/L | Molarities M | pH | Viscosities η: mPas | Specific heats: kJ/(kg K) | PHCl vapor pressures: Pa | Boiling points b.p., °C | Melting points m.p., °C |
|---|---|---|---|---|---|---|---|---|---|
| 10% | 104.80 | 1.048 | 2.87 | –0.5 | 1.16 | 3.47 | 0.527 | 103 | –18 |
| 20% | 219.60 | 1.098 | 6.02 | –0.8 | 1.37 | 2.99 | 27.3 | 108 | –59 |
| 30% | 344.70 | 1.149 | 9.45 | –1.0 | 1.70 | 2.60 | 1,410 | 90 | –52 |
| 32% | 370.88 | 1.159 | 10.17 | –1.0 | 1.80 | 2.55 | 3,130 | 84 | –43 |
| 34% | 397.46 | 1.169 | 10.90 | –1.0 | 1.90 | 2.50 | 6,733 | 71 | –36 |
| 36% | 424.44 | 1.179 | 11.64 | –1.1 | 1.99 | 2.46 | 14,100 | 61 | –30 |
| 38% | 451.82 | 1.189 | 12.39 | –1.1 | 2.10 | 2.43 | 28,000 | 48 | –26 |

**Table 20.23.** *Physical characteristics of hydrochloric acid*

## 20.8. References

Benefield, L., Judkins, J., Weand, B. (1982). *Process Chemistry for Water and Wastewater Treatment*. Prentice Hall, Inc., Englewood Cliffs.

Caldwell, D.H. and Lawrence, W.B. (1953). Water softening and conditioning problems: Solution by chemical equilibrium methods. *Indust. Eng. Chem.*, 45(3), 535–548.

Cardot, C. (1999). *Les traitements de l'eau. Procédés physico-chimiques et biologiques*. Ellipses, Paris.

Chan, P.K. and Rochelle, G.T. (1982). Limestone dissolution: Effects of pH, $CO_2$ and buffers modeled by mass transfer. *ACS Symp. Series*, 97, 75–97.

Coin, L. (1942). Détermination de l'agressivité des eaux naturelles. *Ann. d'Hyg. publ. industrielle et sociale*, XX(5), 145.

Cox, C.R. (1930). The removal of aggressive carbon dioxide by contact beds of limestone or marble. *J. Am. Water Works Assoc.*, 25, 1505–1522.

David, R.L. (1990). *CRC Handbook of Chemistry and Physics*. CRC Press, Ann Arbor/Boston/Boca Raton.

Duguet, J., Bernazeau, F., Clert, D., Gaïd, A., Laplanche, A., Moles, J., Montiel, A., Riou, G., Simon, P. (2006). *Réglementation et traitement des eaux destinées à la consommation humaine*, 1st edition. Association Scientifique et Technique pour l'Eau et l'Environnement, Paris.

Gebbie, P. (2000). Water stability, what does it mean and how do you measure it? In *63rd Annual Water Industry Engineers and Operators' Conference Civic Centre*, Warrnambool.

Geldenhuys, A.J., Maree, J.P., de Beer, M., Hlabela, P. (2003). An integrated limestone/lime process for partial sulphate removal. *J. South Afr. Inst. Min. Metall.*, 7/8, 345–354.

Hallopeau, J. (1949). L'acide carbonique dans l'eau. *T.S.M.*, 1, 1–11.

Hallopeau, J. (1960). Les équilibres carboniques dans les eaux. *Terres et Eaux*, 13(35), 2–31.

Hernández-Suárez, M. (2012). New design for up-flow limestone contactors for large desalination plants. In *Desalination Congress*, Canary Isles.

Hoover, C.P. (1938). Practical application of the Langelier methods. *J. Am. Water Works Assoc.*, 30(11), 1802.

John, J.C. and Mather, A.E. (1992). The system carbon dioxide-water and the Krichevsky-Kasarnovsky equation. *J. Sol. Chem.*, 21, 607–621.

Langelier, W.F. (1936). Analytical control of anticorrosion water treatment. *J. Am. Water Works Assoc.*, 28, 1500.

Langelier, W.F. (1946a). Étude de l'indice de saturation – Influence de la température. *J. Am. Water Works Assoc.*, 38, 179–189.

Langelier, W.F. (1946b). Effect of temperature on the pH of natural waters. *J. Am. Water Works Assoc.*, 38, 169–185.

Langelier, W.F. (1946c). Chemical equilibria in water treatment. *J. Am. Water Works Assoc.*, 38, 169.

Larson, T.E. and Buswell, A.M. (1942). Calcium, carbonate saturation index and alkalinity interpretations. *J. Am. Water Works Assoc.*, 24, 1667.

Legrand, L. and Poirier, G. (1976). *Chimie des Eaux naturelles : agressivité, corrosivité, traitements, applications numériques*. Eyrolles, Paris.

Legrand, L., Poirier, G., Leroy, P. (1981). *Les équilibres carboniques et l'équilibre calcocarbonique dans les eaux naturelles*. Eyrolles, Paris.

Letterman, R.D., Driscoll, C.T., Haddad, M., Hsu, H.A. (1987). Limestone bed contactors for control of corrosion at small water utilities. Report, Syracuse University, Ageement Cr-809979–01–3 with EPA office of R&D, New York.

Mackintosh, G.S., de Souza, P.F., de Villiers, H.A. (2002). Design and operation guidelines for municipal sized surface water limestone contactors. *Water S.A.*, Special edition, 50–55.

Maree, J.P. and Du Plessis, P. (1994). Neutralisation of acid mine water with calcium carbonate. *Water Sci. Tech.*, 26(9), 285–196.

Maree, J.P., Du Plessis, P., Van Der Walt, C.J. (1992). Treatment of acidic effluents with limestone instead of lime. *Water Sci. Tech.*, 26(1–2), 345–355.

Morse, J.W. (1974). Dissolution kinetics of $CaCO_3$. *Am. J. Sci.*, 274, 97–107.

Pichard, T. and Belaubre, C. (2013). Étude relative à la substitution du calcaire marin sur les stations d'eau potable équipées de filtre de reminéralisation, pour l'agence de l'eau. Report, Consultation 2011–035, IRH (Agence Sud-Ouest), Adour-Garonne.

Plummer, L.N. and Pakhurst, D.L. (1979). Critical review of the kinetics of calcite dissolution and precipitation. *ACS Symp. Series*, 93, 538–548.

Robert, C.R., Prausnitz, J.M., Poling, B.E. (1987). *The Properties of Gases & Liquids*, 4th edition. McGraw-Hill, Boston.

Rooklidge, S.J. and Ketchum Jr., L.H. (2002). Corrosion control enhancement from a dolomiteamended slow sand filter. *Water Res.*, 36(11), 2689–2694.

Rossum, J.R. and Merrill, D.T. (1983). An evaluation of the calcium carbonate saturation indexes. *J. Am. Water Works Assoc.*, 75(2), 95–100.

Tillmans, J. and Heunblein, O. (1912). *Gesundh Ing.*, 35, 669.

Tillmans, J. and Hirsch, H. (1931). Der Einflub von hoheren Temperaturen und Salzzusatzen auf das Kalk-Kohlensaure-Gleichgewicht im Wasser un die Kalk-Rost Schutzschich. *Gas und Wasserfach*, 74, 1–9.

Yamauchi, Y., Tanaka, K., Hattori, K., Kondo, M., Ukawa, N. (1987). Remineralisation by limestone dissolution. *Desalination*, 60, 365–383.

# 21

# Disinfection

The goal of disinfection is to distribute water free from pathogenic microorganisms, so not to pose any risk to public health. Water contains a multitude of microorganisms resulting from natural biological activity, some of which are pathogenic, that is to say, they can cause illness.

The main reservoir for the microorganisms responsible for waterborne diseases is the digestive tract of humans and warm-blooded animals. Feces contain an ample variety of pathogenic organisms – bacteria, viruses and parasites – which can cause illness and contaminate healthy carriers. Pathogenic germs can be responsible for mild infections (mild gastroenteritis, diarrhea) or more serious, even fatal infections (cholera, typhoid).

Some bacteria (known as opportunistic bacteria) that are normally present in humans can infect people whose immune defenses are weakened.

The main waterborne diseases are infections due to bacteria, viruses and parasites. Apart from enterobacteriaceae (*Salmonella, Shigella, Escherichia coli, Campylobacter*, *Pseudomonas aeruginosa*, *Vibrio cholerae*), there are viruses (adenovirus, enterovirus, rotavirus and hepatitis), and above all, parasites (*Cryptosporidium* and *Giardia*) whose presence is being increasingly controlled.

## 21.1. Microorganisms present in the water

### 21.1.1. *Bacteria*

With an intermediate size between protozoa and viruses, bacteria measure 1–10 µm long. They are unicellular organisms but, unlike protozoa, their DNA is not enclosed within a nucleus. Bacteria have different shapes (rod-shaped = bacilli, spherical = cocci). They are present in all kinds of environments and present a host of different lifestyles. In water, apart from natural and harmless bacteria, there are pathogenic species that often have a fecal origin. The most frequently encountered belong to the genera *Salmonella* and *Shigella*, as well as species such as *Escherichia coli*, *Vibrio cholerae*, *Yersinia enterocolitica* and *Campylobacter fetus*.

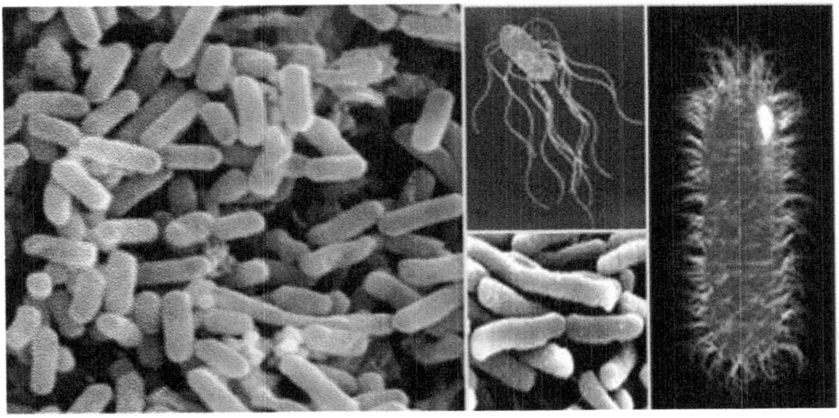

**Figure 21.1.** *Various bacteria species. For a color version of this figure, see www.iste.co.uk/gaid/watertreatment5.zip*

Epidemics caused by bacteria represent the majority of waterborne epidemics, including *Legionella pneumophila* (pneumonia), *Schigellae* (gastrointestinal disorders) and *Salmonellae* (typhoid and paratyphoid fevers, salmonellosis), which are responsible for a variety of syndromes. Frequently associated with food poisoning, they are also present in water.

– *Pathogenic staphylococci*: these germs are the cause of a wide variety of infections: skin conditions, impetigo, septicemia, endocarditis and osteomyelitis.

– *Enteroviruses*: these are abundantly excreted with the feces and risk contaminating the water supply. The main enteroviruses are polioviruses (responsible for poliomyelitis), coxsackie viruses and echoviruses. For the most part, they can induce a vast array of syndromes, including rash, fever, gastroenteritis, myocarditis, meningitis, respiratory conditions and hepatitis.

– *Coliforms*: not only do coliform bacteria exist in the feces, but they can also develop in natural environments: soil, vegetation and natural waterways. The presence of a small number of coliforms in groundwater (1–10 in 100 mL) has only a slight significance in terms of health, provided that it is not accompanied by fecal coliforms.

– *Thermo-tolerant coliforms*: these coliforms are able to grow at 44°C, whereas no growth is observed at this temperature for non-fecal strains. The main coliform bacterium specifically of fecal origin is *Escherichia coli*. This bacterium always appears in large quantities in animal and human excrement, and is only occasionally found in soils and waters which have not been exposed to fecal pollution. Fecal or thermotolerant coliforms are a good indicator of fecal contamination in water.

– *Fecal streptococci*: they are normal hosts in the gut of humans and warm-blooded animals. They are generally not considered pathogenic. Fecal streptococci belong to a group of streptococci, which are not all of fecal origin. However, their survey associated with the search for fecal coliforms constitutes a good index of fecal contamination. Some studies have even shown that streptococci are a better indicator of certain waterborne infectious pathologies than fecal coliforms.

– *Sulfite-reducing bacteria spores*: these are resistant ("sporulated") forms of anaerobic organisms, among which the most frequent and easiest to identify are the clostridia (sulfite-reducing *clostridium*). They are normally present in feces, but in smaller quantities than *E. coli*. They are also present in soils and rivers; their absence in an underground aquifer or an alluvial aquifer is a sign of the efficiency of natural filtration in soils. Similarly, the absence of spores after a filtration treatment is a sign of the treatment's efficiency. Their persistence in disinfected water may signal defects in physicochemical treatments, the lack of efficiency of certain disinfectants or insufficient filtration. In virtue of their behavioral similarities with parasites (formation of resistant cysts), spores are a good indicator of the presence of these microorganisms. This explains why it is not bacteria themselves which are sought after, but their spores.

– *AAHB and atypical colonies*: the survey of aerobic and anaerobic (including facultative) heterotrophic bacteria (AAHB) aims to estimate the general bacterial population in water. The survey of atypical colonies is used to assess the water's bacterial density. The AAHB analyses support and validate the conventional counts of total coliform bacteria per membrane filter.

If they exceed a threshold of 500 AAHB CFU/mL, they enter into competition with total coliform bacteria and mask their presence during the analysis, preventing the development of the typical characteristics of coliform colonies. Similarly, if the colonies of atypical bacteria (which develop on the same culture medium as total coliforms) exceed 200 CFU/100 mL, they may mask the survey of the latter. Not only do these monitoring checks validate the survey of coliforms, but they also provide an indication of the general development of bacteria in the water of distribution networks even before coliforms are detected. This indicator of variations in AAHBs and atypical bacteria over time serves as a management tool for the operator.

AAHB and atypical colonies signal the level of water quality degradation or poor maintenance in the distribution network. In addition, when analyzed in treated water, they are good indicators of the treatment's efficiency for removing bacteria.

### 21.1.2. *Indicator microorganisms or test germs*

It is very difficult and time consuming to control all the pathogenic microorganisms likely to cause waterborne infections in drinking water, for several reasons:

– a certain number of pathogenic microorganisms are not taken into account: some of them are not identifiable in standard laboratory practice;

– in general, pathogenic germs are irregularly present in drinking water. It thus becomes tedious to search for them continuously;

– the delays in obtaining the results of systematic analyses often take too long, hindering an efficient control of the quality of water.

For all these reasons, some indicators are identified so as to reveal the presence or absence of pathogenic germs.

#### 21.1.2.1. *Germs signaling fecal contamination*

The pathogenic germs found in water nearly always have a fecal origin (apart from the case of cyanophyceae algae). But they are not easily identified, because there are also non-pathogenic intestinal germs in the feces of humans and warm-blooded animals, which are easier to identify. Pathogenic germs are really diverse and few in number, in comparison with the flora of an aqueous medium. There is not always a simple and standardized analytical method to search for them.

The assessment of contamination risk is based on the search for test germs, which serve as bacteriological quality indicators.

The criteria for choosing a test germ are as follows:

– it must signal fecal pollution;

– it must be present in large numbers;

– it must be easy to analyze;

– it must simultaneously be present with pathogenic germs;

– it must be harmless during laboratory manipulations.

The indicators used for wastewater are as follows:

– Total coliforms (TC): these are not strictly of fecal origin.

– Fecal coliforms (FC): these belong to the family of Enterobacteriaceae, including *E. coli*. They are heat resistant. They are differentiated from total coliforms by changing the culture medium from 37 to 44°C. Their habitat is the gut of humans or warm-blooded animals.

– Fecal streptococci (FS): they have same origin as fecal coliforms. However, the FS/FC ratio is higher in animals than in human beings. They are particularly useful for identifying the origin of the contamination.

These microorganisms are called fecal contamination "indicators" or test germs: fecal streptococci, as well as *Escherichia coli*, for example, are part of these test germs. Health indicators are essentially fecal contamination indicators, whose presence signals the risk level they pose to health. These indicators are the bacteria *E. coli*, fecal and enterococci coliforms and male-specific coliphage viruses. They are present in the normal intestinal flora of humans and animals in much higher concentrations than enteric pathogenic microorganisms.

Epidemiological studies have shown that the health risk is proportional to the number of test germs surveyed, and that intestinal pathologies can be attributed to water as from the presence of one germ every 100 mL.

If fecal contamination test germs are identified, the water is considered undrinkable and declared non-compliant. On the other hand, the absence of indicator germs does not necessarily mean that water is free from pathogenic microorganisms. One may only conclude that the sample taken is not polluted by feces, and that the water does not pose any of those particular microbiological risks for health.

One should always bear in mind that there are a number of resistant pathogenic organisms (such as viruses or parasites), which can "survive" efficient treatments, namely treatments having resulted in the removal or inactivation of less resistant test germs. This is, for example, the phenomenon that may occur when one finds encysted parasitic species (*Giardia*, *Cryptosporidium*) in apparently properly disinfected water.

The water quality indicators used for evaluating and monitoring the bacterial quality of system water are total coliforms, atypical colonies and AAHB.

### 21.1.3. *Viruses*

Human enteric viruses are likely to be found in almost all surface waters, as well as in vulnerable groundwater. Unlike bacteria and parasites, viruses usually infect only one species.

Although the resistance of viruses to disinfection is variable, it is often greater than that of bacteria. Viruses persist longer in the environment than bacteria. As they are much smaller, they seep more easily through a substrate. As a result, bacterial indicators are not particularly valid for signaling the presence of viruses.

Viral indicators in drinking water monitoring measure somatic coliphages and F-specific coliphages. Coliphages are a variety of bacteriophages, that is to say, viruses which specifically infect a bacterial species. In this concrete case, they are coliform bacteria, and mainly *E. coli*. Coliphages only replicate by infecting the target bacterium, forcing it to produce a few hundred viruses, leading to the bursting of the cell and to the release of these infectious viruses into the medium. Coliphages are often found in domestic wastewater and are present in greater quantities than enteric viruses in the feces.

Somatic coliphages are not very good indicators of fecal contamination, but rather point to the vulnerability of groundwater to viral contamination. In contrast, male-specific coliphages are primarily infective for the bacterium *E. coli*. These coliphages are very good indicators of fecal contamination.

Coliphages signal fecal contamination in groundwater, and as well as the possible presence of human enteric viruses. These microorganisms are more resistant than bacteria. They can exist in the vegetative state, or be active on living cells:

– hepatitis A virus;

– poliomyelitis virus;

– conjunctivitis virus.

Fecal bacteriophages are the fecal bacteria-specific viruses corresponding to them. They are found in water contaminated with feces. They are a good indicator of this type of contamination.

Their presence may indicate old or intermittent contamination. Thanks to their resistance to chlorine treatments, they can be detected in the network and serve as the only trace of previous contamination.

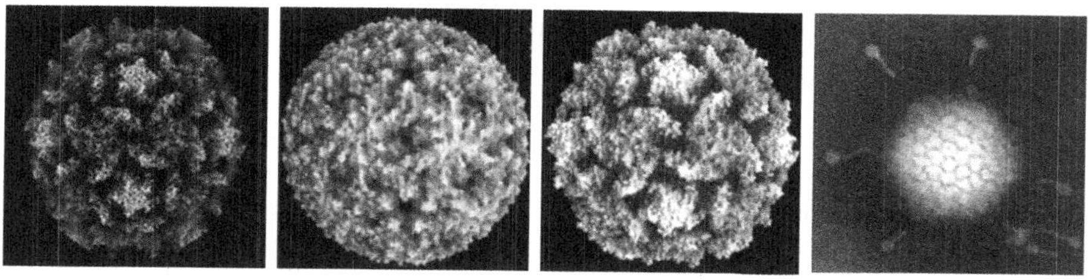


**Figure 21.2.** *Some surface water viruses. For a color version of this figure, see www.iste.co.uk/gaid/watertreatment5.zip*

With a size range between 10 and 300 nm, invisible under an optical microscope, viruses do not have a cellular structure but somehow constitute the reservoir of genetic material. Incapable of an autonomous life, viruses can only develop in living cell parasites (bacteria, protozoa, higher organisms, etc.). They exploit the biochemical machinery of their host by deflecting it for their benefit.

The presence of viruses in water is not uncommon and depends on their constitution: some have a coat (called a capsid) that makes them very resistant in an aquatic environment. This is particularly the case for the viruses in the human digestive tract responsible for poliomyelitis and hepatitis A, which are common in natural waterways.

Contaminated groundwater or surface water requires disinfection to remove at least 99.99% of the viruses.

### 21.1.4. *Parasites*

The parasites *Cryptosporidium* and *Giardia* are subject to regulation. They are frequently present in surface waters in an encysted form. They have greater resistance to chlorine disinfection than bacteria and viruses. In addition, they are far more persistent in the environment than bacteria. They multiply in the gut and their main reservoirs are domestic animals, wild animals (muskrats and beavers in particular for *Giardia*) and humans.

The main parasites are protozoa and helminths (worms). Among the various helminths which can contaminate water supplies, one of the best known is the one responsible for schistosomiasis, a serious disease with urinary, digestive or arterio-venous localization, frequent in developing countries. These worms and their eggs are not destroyed by conventional disinfection, although they are efficiently removed by filtration.

Protozoa are motile unicellular organisms whose DNA is contained in a nucleus surrounded by a membrane. Their cysts are among the most resistant forms to drinking water treatments, including chemical disinfection treatments. Among the most common, the cysts of *Giardia* and *Cryptosporidium* are clearly identified. The size of *Cryptosporidium* ranges between 4 and 6 µm, and the size of *Giardia* between 8 and 10 µm.

Protozoa feed on suspended organic particles or bacteria and most often reproduce by asexual reproduction (the division of a "mother" cell giving rise to two "daughter" cells). There are four main classes differentiated by the nature of their musculoskeletal system and the characteristics of their biological cycle. These are the flagellates, which have a flagellum (*Giardia lamblia*, etc.), rhizopods (amoeba *Entomoeba histolytica*, etc.), ciliates and sporozoa.

The analysis of aerobic spore-forming bacteria (ASFB) is an interesting indicator to assess the parasite-removal performance of a treatment station. ASFBs do not pose any health risk and are abundantly present in the soil and all surface waters. They are analyzed using simple and rapid methods. After conventional filtration, the removal rate of bacterial spores has the same order of magnitude as for parasites. Also the resistance of spores to chlorination is generally higher than that of *Giardia* cysts.

The monitoring of small particles (a few micrometers) is also an interesting tool for evaluating the performance of each filtration stage when it comes to the removal of *Giardia* and *Cryptosporidium* cysts and oocysts.

More often than not, the epidemics listed for *Cryptosporidium* appear in full treatment systems having demonstrated a deficiency in their filtration system.

The UV disinfection treatment is particularly efficient against these parasites. Some operators have observed that ozonation has proven to be less effective in the winter period. Since bacterial indicators are not useful for detecting the presence of parasites, the treatment must ensure their safe removal at all times. *Cryptosporidium* is a rather small parasite and one of the most resistant to chlorination. It is assumed that an efficient treatment to safely remove this parasite will be sufficiently potent against other parasites.

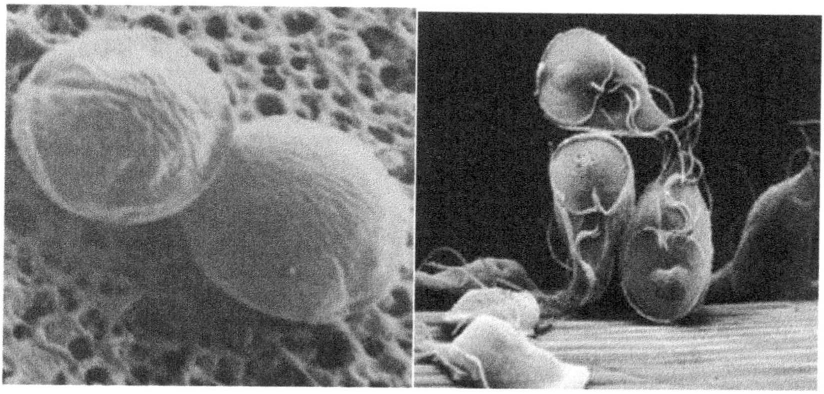

a) b)

**Figure 21.3.** *(a) Cryptosporidium and (b) Giardia*

| Microorganisms | Shapes | Dimensions (μm) |
|---|---|---|
| | **Bacteria** | |
| ***Bacilli*** | Straight | 0.3–1.5 Φ × 1–10 L |
| ***Bacillus (E. coli)*** | Straight | 0.6–1.2 Φ × 2–3 L |
| ***Cocci*** | Spherical | 0.5 × 4 |
| ***Spirilla*** | Spiral | 0.6–2 Φ × 20–50 L |
| ***Vibrio*** | Straight, curved | 0.4–2 Φ × 1–10 L |
| | **Protozoa** | |
| ***Cryptosporidium Oocysts*** | Spherical | 3–6 |
| ***Giardia Lamblia Cysts*** | Oval | 6–8 w × 8–14 L |
| | **Virus** | |
| **MS2** | Spherical | 0.022–0.026 |
| **Enterovirus** | Spherical | 0.020–0.030 |
| **Polio** | Spherical | 0.025–0.030 |
| **Rotavirus** | Spherical | 0.070–0.080 |

Φ: diameter, L: length, w: width.

**Table 21.1.** *Shapes and dimensions of various microorganisms analyzed in raw water*

### 21.1.5. *Micro-algae*

Algae play a positive role in the natural equilibrium of aquatic environments, notably by ensuring the oxygenation of water through photosynthesis. From a sanitary point of view, they do not awaken great interest, except for the fact that they can harbor pathogenic germs and protect them from disinfection.

Some algae are nonetheless toxic to humans: these are blue algae, also called Cyanophyceae or Cyanobacteria, which excrete the toxins responsible for skin and respiratory allergies, nausea, gastroenteritis or dysentery.

## 21.2. Quality of potable water

Disinfection is an essential stage at the treatment station because it ensures the inactivation of pathogenic microorganisms present in the water. It contributes to the good bacteriological quality of distributed water, which is the main criterion for assessing the microbiological potability of water.

Water must be disinfected at the station's outlet and contain a bacteriostatic agent to preserve its quality in the distribution network until it reaches the point-of-use. The disinfectant reagents used at the station and for obtaining a residual for distribution may be the same or different.

### 21.2.1. *French regulations*

The text presented below is drawn from the French government resolution 2007-49 of January 11, 2007, which sets the quality limits and references for potable water:

> Potable water must be free from any number or concentration of microorganisms and parasites or from any other substance constituting a potential danger to human health. Water must be free of pathogenic organisms, in particular Salmonella in a 5 L water sample, pathogenic Staphylococci in a 100 mL water sample, fecal bacteriophages in a 50 mL water sample and enteroviruses in a 10 liter water volume. At least 95% of the samples drawn must not contain coliforms in a 100 mL water sample. Water should be free of thermo-tolerant coliforms and fecal Streptococci in a 100 mL water sample. Water should not contain more than one spore of sulfite-reducing bacteria in a 20 mL water sample.

| Parameters | Quality limits (number/100 mL) | Quality reference values (number/100 mL) |
|---|---|---|
| ***Escherichia coli* (*E. coli*)** | 0 | – |
| **Enterococci** | 0 | – |
| **Coliform bacteria** | – | 0 |
| **Sulfite-reducing bacteria (including spores)** | – | 0 |
| **Count of viable aerobic germs at 22°C and 37°C** | – | Variation in a ratio of 10 compared to the usual value |

**Table 21.2.** *Potable water quality limits: bacteriological parameters (French regulations)*

The quality limits are set for microbiological and chemical health parameters. The reference qualities are defined for the operating indicator parameters of drinking water production and distributions facilities.

In all cases, a rise in temperature increases the efficiency of disinfection because it leads to faster inactivation kinetics. A 10°C temperature rise, for example, multiplies the microbicidal effect of chlorine by a factor of 2. Likewise, the solubility and stability of the chemical oxidant can also decrease with temperature.

| Indicators | Sanitary risks | Fecal contamination | Treatment efficiency monitoring | Network qualities | Comments |
|---|---|---|---|---|---|
| ***Escherichia coli*** | +++ | ++++ | + | + | Sanitary risks<br>Fecal contamination |
| **Fecal coliforms** | ++ | +++ | + | + | Fecal contamination |
| **Total coliforms** | – | – | ++ | ++ | Indicator of the treatment's efficiency and of the quality of distributed water |
| **Aerobic and anaerobic heterotrophic bacteria (AAHB)** | – | – | +++ | ++++ | Indicator of the treatment's efficiency and the quality of distributed and network water |

| Indicators | Sanitary risks | Fecal contamination | Treatment efficiency monitoring | Network qualities | Comments |
|---|---|---|---|---|---|
| **Atypical colonies** | | | ++ | +++ | Indicator of the treatment's efficiency and the quality of distributed and network water |
| **Enterococci** | ++ | +++ | + | + | Fecal contamination (in relation to *E. coli*) |
| **Coliphages** | + | +++ | + | | Fecal contamination Highly probable presence of pathogenic viruses is fecal contamination of human origin |
| **Anaerobic sulfite-reducing bacteria (SRB)** | | | +++ | | Indicator of the treatment's efficiency (filtration) and parasite inactivation |
| **Turbidity** | ++ | | ++++ | +++ | Associated with the overall health risk for contaminated raw water |
| **Disinfectant residual** | + | | ++ | ++++ | With SRB, distributed water quality |

**Table 21.3.** *Main indicators of contamination and the quality of affected water: + low indicator; ++ medium indicator; +++ high indicator; ++++ very high indicator*

## 21.3. General rules of chemical disinfection

### 21.3.1. *Disinfection mechanisms*

Water disinfection involves two important phases, which correspond to the two different effects of a disinfectant:

– a germicidal effect (bactericidal and virulicidal): the ability of disinfectants to destroy germs at a certain stage of the water treatment chain;

– a residual effect: the disinfectant must remain active in the distribution network in order to guarantee the microbiological quality of the water transported in the pipes. It both has a bacteriostatic effect against bacterial reviviscences and a germicidal effect against the weak and occasional forms of pollution, which may arise in the network.

### 21.3.2. *The mode of action of chemical disinfectants*

The various disinfection techniques include:

– chlorine disinfection:

- sodium or calcium hypochlorite,
- gaseous chlorine;

– chlorine dioxide disinfection;

– UV disinfection generally associated with postchlorination disinfection;

– ozone disinfection combined with postchlorination disinfection;

– chloramination disinfection.

A disinfection treatment is only efficient provided that it is applied to water free from suspended solids, likely to protect germs from the action of disinfecting agents.

The disinfectants commonly used in water treatment (chlorine and its derivatives, ozone) have an unparalleled germicidal power, essentially based on chemical redox mechanisms. A disinfectant's efficiency is directly related to its oxidizing power. The following reagents can thus be classified in descending order of efficiency, depending on their redox potential (Table 21.4).

| **Disinfectants** | **Oxidation–reduction potential, V** |
|---|---|
| Ozone in an acid medium $O_3/O_2$ ($H^+$) | 2.07 |
| Hypochlorous acid $HClO/ClO^-$ | 1.49 |
| Chlorine $Cl_2/Cl^-$ | 1.36 |
| Ozone in a basic medium $O_3/O_2$ ($OH^-$) | 1.24 |
| Chlorine dioxide $ClO_2/ClO_2^-$ | 0.95 |
| Hypochlorite $ClO^-/Cl^-$ | 0.90 |

**Table 21.4.** *Oxidation–reduction potential (ORP) of various disinfectants*

The action sites of the disinfecting agent depend on the nature of the target microorganism and its chemical structure. In the case of viruses, the oxidant alters the complex macromolecules of nucleic acids (DNA, RNA), hindering the microorganism from reproducing, and therefore reducing its infectivity. In the case of bacteria, the oxidant penetrates the cytoplasmic membrane and oxidizes the enzymes' sulfur bonds, altering them irreversibly, causing the microorganism's death. Higher oxidant doses also lead to the degradation of the bacteria's nucleic acids.

As viruses are naturally more resistant than bacteria, regardless of the disinfectant used, virucidal treatment conditions are more severe than bactericidal treatments.

As a matter of fact, some microorganisms can adapt to be very resistant to chemical attacks. Indeed, bacteria that have survived limiting conditions (deprivation of nutrients, repeated oxidation or temperature) show an increased resistance to the action of oxidants, in comparison with the same strains subjected to optimal growth conditions. Sporulation mechanisms (the microorganism's secretion of a protective polysaccharide "coat") help certain germs to withstand limiting conditions by adopting a decelerated life cycle.

UV disinfection generally involves the microorganism's absorption of the UV radiation spectrum between 200 and 300 nm. The maximum absorption level, corresponding to the highest biocidal action, is obtained for a 253.7 nm wavelength.

UV radiation acts on the microorganisms' DNA and RNA by blocking their replications, which results in a halt in cell division. Prolonged irradiation provokes an expansion of the cell's nucleus, ultimately causing the cell to burst.

### 21.3.3. *Inactivation kinetics*

The destruction – or inactivation – of microorganisms during the disinfection treatment is not instantaneous. It is implemented following a rhythm which responds to statistical laws. The germ mortality curve is similar to the exponential growth curve, although it is expressed in the opposite sense because the number of surviving germs gradually decreases as a function of time.

Depending on the microorganisms, there are generally several suppression phases over time. For relatively vulnerable pathogenic bacteria, as well as for viruses, there is an almost instantaneous inactivation followed by a slower inactivation. Given the great resistance in protozoa (the case of parasites), inactivation is preceded by a latency phase.

#### 21.3.3.1. *The "log"-reduction principle in the number of germ-units*

The concept of log reduction unit is calculated using the following formula:

$$\text{Log reduction number} = \log\frac{\text{Initial concentration}}{\text{Final concentration}}$$

$$\text{reduction }\% = \frac{\text{Initial concentration} - \text{Final concentration}}{\text{Initial concentration}}$$

For example, if the microorganism concentration is reduced by a factor of 100 following a treatment: the percentage reduction is equal to 99%, and the reduction in log unit is equal to 2.

In a similar manner, a 3 log reduction corresponds to a 99.9% percentage, 4 log corresponds to 99.99%, etc. This reduction measure assesses the impact of the inactivation resulting from a certain type of treatment, comparing the respective efficiencies of several systems. First applied to physical retention processes, this unit is now used by extension to measure the efficiency of chemical disinfection treatments (this is known as rate of inactivation).

### 21.3.4. *The notion of Ct*

The chemical disinfection of potable water relies on the use of the oxidizing properties of various reagents. At present, the most widely used reagents are chlorinated compounds (gaseous chlorine, sodium and calcium hypochlorites), chlorine dioxide and ozone. The bactericidal and virucidal efficacy of these multiple reagents is variable.

Different microorganisms have different sensitivities to disinfectants, which themselves have different oxidizing powers (Table 21.2). For a given microorganism, the efficiency of disinfection is mainly affected by the disinfectant's concentration, contact time, pH, temperature as well as by the water's disinfection requirements.

The product of the disinfectant's concentration (C in $mg \cdot L^{-1}$, measured at the outlet of the contact tank) by the contact time (t in minutes) is called Ct (in $mg \cdot L^{-1} \cdot min^{-1}$). Ct values are useful for comparing the efficiency of disinfectants. The concentration or contact time, or both parameters, can be modified to reach the most efficient Ct value. The higher the Ct value, the more effective the disinfection. But the lower the Ct required to achieve a certain microorganism rate of inactivation, the better the disinfectant used is in comparison with another disinfectant requiring a higher Ct for the same reduction. The Ct value is often used to rank the relative susceptibility of different microorganisms. The higher the Ct value required to attain a given rate of inactivation, the more resistant the microorganism.

The efficiency of disinfection can be written out as an equation, following Chick's law:

$$\frac{Nt}{No} = Exp\,(-kt)$$

and:

$$k = \lambda C^{\eta}$$

where:

– No: number of microorganisms initially present;

– Nt: number of surviving microorganisms at time t in the presence of concentration C;

– C: disinfectant concentration (assumed to be constant over time);

– k: rate constant, depending on the microorganism, the disinfectant and the operating conditions (temperature, pH, medium's ionic strength);

– t: contact time with the disinfectant;

– λ: lethality coefficient of the microorganism in the presence of the disinfectant;

– n: a constant called "the dilution coefficient", close to 1.

This law reflects kinetics of order 1 in relation to time and order n in relation to the oxidant concentration for a given microorganism, under determined operating conditions and for a constant oxidant concentration. Doubling the contact time should lead to an additional logarithmic scale in germ reduction. This law stresses the importance of the product Ct.

Disinfection is considered to be efficient when a residual of oxidant C remains for a time lapse t on the water to be treated. For a given disinfectant, the Ct required for the inactivation of n log of germs depends on the target microorganism, temperature and pH.

Table 21.5 shows comparative Ct values for different disinfectants and various target microorganisms (99.9% inactivation, or 3 logarithmic scales). In particular, the Ct values recorded in Table 21.5 highlight the low biocidal power of monochloramine, as well as the superiority of ozone for the inactivation of encysted parasites (*Giardia Lamblia* and *Muris*).

| | **Disinfectant Ct (mg·L$^{-1}$·min$^{-1}$)** | | | |
|---|---|---|---|---|
| **Target microorganisms** | **Chlorine pH 6–7** | **Chlorine dioxide pH 6–7** | **Ozone pH 6–7** | **Monochloramine pH 8–9** |
| ***E. coli*** | 0.034–0.050 | 0.4–0.75 | 0.02 | 95–180 |
| **Poliovirus 1** | 1.1–2.5 | 0.2–6.7 | 0.1–0.2 | 768–3,740 |
| **Rotavirus** | 0.01–0.05 | 0.2–2.1 | 0.006–0.06 | 3,806–6,476 |
| ***Giardia lamblia* (cyst)** | 47 > 150 | – | 0.5–0.6 | – |
| ***Giardia muris* (cyst)** | 30–630 | 120–180 | 1.8–2.0 | – |

**Table 21.5.** *Required Ct values (mg·L$^{-1}$·min$^{-1}$) for a 3 log inactivation at 20°C*

Ct can be calculated from the following ratio:

$$Ct\ (\text{mg}\cdot\text{L}^{-1}\cdot\text{min}^{-1}) = \left(b\,\frac{Cout}{Cin}\right)^{-\frac{1}{n}}$$

In this equation, Cin and Cout are the microorganism concentrations before and after disinfection. Also, n and b are the coefficients related to each microorganism and depend on the inactivation kinetics associated with each of them, as well as on the type of disinfectant. For example, for chlorine disinfection, the coefficients b and n are, respectively, 4 and 2.65 for total and fecal coliforms.

For the calculation of Ct, C is expressed in mg·L$^{-1}$ and t in minutes. Table 21.6 shows the estimated Ct values for different disinfectants.

In practice, C is assimilated to the residual disinfectant concentration at the outlet of the disinfection tank. The accurate calculation of t can only be done using a tracer. The value $t_{10}$ is often proposed for calculating Ct. This corresponds to the time when 10% of the flow has left the reactor (in other words, a minimum transit time of 90% of the water flow). In practice, t represents around 50–60% of the average retention time, obtained by dividing the tank's volume by the flow rate through:

$$t = \frac{\text{Disinfection tank volume}}{\text{Flow}}$$

| Actions | Parameters of use | Gaseous chlorine and hypochlorite | Chlorine dioxide | Ozone |
|---|---|---|---|---|
| Bactericidal | Concentration | 0.3–0.5 mg·L$^{-1}$ | 0.3–0.5 mg·L$^{-1}$ | 0.4 mg·L$^{-1}$ |
| | Contact time | 30 min | 20 min | 4 min |
| Virulicidal | Concentration | 0.3–0.5 mg·L$^{-1}$ | 0.3–0.5 mg·L$^{-1}$ | 0.4 mg·L$^{-1}$ |
| | Contact time | 30–45 min | 30 min | 4 min |
| Ct scale in (mg·L$^{-1}$·min$^{-1}$) | – | 9–22.5 | 6–15 | 1.6 |

**Table 21.6.** *Virulicidal and bactericidal Ct values for different disinfectants (10°C, pH 6.5–7.5 and >3 log reduction)*

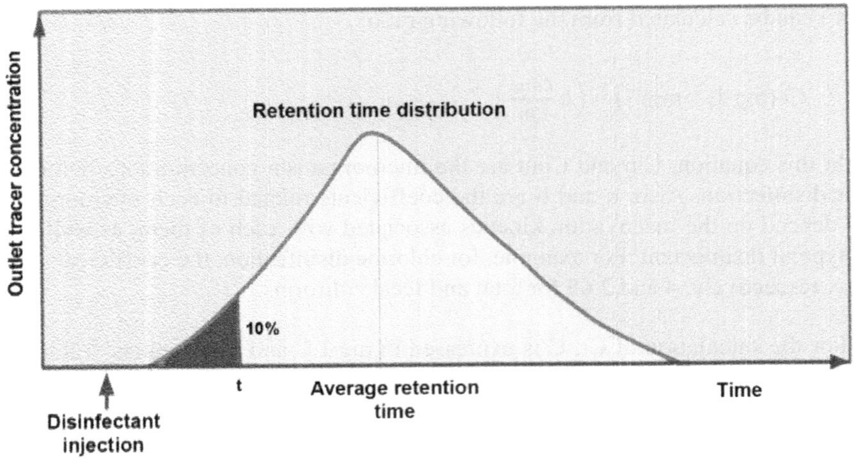


**Figure 21.4.** *Distribution of retention times in a disinfection tank*

The chlorination contact time is always calculated on the basis of the peak flow. The chlorine dose generally varies from 1 to 3 mg·L$^{-1}$. There are several other factors to ensure optimal disinfection conditions, such as temperature, suspended solids, turbidity, organic matter, pH and concentration of oxidizable compounds ($NH_4^+$, $NO_2^-$, $Fe^{2+}$, $Mn^{2+}$, color, etc).

## 21.4. Factors affecting the efficiency of chemical disinfection

The following factors influence the performances of chemical disinfection.

### 21.4.1. *Contact time*

Contact time is one of the most important parameters to be kept under control in order to ensure an efficient disinfection. According to Chick's law, it is expressed as:

$$\frac{Nt}{No} = e^{-kt}$$

where:

– No and Nt: microorganism concentrations at time t = 0 and at time t;

– k: inactivation constant ($min^{-1}$);

– t: contact time (min).

The modified Chick–Watson equation involves the disinfectant's concentration (C in $mg \cdot L^{-1}$), $k'$ being the mortality constant:

$$Ln\frac{Nt}{No} = -k'C^{n}t$$

For *E. coli* and in the presence of chlorine (pH 8.5, T = 5°C), this leads to:

$$Ln\frac{Nt}{No} = -10.48\ C^{1.2}\ t$$

The effect of temperature follows the Van't Hoff–Arrhenius equation where:

$$Ln\frac{t1}{t2} = \frac{E(T2 - T1)}{RT1T2}$$

where:

– E: activation energy ($J \cdot mol^{-1}$);

– T: temperature in Kelvin;

– R: 8.3144 $J \cdot mol^{-1} \cdot K^{-1}$.

The activation energy of the disinfectant is particularly related to the pH. Table 21.7 shows the values of E (activation energy) for chlorine.

| pH | Activation energy ($J \cdot mol^{-1}$) |
|---|---|
| 7.0 | 34,340 |
| 8.5 | 26,800 |
| 9.5 | 50,250 |

**Table 21.7.** *Activation energy for chlorine at different pH values*

### 21.4.2. *Turbidity*

The turbidity in treated water before disinfection must be as low as possible for the following reasons: colloidal turbidity and suspended solids constitute support media for germ adhesion ("shelters"), which reduces the efficiency of disinfectants. In the case of disinfection by UV radiation, turbidity must be as low as possible because it conditions the UV absorption coefficient, thus affecting the efficiency of radiation over germs. In short, turbidity has the effect of protecting microorganisms against the action of disinfectants.

#### 21.4.2.1. *Turbidity and parasite risk*

In a conventional treatment station, the clarification stage is likely to lead to a high reduction in parasites. Among the disinfection processes, UV and ozone (to a lesser extent) are the most effective techniques. On the other hand, chlorine dioxide and chlorine treatments are not particularly efficient under the usual Ct conditions. Before disinfection, the conditions to obtain the lowest parasite concentration require a maximum removal of particles and the low turbidity of filtered water. This goal must be sought under all circumstances, given the health hazards associated with parasites.

### 21.4.3. *Presence of oxidizable matter*

Whether organic or mineral in nature, oxidizable compounds or matter play an important role in maintaining the oxidant concentration in water. They largely contribute to the oxidant requirement and condition the concentration of disinfectants needed for the treatment and to maintain a sufficient residual in the network. It is necessary to be vigilant against the presence of these compounds and to actively work to reduce their concentration. The organic matter content in the water to be disinfected must be limited, on the one hand, to minimize the formation of by-products with chlorine, and more generally, to limit competitive reactions.

Indeed, the presence of organic matter in the water to be treated makes it difficult to maintain a bacteriostatic residual, favoring a possible bacterial revival in the network.

In general, physicochemical disinfection will act more efficiently if clarification has been optimized upstream.

### 21.4.4. *pH*

The pH of the water to be treated is also a factor to consider, especially in the case of chlorine disinfection. Chlorine loses its effectiveness at basic pH (its disinfecting power becomes very weak beyond a pH of 8.5). Hypochlorous acid HClO, which is the active form of chlorine (more efficient than hypochlorite $ClO^-$), is predominant at acidic pH.

### 21.4.5. *Injection mode and injection point*

The optimum efficiency is obtained when the disinfectant is intimately and quickly mixed with the water to be treated. To ensure disinfection, the reagent has to be injected upstream of a tank or a reservoir for it to have sufficient contact time.

It is advisable not to inject chlorine in water containing residual ozone because there is competition between the two oxidants, causing a reagent overconsumption. For the same reason, chlorine dioxide should not be injected into water containing residual ozone. The reaction between ozone and chlorine dioxide also leads to the formation of chlorates, which are undesirable from a health perspective.

The solution then is to inject chlorine or chlorine dioxide as far downstream as possible from the ozonation towers or, if necessary, to neutralize residual ozone.

### 21.4.6. *Design of the contact tank*

From a hydraulic point of view, the contact tank must be designed to limit the presence of short circuits, whereby part of the water would escape and fail to achieve the minimum contact time required. Circulation circuits limiting the tank's usable volume must be restricted. The interior layout of contact tanks and reservoirs must be optimized with spillways and baffles to create the turbulence needed for mixing as well as to prevent hydraulic short circuits.

## 21.5. Qualities of a good disinfectant

The ideal disinfectant should have the following qualities:

– remove pathogenic germs, including the most resistant ones (cysts, spores);

– have a residual effect;

– be stable (to prevent self-decomposition phenomena);

– do not generate undesirable by-products;

– inhibit biofilm growth;

– be simple to implement.

In practice, each disinfection technique has its own advantages and drawbacks:

– *Ozone* is a powerful oxidant and an excellent germicide. On the other hand, it has a low stability in water, making any residual action impossible (ozone is considered to be stable in water for about 20 min). Ozone degrades organic compounds into small molecules, which can be used as carbonaceous substrate for microorganisms (biodegradable dissolved organic carbon [BDOC]), especially in distribution networks. Regarding the toxic – or allegedly toxic – by-products associated with the use of ozone, it is important to raise the question of aldehyde formation and the appearance of bromates after ozonation in water containing bromides.

– *Chlorine* is the most widely used disinfectant because it provides both a germicidal effect and a residual effect. Its implementation conditions are well known and controlled. Chlorine participates in competitive reactions with organic matter. By reacting with humic acids, it gives rise to halogenated compounds, or "haloforms", which are suspected carcinogens (trihalomethanes [THM] or volatile organic compounds). The action of chlorine on certain organic molecules can also produce undesirable sapid products (e.g. with phenols or with planktonic micro-algae).

– *Chlorine dioxide* has higher disinfectant qualities than chlorine, including on sporulated microorganisms. It produces no sapid compounds and does not generate haloforms. On the other hand, chlorine dioxide contains chlorates which can be reduced to chlorites and which transmit an unpleasant metallic flavor to water, which is why their content in tap water is highly regulated. The implementation of chlorine dioxide is quite delicate, which sometimes constitutes an obstacle to its use. Due to its instability, dioxide cannot be transported and must be prepared in situ before use.

– Among all the chlorine-based disinfectants, *monochloramine* (obtained by the joint use of ammonia and chlorine) has the lowest disinfecting power. Indeed, it is approximately 50 times less effective than chlorine and chlorine dioxide (for comparable contact times). On the other hand, it has the advantage of reducing the formation of haloforms (THMs). One of the appreciable qualities of monochloramine is its stability, both over time and in relation to heat (water at 25°C or over), which gives it an interesting residual power.

Table 21.8 summarizes the qualities of the main disinfectants used in water treatments.

To optimize disinfection, oxidant combinations are implemented, for example, a disinfection with ozone (germicidal effect) followed by additional chlorination, before delivery into the network (residual effect).

| | **Ozone** | **Chlorine** | **Chlorine dioxide** | **Chloramines** | **UV** |
|---|---|---|---|---|---|
| **Germicidal effect** | ++++ | ++ | +++ | + | ++ |
| **Residual effect** | 0 | ++ | ++ | +++ | 0 |

**Table 21.8.** *Qualities of the main disinfectants*

## 21.6. Chlorine disinfection

Gaseous chlorine and chlorinated compounds are widely used for the disinfection of potable water. In solution, chlorine can exist in different states and react with certain organic or mineral molecules.

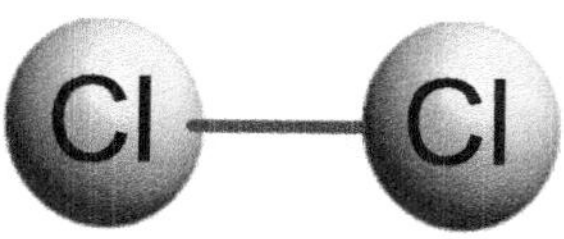


**Figure 21.5.** *Structural formula of chlorine*

The action of chlorine on microorganisms takes place on the surface and by slow diffusion. When microorganisms form clusters or are attached to suspended matter, the action of chlorine is much less effective because it does not completely diffuse inside the formed clusters. In addition, microorganisms can release polysaccharides, which create a kind of barrier which can reduce the effectiveness of chlorination:

chlorination reagents: gaseous chlorine – bleach

For disinfection purposes, chlorine can be used in three forms:

– gaseous chlorine $Cl_2$, stored under pressure in the liquid phase (cylinders, tanks);

– sodium hypochlorite (or bleach), NaOCl (liquid);

– calcium hypochlorite, $Ca(ClO)_2 \cdot 2H_2O$ (solid).

The first two forms, gaseous chlorine and bleach, are extremely common in water treatment. In all three cases, the disinfectant resulting from the reagent's contact with the water to be treated is the same: hypochlorous acid HClO and hypochlorite ($ClO^-$). In general, the effectiveness of chlorine strongly depends on water pH due to the chemical equilibria between the different dissolved forms of chlorine (Table 21.9). For example, between a pH of 7.5 and 7.9, chlorine loses half of its effectiveness.

| **Residual chlorine content** | | | **Composition** |
|---|---|---|---|
| Total chlorine | Free chlorine | Active free chlorine | Elemental chlorine $Cl_2$ Hypochlorous acid (HClO) |
| | | Potential free chlorine | Hypochlorites $ClO^-$ Chlorocyanurates |
| | Combined chlorine | – | Mono-, di- and trichloramines |

**Table 21.9.** *Different forms of chlorine*

### 21.6.1. *Gaseous chlorine*

Gaseous chlorine is dissolved in utility water, using a hydro-ejection device, at an average concentration of 1–2 $g \cdot L^{-1}$. The solution obtained is chlorinated water or chlorine water and used for dosing the disinfectant. Chlorine is the most widely used chemical disinfectant. The term chlorination is often used as a synonym for disinfection. Chlorine can be used as a gas ($Cl_2$), as sodium hypochlorite (NaOCl) which is also known as bleach, or as calcium hypochlorite [$Ca(OCl)_2$], commonly known as HTH or as chlorinated lime ($CaOCl_2$).

The most active form in disinfection is hypochlorous acid HClO, called active chlorine. Depending on pH, hypochlorous acid is in equilibrium with the hypochlorite ion ($ClO^-$) following the reaction:

$Cl_2 + H_2O \rightarrow$ HClO (hypochlorous acid) + HCl

This reaction is accompanied by an acidification of the solution. In a diluted solution and at pH values above 1.0, the equilibrium shifts to the right and very little $Cl_2$ exists in the solution.

As Figure 21.6 shows, the proportion of hypochlorous acid is lower as pH decreases. The phrase "active chlorine" conveys the concentration of a potentially reactive chlorine-based compound in terms of $Cl_2$. For example, 1 g NaOCl corresponds to 1/74.5 = 0.0134 moles of NaOCl, or 1 × 0.0134 × 71 = 0.95 g of active chlorine. Unlike combined chlorine, free chlorine results from an analysis. It corresponds to the sum of hypochlorous acid (HClO) and hypochlorite ($ClO^-$).

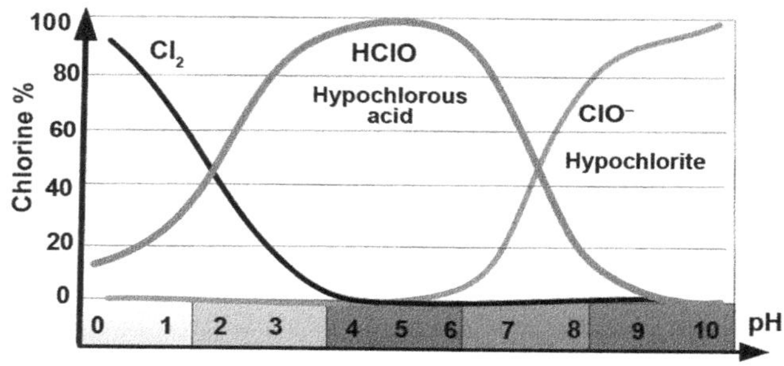


**Figure 21.6.** *Chlorine dissociation in function of pH. For a color version of this figure, see www.iste.co.uk/gaid/watertreatment5.zip*

Given a certain free chlorine content, the percentage of active free chlorine formation is all the higher as pH and temperature decrease (it is optimum at pH 5). The pH of the water to be disinfected is an essential parameter in a chlorine treatment, insofar as the pH range corresponding to high HClO levels is situated in the acidic to neutral pH, rather than in the alkaline pH. In the case of natural waterways with low mineral content, pH values in equilibrium are relatively high. In these conditions, chlorine is less active. At pH 8.0, for example, the percentage of HClO is only about 20% of the free chlorine measured.

The ratio between hypochlorous acid and hypochlorite is given by the equation:

$$\text{Log}\ ([HClO]/[ClO^-]) = 7.6 - pH$$

It is then hypochlorous acid that stands out as the real disinfectant agent, with a bactericidal power about 100 times higher than the hypochlorite ion, called potential chlorine. The sum of the active chlorine and potential chlorine concentrations is called free chlorine. Chlorine gas has the following advantages:

– better effectiveness as a disinfecting agent, compared to bleach;

– homogeneous and immediately effective mixture for the destruction of germs. All the chlorine injected is transformed into active free chlorine, against only 50% for bleach;

– concentrated product reducing storage volume (storage amounting to tons, and not hundreds of tons like bleach);

– good preservation, unlike bleach;

– diversification and security.

Chlorine must be kept under appropriate safety measures and the industrial risk must be carefully monitored:

– highly protected and supervised storage;

– distribution under pressure;

– automatic neutralization unit in the event of a leak;

– setting up superchlorination stations that can be activated in times of risk.

**Figure 21.7.** *Gaseous chlorine facility. For a color version of this figure, see www.iste.co.uk/gaid/watertreatment5.zip*

### 21.6.2. *Hypochlorite*

There are three types of hypochlorites: sodium hypochlorite, calcium hypochlorite and commercial bleach:

– sodium hypochlorite (NaOCl) is available in the liquid form containing up to 12% chlorine;

– calcium hypochlorite ($Ca(OCl)_2$), also called HTH, is a solid which, after dissolution, results in a 65–70% concentration;

– commercial bleach is a sodium hypochlorite solution whose concentration varies between 3 and 5%.

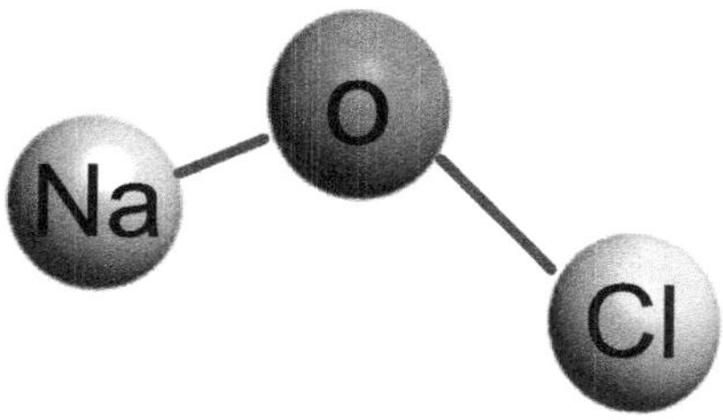


**Figure 21.8.** *Structural formula of sodium hypochlorite*

Hypochlorites react with water to form hypochlorous acid. The reactions are the following:

$$NaOCl + H_2O \rightarrow HClO + NaOH$$

Sodium hypochlorite + water → hypochlorous acid + sodium hydroxide

and:

$$Ca(OCl)_2 + 2H_2O \rightarrow 2HClO + Ca(OH)_2$$

Calcium hypochlorite + water → hypochlorous acid + lime

Hypochlorous acid is a weak acid and dissociates poorly at pH levels below around 6. Between pH 6.0 and 8.5, there is a clear change from undissociated HOCl to an almost complete dissociation:

$$HOCl \rightarrow H^+ + OCl^-$$

Chlorine exists primarily as HOCl at pH levels between 4.0 and 6.0.

Below pH 1.0, and depending on the chloride concentration, HOCl reverts to $Cl_2$ as illustrated by the previous equation. At 20°C, above a pH of 7.5 and at 0°C, above a pH of 7.8, hypochlorite ions (OCl) are predominant. Hypochlorite ions exist almost exclusively at pH levels around 9 and above. The chlorine existing in the HOCl and/or $OCl^-$ form is defined as free chlorine.

For example, if 3 mg·L$^{-1}$ of HOCl is injected into the water during the disinfection stage and the pH is 7, the portion of HOCl which is not dissociated can be deduced as a percentage. The temperature is 25°C.

The dissociation reaction of HOCl is:

$HOCl \rightarrow H^+ + OCl^-$ pKa: 7.54 at 25°C

where:

$$Ka = \frac{[H^+][OCl^-]}{[HOCl]} = 10^{-7.54} = 2.88 \times 10^{-8}$$

$$2.88 \times 10^{-8} = \frac{[10^{-7}][OCl^-]}{[HOCl]}$$

$[HOCl] = 3.47\ [OCl^-]$

The mass conservation law shows that:

$[HOCl] + [OCl^-] = 100\%$

$3.47\ [OCl^-] + [OCl^-] = 100\%$

Therefore, $[OCl^-] = 22.37\%$ and $[HOCl] = 77.63\%$.

This reaction is accompanied by a rise in pH, which can lead to scaling phenomena if the dissolution water is hard.

Regardless of the reagent producing it, hypochlorous acid is a weak acid (pKa 7.6 at 25°C), which is partially dissociated in water, and is in equilibrium with the hypochlorite ion, following the reaction:

$HClO \rightarrow ClO^- + H^+$

The influence of pH on the division between hypochlorous acid and the hypochlorite ion is shown in Table 21.10 and illustrated in Figure 21.9. Although hypochlorite has germicidal properties, the true disinfecting agent is hypochlorous acid. One explanation is that bacterial membranes are lipid media in which ($ClO^-$) ions are poorly diffused, whereas the chemical structure of hypochlorous acid enables it to penetrate the cytoplasmic membranes.

| pH | Total chlorine % in the HClO form | Total chlorine % in the $ClO^-$ form |
|---|---|---|
| 6.0 | 96.5 | 3.5 |
| 7.0 | 72.5 | 27.5 |
| 7.2 | 63.5 | 36.5 |
| 7.3 | 58.6 | 41.4 |
| 7.4 | 53.3 | 46.7 |
| 7.5 | 47.0 | 53.0 |
| 7.6 | 41.1 | 58.9 |
| 7.7 | 35.5 | 64.5 |
| 7.8 | 30.0 | 70.0 |
| 7.9 | 24.5 | 75.5 |
| 8.0 | 22.0 | 78.0 |
| 8.2 | 17.0 | 83.0 |

**Table 21.10.** *Influence of pH on the dissociation of hypochlorous acid at 20°C*

Depending on the form in which chlorine is injected into the water ($Cl_2$, NaOCl or $Ca(OCl)_2$, the different species of hypochlorous acid, hypochlorite and molecular chlorine are present in accordance with the reactions previously described.

The only difference is that gaseous chlorine produces an acidic reaction, whereas sodium and calcium hypochlorites lead to basic reactions. However, this increase or decrease in pH is very low due to the low dosage of additional oxidants. At pH > 5, elemental chlorine $Cl_2$ is found in low concentrations in water.

The sum of the three species $Cl_2$, HClO and $ClO^-$ is referred to as free available chlorine, and the concentration of each of the species is expressed in mg $Cl_2 \cdot L^{-1}$.

Hypochlorous acid is referred to as active chlorine as opposed to free chlorine in the $ClO^-$ hypochlorite form and to combined chlorine which comprises chloramines (mono-, di- and trichloramines resulting from the action of chlorine with ammonia, and the organic chloramines formed by the action of chlorine on certain organic molecules). The sum of free chlorine and combined chlorine is called total chlorine.

At 20°C, hypochlorous acid is the predominant species for pH < 7.6, whereas the hypochlorite ion is predominant for pH > 7.6. These two species are found in equal concentrations at pH 7.6 and 20°C.

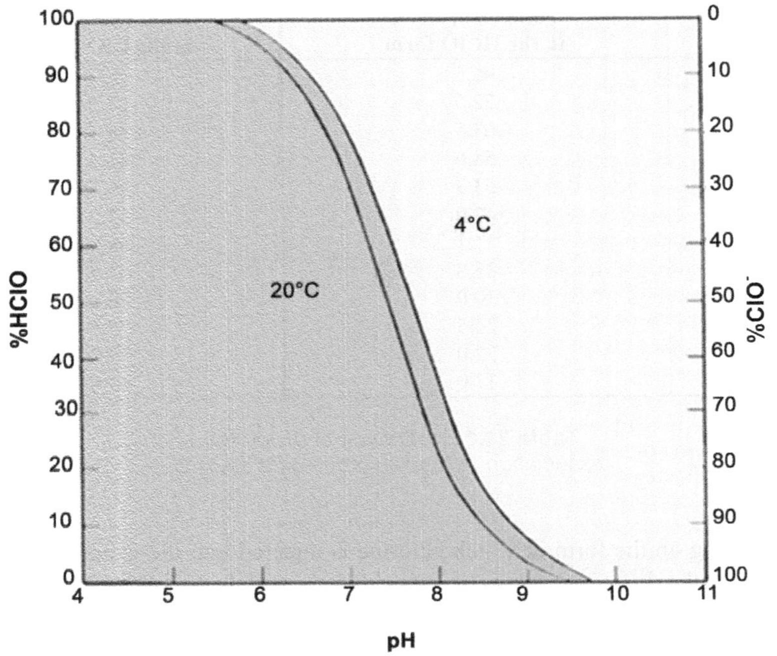


**Figure 21.9.** *Distribution of hypochlorous acid HClO and hypochlorite ion ClO⁻ depending on pH and temperature*

### 21.6.3. *Chlorine application points*

The chlorine application points for disinfection are located upstream and downstream of filtration and upstream of the tanks.

#### 21.6.3.1. *Interchlorination*

The intermediate point which is located upstream of filtration is often used as the first barrier, especially in hot countries where it prevents the development of algae and oxidizes the residual elements (Fe, As). It is injected in-line and the reaction takes place in the volume of water above the media. The concentrations oscillate around 0.3–0.6 $mg \cdot L^{-1}$. This solution does not lead to the formation of THMs because organic matter was removed in the previous stages. But it should immediately be avoided in the event of filter nitrification.

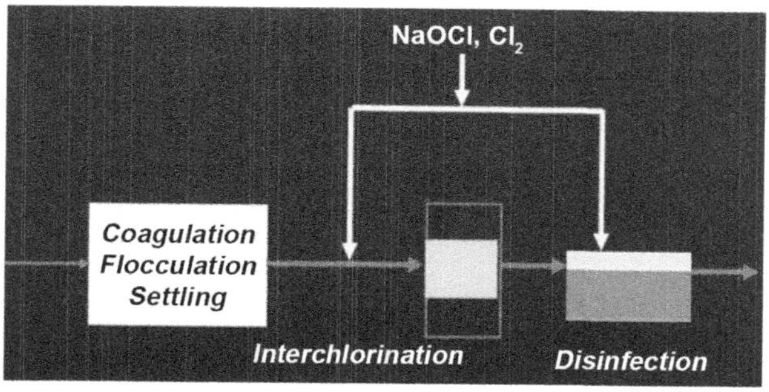


**Figure 21.10.** *Intermediate and final chlorine injection points. For a color version of this figure, see www.iste.co.uk/gaid/watertreatment5.zip*

### 21.6.3.2. *Final chlorination and disinfection tank configuration*

Bacteriostatic chlorination is imperatively scheduled at the end of the treatment chain at a rate ensuring a free residual chlorine of at least 0.4 $mg \cdot L^{-1}$. Depending on the network's configuration, it may be that at points far from the production station, it is only feasible to obtain a sufficient residual at the cost of a large residual at the station's outlet. In this case, intermediate chlorination at various network points can be considered.

Bactericidal conditions correspond to maintaining a free chlorine residual between 0.2 and 0.3 $mg \cdot L^{-1}$ for 15 min (at $pH < 8$).

On the other hand, virulicidal conditions correspond to 0.3–0.5 $mg \cdot L^{-1}$ for 30 min (at $pH < 8$). Water circulation in the disinfection tanks directly influences the water–chlorine contact time, thus conditioning the quality of disinfection. Retention time depends on the geometry of the tanks and their hydraulic loading rate, which has a prominent influence on the quality of the treatment. In order to obtain the optimum retention time, one of the approaches is to simulate hydraulics using fluent-type computational fluid dynamics (CFD) models. This shows different scenarios in terms of flow rates, water heights and interior layouts. From a hydraulic point of view, it is recommended to provide internal baffles to ensure a piston flow. The purpose is to obtain a tank where water moves uniformly from inlet to outlet, limiting recirculation phenomena.

The distribution of the optimum retention time is characterized by time parameters such as $t_{10}$, $t_{90}$, which correspond to 10% and 90% of the input and output volume, whereas tm is the average retention time. The divergence from a piston flow is given by the $t_{10}$/tm ratio, which is also used in the Ct calculation.

Deviation from 1 signals a short circuit level. On the other hand, the $t_{90}$/tm ratio characterizes the recirculation rate inside the tank, quantifying the divergence in relation to the piston flow rate. The closer these ratios are to 1, the more the flow respects a piston flow.

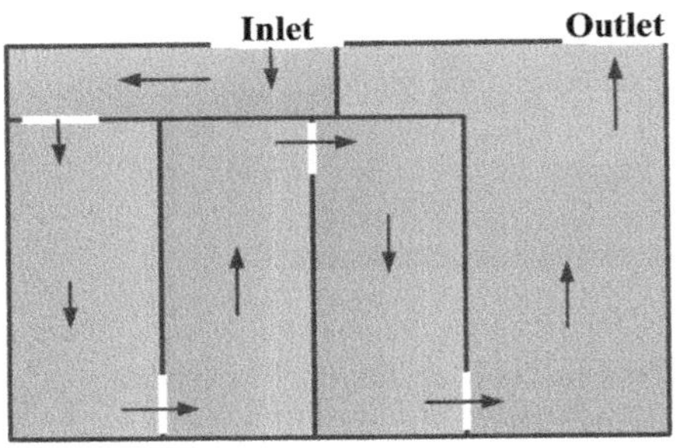


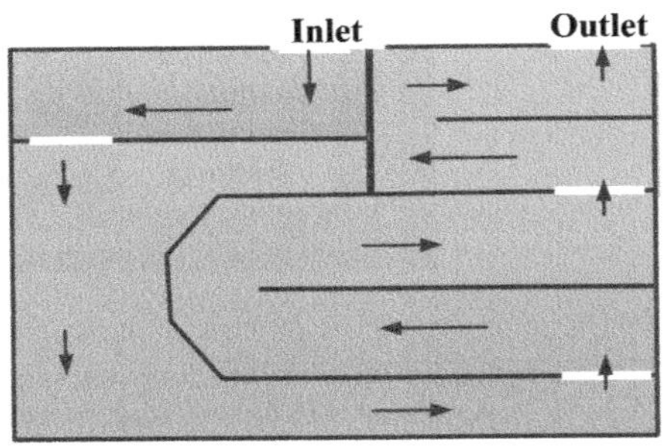


**Figure 21.11.** *Types of configuration of disinfection tanks enabling an optimum contact time*

The transverse baffles, characterized by a low slenderness ratio (compartment length/width), promote recirculation areas in the zones located behind each baffle. The results obtained (Figure 21.11b) show an improvement in hydraulic efficiency by a factor of 1.5 with a $t_{10}$/tm ratio of 0.6.

The same recirculation level is observed for nominal and maximum flows. This means that 90% of the fluid passes through the tank in 1.5 times the average retention time. The longitudinal baffle arrangement is useful for monitoring the contact times along the entire flow rate range.

In terms of water quality, the use of longitudinal baffles instead of transverse baffles improves the Ct value by approximately 50% at nominal flow. At low temperature (5°C), the additional inactivation of *Giardia* due to the slenderness ratio reaches the value of 0.9 log at 25°C. The longitudinal layout of the deflector thus enables a higher level of *Giardia* inactivation because of its high hydraulic efficiency.

21.6.3.2.1. Head loss in disinfection tanks

$$H = \frac{1}{2g}\left(\frac{Q}{Cna}\right)^2$$

where:

– $H$: head loss (m);

– $g$: acceleration due to gravity (9.81 $m \cdot s^{-2}$);

– $Q$: flow rate through ($m^3 \cdot s^{-1}$);

– $C$: transfer coefficient (0.8);

– $n$: number of corridors;

– $a$: surface area of each corridor ($m^2$).

### 21.6.3.2.2. Designing a chlorine disinfection tank

The dispersion coefficient $D$ is:

$$D = 1.01\ \eta\ (Re)^{0.875}$$

where $\eta$ is the kinematic viscosity at T°C ($m^2 \cdot s^{-1}$):

$$Re = \frac{4vrh}{\eta}$$

where:

– $rh$: hydraulic radius: surface area/perimeter (m);

– $v$: flow velocity ($m \cdot s^{-1}$):

$$rh = \frac{H * l}{2\ (H + l)}$$

Dispersion number $d$ (must be $< 0.015$):

$$d = \frac{Dt}{L^2}$$

For example, the dimensions of the disinfection tank for chlorine water can be calculated as follows:

– flow rate: 10,000 $m^3 \cdot j^{-1}$, contact time: 30 min, kinematic viscosity: 1.003 $10^{-6}\ m^2 \cdot s^{-1}$;

– number of channels: n = 2;

– tank volume ($m^3$):

$$V = 10{,}000(\mathrm{m}^3 \cdot \mathrm{j}^{-1}) * \frac{30(\mathrm{min}) * 60\ (s)}{86{,}400} = 208.3\ \mathrm{m}^3$$

– with a depth H = 3 m and a width l = 3 m;

– unit volume for each channel: V ($m^3$) = 104.15;

– the total surface area is $S$ ($m^2$): $S = \frac{208.3}{3} = 69.4\ m^2$;

– unit area: $Su = \frac{69.4}{2} = 34.7\ m^2$;

– total length: $L = \frac{69.4}{3} = 23.14\ m^2$;

– flow velocity:

$$v = \frac{Q}{H * l} = \frac{10{,}000}{86{,}400 * 3 * 3} = 0.0128\ m \cdot s^{-1}$$

– hydraulic radius: $rh = \frac{(3)(3)}{2(3+3)} = 0.75\ m$;

– Reynolds $Re = \frac{(4)(0.0128)(0.75)}{1.003 \times 10^{-6}} = 38{,}285$;

– dispersion coefficient $D$:

$$D = (1.01)(1.003 \times 10^{-6})(38{,}285)^{0.875} = 0.0104\ m^2 \cdot s^{-1}$$

– dispersion number $d$:

$$d = \frac{(0.0104)(30\ x\ 60)}{(23.14)^2} = 0.0135$$

### 21.6.4. *Oxidant demand*

Many inorganic and organic molecules that occur in raw water exert a “demand”, that is to say, a capacity to react with and consume the disinfectant. Therefore, higher-demand waters require a greater disinfectant dose to achieve a specific concentration of the active species for the inactivation of microorganisms. This demand must be satisfied to ensure adequate biocidal treatment. Ferrous ions, nitrites, hydrogen sulfide and various organic molecules exert a demand for oxidizing disinfectants such as chlorine, ozone or chlorine dioxide.

With a halogen disinfectant, a given dosage should theoretically impart a residual concentration of the active agent in the water. From a practical point of view, most natural waters exert a demand for the disinfectant so that the residual in the water is less than the calculated amount based on the dosage. In most cases, the decrease in residual, which is caused by the demand, is rapid while still at the station, but it may be prolonged in the distribution network, until the residual eventually disappears. It is then necessary to incorporate an additional dosage to renew the residual.

The chemical agent may decompose spontaneously, thereby yielding substances having little or no disinfection ability and exerting no measurable residual. For example, not only does ozone react with substances in water exerting a demand, but also decomposes rapidly.

To achieve microbial inactivation with a chemical agent, a residual must be present for a specific time. The nature and level of the residual, together with time of exposure, are important in achieving disinfection or microbial inactivation. Given the fact that the nature of the dosage-residual relationship for natural waters has not been and cannot be reliably defined, the efficacy of disinfection with a chemical agent must be based on a residual concentration/time of exposure relationship. Residual measurements are important and useful in controlling the disinfection process. By knowing the residual–time relationship required to inactivate pathogenic or infectious agents, one can adjust the dosage of the disinfecting agent to achieve the residual required for a given contact time. Thus, the effectiveness of the disinfection process can be controlled and/or judged by monitoring the residual.

#### 21.6.4.1. *Chlorine demand*

Humic substances react with chlorine, producing chloroform ($CHCl_3$) and other THMs. Ammonia and amines in raw water react with chlorine yielding chloramines which do have some biocidal activity, unlike most products of these side reactions. If, in the presence of ammonia, chlorination progresses to the breakpoint, that is to say, to a free-chlorine residual, these chloramines will be oxidized. This causes an even higher chlorine demand. This phenomenon will be examined in further detail. The nature of the disinfectant demand reactions varies with the composition of the water and the type of disinfectant.

Chlorination comes down to satisfying the chlorine demand in the water to be treated, namely by injecting the sufficient quantity of chlorine (or bleach) to detect the appearance of the free-chlorine required for the biocidal action. After the chlorine demand is met, the remaining chlorine is called total chlorine, which is further divided into:

1) combined chlorine, which is the amount of chlorine having reacted with mineral and organic molecules;

2) free-chlorine, chlorine that is left over and is available to inactivate disease-causing organisms. Total chlorine is equal to the sum of combined chlorine and free-chlorine.

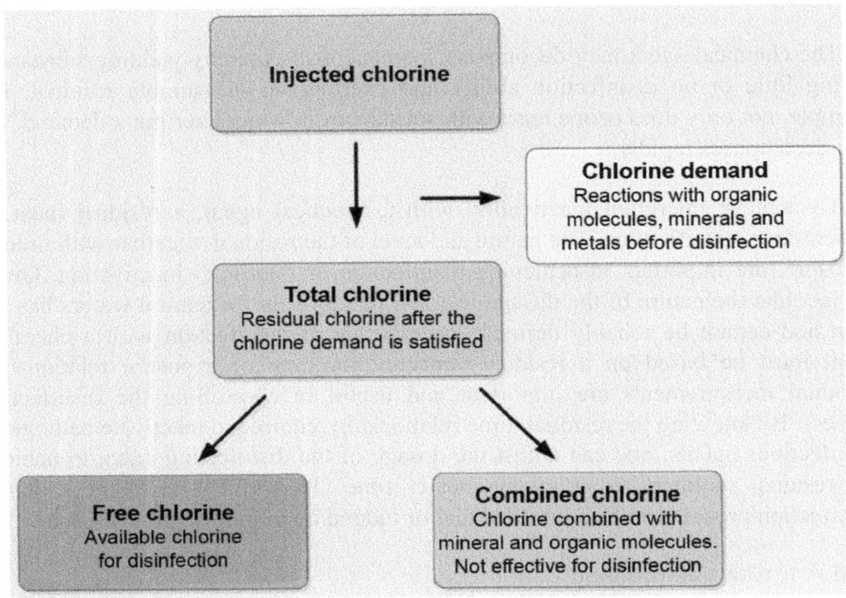


**Figure 21.12.** *Distribution of chlorine fractions depending on water composition*

Since hypochlorous acid has good chemical reactivity, it gives rise to oxidation-reduction reactions (with $Fe^{2+}$ or $NO_2^-$, for example), as well as many substitution reactions (with ammonia and certain organic molecules such as amines, proteins, phenols). The disinfectant concentration to be considered is therefore the residual concentration remaining after all these chemical reactions.

In clean water with no contaminants, the chlorine demand will be zero, and as there will be no inorganic or organic matter, no combined chlorine will be present. Thus, the free-chlorine concentration is equal to the chlorine concentration initially added. In natural waters, especially surface waters, organic matter and metals such as $Fe^{2+}$ are often present, exerting a chlorine demand, and organic compounds like amines and proteins will form combined chlorine. Thus, the concentration of free chlorine will be lower than the concentration of chlorine initially added (free chlorine = total chlorine measurement – combined chlorine measurement).

The chlorine demand of chemical origin is rapid in the minutes following the introduction of chlorine, then becomes slower after a few hours but does not disappear, which makes the notion of residual disinfectant sometimes uncertain. These slow kinetic reactions partly explain the evolution of the chlorine residual in the distribution networks.

It is usually considered that the disinfection treatment has achieved its goals (virulicidal conditions) when the remaining free-chlorine residual equals 0.3–0.5 $mg \cdot L^{-1}$ after a 30–45 min contact time.

In the absence of ammonia or organic nitrogen compounds, water chlorination is very simple. Total chlorine is always equal to free available chlorine.

The chlorine demand by organic substances can be calculated from the following empirical equation:

$$D\ mg \cdot L^{-1} = 0.15\ (TOC)\ (0.3\ e^{1.25t} + 0.7e^{0.017t})$$

where:

– TOC: total organic carbon in $mg \cdot L^{-1}$;

– t: contact time in h.

### 21.6.4.2. *Chlorine demand by the water's mineral elements*

The chlorine demand is defined as the difference between the quantity of chlorine added to the water (chlorine dose) and the remaining quantity of chlorine (residual chlorine) at the end of a given contact time. Inorganic ions ($Fe^{2+}$, $HS^-$, $CN^-$, $NO_2^-$, etc.) react quickly with chlorine, causing an initial rapid chlorine demand.

| **Elements** | **Reactions with chlorine** | **$Cl_2$ demand ($mg \cdot mg^{-1}$)** |
|---|---|---|
| $Fe^{2+}$ | $2\ Fe^{2+} + HOCI + 5\ H_2O \rightarrow 2\ Fe(OH)_3(s) + 5\ H^+ + Cl^-$ | 0.64 |
| $HS^-$ | $HS^- + HOCI \rightarrow S(s) + H_2O + Cl^-$<br>$HS^- + 4HOCI \rightarrow S0_4^{2-} + 5H^+ + 4Cl^-$ | 2.2 ($HS^-$ in S)<br>8.8 ($HS^-$ in S) |
| $NO_2^-$ | $NO_2^- + HOCI \rightarrow NO_3^- + H^+ + Cl^-$ | 5 ($NO_2^-$ in N) |
| $CN^-$ | $CN^- + HOCI \rightarrow CNO- + H^+ + Cl^-$<br>$2CN^- + 5HOCl + H_2O \rightarrow N2(g) + 2\ HCO_3^- + 5Cl^- + 5H^+$ | 2.73<br>6.82 |
| $NO_2^-$ | $NO_2^- + HOCl + H_2O \rightarrow MnO_2(s) + 3\ H^+ + Cl^-$<br>(pH > 8) | 1.3 |

**Table 21.11.** *$Cl_2$/element stoichiometric ratios*

### 21.6.4.3. *Breakpoint*

In the presence of ammonia, chlorine reacts to form a class of compounds called chloramines. Chloramines are characterized by the presence of one or more N-Cl bonds. Three types of chloramines can be formed from the reactions of chlorine with ammoniacal nitrogen:

$$HOCl + NH_3 \rightarrow NH_2Cl \text{ (monochloramine)} + H_2O$$

$$HOCl + NH_2Cl \rightarrow NHCl_2 \text{ (dichloramine)} + H_2O$$

$$HOCl + NHCl_2 \rightarrow NCl_3 \text{ (trichloramine)} + H_2O$$

In the presence of excess chlorine, one obtains the total oxidation of ammoniacal nitrogen into gaseous nitrogen:

$$2NH_3 + 3HClO \rightarrow N_2 + 3H_2O + 3H^+ + 3Cl^-$$

The competition between the reactions depends on water pH, temperature, contact time and, above all, the initial chlorine/ammonia ratio. Monochloramine and dichloramine are the two dominant species in most cases. Chlorine reacts with certain organic nitrogen compounds to form organic chloramines. The sum of all chloramines is called combined available chlorine. The sum of combined chlorine and free-chlorine is called total residual chlorine.

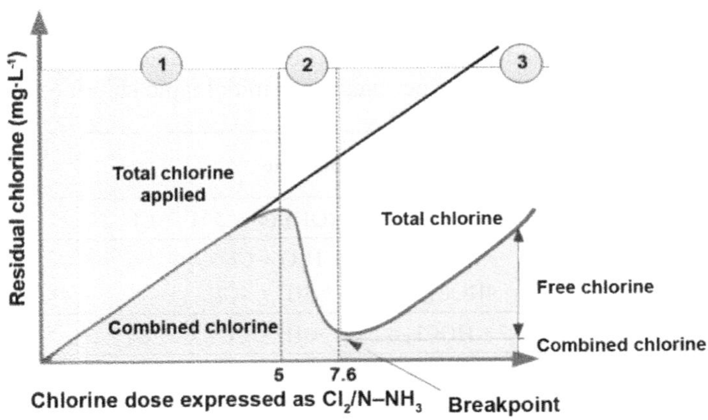


**Figure 21.13.** *Break-point chlorination. For a color version of this figure, see www.iste.co.uk/gaid/watertreatment5.zip*

COMMENT ON FIGURE 21.13.–

*– Zone 1: chlorine destruction by mineral compounds.*

*– Zone 2: formation followed by the destruction of chloramines by adding supplementary chlorine.*

*– Zone 3: production of free active chlorine.*

For water containing ammonia, disinfection must be carried out by chlorination above the breakpoint. The estimation of breakpoint is done with a beaker test, by adding chlorine to water in increasing quantities, and then measuring the residual chlorine after a given contact time (generally 30 min).

The shape of the residual chlorine curve = f (chlorine dose added) is presented and commented in Figure 21.13. Several characteristic areas are identified. The dominant reaction in zone 1 is the reaction between chlorine and ammonia.

The maximum total chlorine theoretically occurs when the ratio of chlorine to ammonia is at 5:1 by weight (1:1 molar ratio), with the formation of monochloramine. The addition of chlorine imparts an increase in the previous molar ratio, provoking the formation of dichloramine. Dichloramine is unstable and the further addition of chlorine quickly results in the formation of trichloramines. The additional dosages lead to a complete oxidation of ammoniacal nitrogen. Residual chlorine is at its lowest level. Breakpoint is then reached.

The complete ammonia oxidation equation indicates that it takes 7.6 mg of chlorine ($Cl_2$) to oxidize 1 mg of ammoniacal nitrogen ($N–NH_4$). In practice, taking into account the competitive chemical reactions, the dose of chlorine required to reach breakpoint is of around 9–10 mg per mg of ammoniacal nitrogen, depending of the presence of organic compounds and/or ions in the water. At this point, the chloramines are oxidized and the dominant residue existing in zone 3 water is free chlorine. An increase in the chlorine dose in this zone leads to a corresponding increase in the free available chlorine.

### 21.6.5. *The implementation of chlorination*

Regardless of whether chlorine is introduced into the water in the form of gaseous chlorine or hypochlorite, only the medium's pH and temperature will determine the nature and proportion of the species present. In practice, the most frequently used expressions to convey the chlorine concentration are as follows: the chlorometric degree (of a hypochlorite solution), which corresponds to the number of liters of gaseous chlorine released by 1 L of hypochlorite solution (under normal temperature and pressure conditions). One chlorometric degree is equivalent to

3.17 $g \cdot L^{-1}$ of free chlorine. The hypochlorite concentration is often expressed as free chlorine: 1 g of sodium hypochlorite (NaOCl) equals 71/74.5 = 0.96 g of free chlorine ($Cl_2$).

#### 21.6.5.1. *Bleach*

Hypochlorites are simpler to implement. They are available as sodium or calcium salt. Sodium hypochlorite NaOCl is formed by the dissolution of gaseous chlorine into an excess soda solution:

$$2NaOH + Cl_2 \rightarrow 2NaOCl + H_2O$$

The commercial aqueous solution, commonly known as bleach, is a greenish-yellow liquid with a pH of 11. Industrial bleach "47/50" titers at 47–50 chlorometric degrees, that is to say, approximately 150 g of active chlorine per liter (1 chlorometric degree is equivalent to 3.17 g of active chlorine). This reagent is injected directly into the water to be treated, generally using dosing pumps. For small facilities, or medium-sized facilities, it is less expensive to use bleach (compared to gaseous chlorine), in which case the equipment is reduced to a storage area and one or more metering pumps.

The industrial bleach solution can be delivered in 20–30 L canisters, in 2,400 L containers, or by tank truck. The quantity of injected hypochlorite must be regularly monitored, because the title of the storage solution decreases over time.

Acids (such as hydrochloric acid or sulfuric acid) should never be stored in the same room as bleach in order to prevent any risk of contact between the two products (violent reaction with chlorine release). Bleach is corrosive and requires the implementation of safety measures, in particular, for the staff. Bleach must be stored in plastic tanks (PVC, polyethylene, polyester resin, etc.) or ebonite steel, inside a watertight container and protected from light, to prevent its decomposition. As a result of chlorine degassing, to prevent the dosing pump from becoming unprimed, vents should be provided on the suction and discharge pipes.

As beach is a basic product, it is necessary to ensure that the service water used for its dilution is not very hard, because there is a risk of progressive clogging of the dosing pump elements as a result of carbonate deposits. In such a situation, either dilute the bleach with softened water, or add an anti-scaling to the dilution water (e.g. sodium hexametaphosphate, with a ration of 50 g/100 L of water).

The chlorometric title of bleach decreases when stored for long (40% decrease after a few months of storage or over when the outside temperature is high). The solution is relatively unstable. It forms chlorates and chlorides. The decomposition reactions are as follows:

$$2\ ClO^- \rightarrow ClO_2^- + Cl^-$$

$$ClO^- + ClO_2^- \rightarrow ClO_3^- + Cl^-$$

The half-life of hypochlorite as a function of temperature is shown in Table 21.12.

| **Temperatures (°C)** | **Half-life (d)** |
|---|---|
| 5 | 3,300 |
| 15 | 890 |
| 25 | 240 |
| 35 | 65 |
| 45 | 18 |

**Table 21.12.** *Decomposition of hypochlorite as a function of temperature (°C)*

#### 21.6.5.2. *Gaseous chlorine*

For large facilities, a treatment with gaseous chlorine is preferable (compared to bleach), because the transport of water – which is always expensive – is avoided and the storage volumes are reduced.

Chlorine is a corrosive and irritating gas, greenish yellow, two and a half times heavier than air (d = 2.491). Its liquefaction temperature under 1 bar is –34°C. Its solubility in water at 20°C under 1 bar pressure is 7–407 $mg \cdot L^{-1}$ of pure gas. It has a high redox potential (1.36 V at 25°C). Gaseous chlorine is commercially available in liquid form at 5–6 bar in 50 kg cylinders or 1,000 kg tanks. Distribution is ensured using chlorometers mounted directly on the containers and operating in a vacuum, which ensures automatic closure of the container in the event of a leak on the pipe carrying the gaseous chlorine.

Since the chlorine is stored in the liquid state and withdrawn in the gaseous state, for a temperature of approximately 20°C, the *maximum* flow rates which can be withdrawn are 1 $kg \cdot h^{-1}$ for a cylinder of 50 kg, and 20 $kg \cdot h^{-1}$ for a 1,000 kg tank.

Gaseous chlorine can also be used on medium-sized facilities. The implementation of disinfection using gaseous chlorine requires the following equipment:

– a liquified chlorine storage device: in 50 kg cylinders for medium-sized facilities or in 1,000 kg tanks for larger facilities;

– a chlorometer ensuring the distribution of gaseous chlorine.

Chlorometers are mounted directly on the chlorine tank or on the wall. They operate in a vacuum and are controlled by an automatic closing of the tank in the event of a leak on the pipe transporting the gaseous chlorine.

As chlorine is stored in the liquid form and withdrawn in gaseous state, the *maximum* flow rates that can be drawn (at 15–20°C) are 1–2 $kg \cdot h^{-1}$ for one cylinder, and 10 $kg \cdot h^{-1}$ for a tank.

For large facilities, several tanks can be commissioned simultaneously and in parallel. If the water flow to be treated is very high, and the number of chlorine tanks to be connected in parallel becomes prohibitive (>6), it is preferable to use an evaporator supplied with liquid chlorine which extracts approximately 25% of the tank's total capacity per hour.

One or more spirometer(s) set with a flow adjustment valve, and one or more hydroejector(s), can help in the preparation of chlorine water by dissolving chlorine gas into service water.

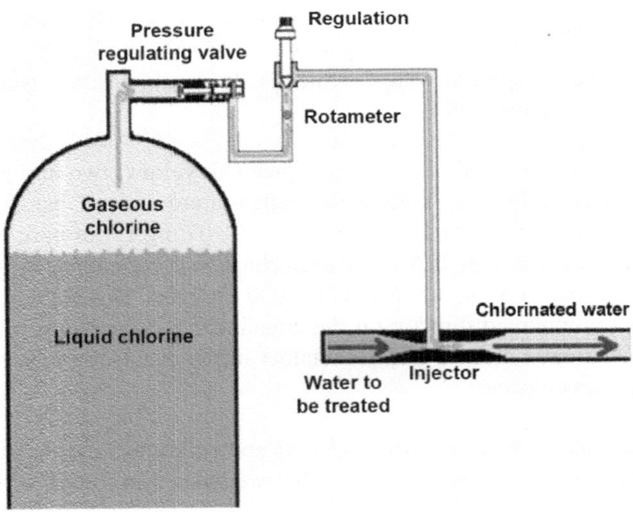


**Figure 21.14.** *Gas chlorination systems. For a color version of this figure, see www.iste.co.uk/gaid/watertreatment5.zip*

In addition, the gaseous chlorine facility can include equipment such as an automatic diverter, which ensures the passage from the empty chlorine tank to the reserve one, without chlorination being interrupted (in the case of stations where continuous monitoring cannot be granted). And a regulation set to interlock the chlorine flow with the water to be treated and/or the residual chlorine.

Chlorine is commercially available in liquid form contained in cylinders at a pressure of 15 bar and an average capacity equal to 40 kg of chlorine. Since pure, dry chlorine is not corrosive, containers are made of ordinary steel. They include a balancing gas space to compensate for pressure variations depending on the outside temperature, which must be between 10 and 40°C during storage. Chlorine storage is regulated under the legislation on "facilities classified for the protection of the environment".

From storage containers, chlorine is distributed to the dosage unit (chlorometer), either in the gaseous form or the liquid form. Direct sampling in the gaseous phase by expansion is limited due to evaporation, which is endothermic. For safety reasons, the direct heating of containers is forbidden. In this case, it is the ambient storage temperature (maintained by air circulation around the confined enclosures), which will determine the gas flow withdrawn. It reaches 10 $kg \cdot h^{-1}$ per 1,000 kg tank in a room at a temperature of 20°C. For higher consumption, chlorine withdrawal is done by liquid phase sampling and forced evaporation. The evaporator is a special bath thermostated at 70–80°C. Its sampling capacity is increased by a factor of 5–10 in comparison with the gas phase sampling system.

The transfer line downstream of the sampling system is thermostatically controlled at a temperature of 3–5°C higher than the temperature of the storage room to prevent any risk of partial condensation. Gaseous chlorine is dosed and expanded to operating pressure in the chlorometer. The underlying principle is to prompt the dissolution of gaseous chlorine using a hydro-ejector, which creates a variable depression in function of the chlorine flow. The capacity of chlorometers can reach 200 $kg \cdot h^{-1}$.

The concentrated aqueous solution from the chlorometer is mixed with the water to be treated by an immersed system designed in such a way that the ejection velocity grants a good mixing quality for the intended application.

In the event of a leak, gaseous chlorine, about 2.5 times denser than air, stagnates at ground level. For chlorine storage of a certain capacity, French legislation requires the installation of a detector for possible chlorine leaks and a neutralization system in a separate room. This room is supplied with air ventilation extending from the storage room's ceiling right down to the floor. In the event of a leak confirmed by the detector, the gas is treated in a neutralization tower by adding a mixture of soda and sodium hyposulfite.

#### 21.6.5.3. *Electrochlorination*

Electrochlorination involves the manufacture of sodium hypochlorite in situ by electrolysis of a sodium chloride solution (brine).

The reactions involved are as follows:

– at the cathode: $Na^+ + e^- \rightarrow Na$;

– in water: $Na + 2H_2O \rightarrow 2NaOH + H_2$;

– at the anode: $2Cl^- \rightarrow Cl_2 + 2e^-$;

– in water: $Cl_2 + H_2O \rightarrow HCl + HClO$.

The hypochlorous acid HClO formed is neutralized with sodium hydroxide NaOH, producing sodium hypochlorite.

An electrochlorination facility includes the following equipment:

– salt saturation: brine preparation;

– brine transfer pump to the electrolyzers;

– softeners to produce softened water for saturated brine, diluting this brine at the electrolyzer (and thus avoid scaling problems at the cathode);

– brine dilution pump;

– electrolysis cells powered by a rectifier;

– storage tank for the hypochlorite produced, equipped with a ventilation system for diluting and discharging the hydrogen formed during electrolysis;

– dosing pump for dosing hypochlorite.

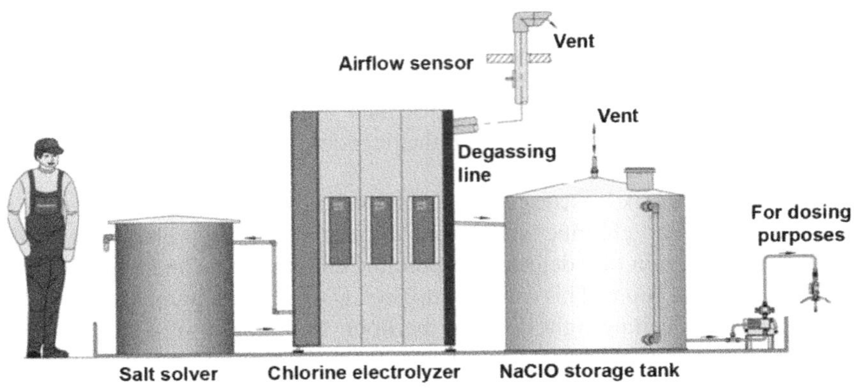


**Figure 21.15.** *Electrochlorination facility. For a color version of this figure, see www.iste.co.uk/gaid/watertreatment5.zip*

The optimum performance of an electrochlorinator is at temperatures between 15 and 20°C.

The announced energy consumption varies between 4.5 and 5.8 kW per kg $Cl_2$, and the salt consumption varies between 3.1 and 3.5 kg NaCl per kg $Cl_2$. The salt used must be food-grade.

The cost of producing sodium hypochlorite by electrolysis depends on the salt concentrations used and the desired hypochlorite concentration. The cost of producing softened water for the generation of brine should also be contemplated. At the industrial level, it is therefore necessary to find a compromise between the salt concentration used and the desired hypochlorite concentration, so as to produce the hypochlorite under the best possible economic conditions (electricity consumption and salt consumption).

Electrochlorination is an interesting alternative to the use of gaseous chlorine (classified facilities, strict regulations) or bleach (degradation over time of the stored industrial product due to its high concentration).

#### 21.6.5.4. *Dechlorination*

When the quantities of chlorine implemented are significant – involving too high a residual for distribution – a dechlorination treatment by injection is a possibility, or the use of a reducing chemical reagent. Dechlorination can be obtained by percolation onto granular activated carbon or by the use of sodium bisulfite ($NaHSO_3$), sulfur dioxide ($SO_2$) or sodium metabisulfite ($Na_2S_2O_5$).

The chemical reactions related to dechlorination involving the use of reducing agents are as follows:

– With *sulfurous gas* ($SO_2$):

$SO_2 + H_2O \rightarrow H_2SO_3$ (sulfurous acid)

$HOCl + H_2SO_3 \rightarrow HCl + H_2SO_4$

that is, $SO_2 + HOCl + H_2O \rightarrow HCl + H_2SO_4$.

or 1.21 mg $SO_2$/mg HOCl or 0.90 mg $SO_2$/mg $Cl_2$.

– With *sodium sulfite* ($Na_2SO_3$):

$Na_2SO_3 + Cl_2 + H_2O \rightarrow Na_2SO_4 + 2HCl$
*sodium sulfite, chlorine, sodium bisulfate, hydrochloric acid*

that is, 1.77 mg sodium sulfite ($Na_2SO_3$)/mg $Cl_2$.

$Na_2SO_3 + HOCl \rightarrow Na_2SO_4 + HCl$
*sodium sulfite, hypochlorous acid, sodium bisulfate, hydrochloric acid*

that is, 2.4 mg sodium sulfite ($Na_2SO_3$)/mg HOCl.

– With *sodium bisulfite* ($NaHSO_3$):

$NaHSO_3 + Cl_2 + H_2O \rightarrow NaHSO_4 + 2HCl$
*sodium bisulfite, chlorine, sodium bisulfate, hydrochloric acid*

that is, 1.46 mg of sodium bisulfite ($NaHSO_3$)/mg $Cl_2$.

$NaHSO_3 + HOCl \rightarrow NaHSO_4 + HCl$
*sodium bisulfite, hypochlorous acid, sodium bisulfate, hydrochloric acid*

that is, 1.98 mg of sodium bisulfite ($NaHSO_3$)/mg HOCl.

– With sodium metabisulfite ($Na_2S_2O_5$):

$Na_2S_2O_5 + HOCl + H_2O \rightarrow Na_2SO_4 + SO_2 + HCl$

$SO_2 + HOCl + H_2O \rightarrow HCl + H_2SO_4$

$Na_2S_2O_5 + 2HOCl + H_2O \rightarrow Na_2SO_4 + 2HCl + H_2SO_4$
*sodium metabisulfite, hypochlorous acid, sodium sulfate, hydrochloric acid, sulfuric acid*

that is, 1.81 mg sodium metabisulfite ($Na_2S_2O_5$)/mg HOCl.

These reactions take place within seconds. For each milligram of neutralized hypochlorite, 2.8 $mg \cdot L^{-1}$ in alkalinity ($CaCO_3$) are consumed. But, as the additional $SO_2$ concentrations are weak (2–5 $mg \cdot L^{-1}$), the pH variation is minor.

– With *sodium thiosulfate* ($Na_2S_2O_3$), there are several possible reactions, depending on water pH:

$Na_2S_2O_3 + 4HOCl + H_2O \rightarrow 2NaHSO_4 + 4HCl$
*sodium thiosulfate, hypochlorous acid, sodium bisulfate, hydrochloric acid*

$Na_2S_2O_3 + HOCl \rightarrow Na_2SO_4 + S + HCl$

$2Na_2S_2O_3 + HOCl \rightarrow Na_2S_4O_6 + NaCl + NaOH$
*sodium thiosulfate, hypochlorous acid, tetrathionate sodium, chloride sodium, caustic soda*

Depending on the reactions, the stoichiometric quantities vary between 0.75 and 6 mg Na2S2$O_3$/mg HOCl.

– With calcium thiosulfate ($CaS_2O_3$):

$$CaS_2O_3 + 4HOCl + H_2O \rightarrow CaSO_4 + 4HCl + H_2SO_4$$

*calcium thiosulfate, hypochlorous acid, calcium sulfate, hydrochloric acid, sulfuric acid*

that is, 0.72 mg calcium metabisulfite ($CaS_2O_3$)/mg HOCl or 0.53 mg $CaS_2O_3$/mg$Cl_2$.

– With calcium metabisulfite ($Ca_2S_2O_5$):

$$Ca_2S_2O_5 + HOCl + H_2O \rightarrow 2CaSO_4 + SO_2 + HCl$$

$$SO_2 + HOCl + H_2O \rightarrow HCl + H_2SO_4$$

$$Ca_2S_2O_5 + 2HOCl + H_2O \rightarrow 2CaSO_4 + 2HCl + H_2SO_4$$

*calcium metabisulfite, hypochlorous acid, calcium sulfate, hydrochloric acid, sulfuric acid*

that is, 2.13 mg calcium metabisulfite ($Ca_2S_2O_5$)/mg HOCl or 1.57 mg $Ca_2S_2O_5$/mg$Cl_2$.

The chemical reactions take place in less than a minute. Dechlorination using these reducing agents can reduce the dissolved oxygen concentration. Furthermore, the formation of sulfuric acid and hydrochloric acid can change the pH. However, as the concentrations used are low, the impacts on $O_2$ and pH are minor, though they need to be monitored.

Reducing agent solutions (sodium metabisulfite, sodium thiosulfate, sodium sulfite, calcium thiosulfate, etc.) are generally prepared in concentrations at 1% and injected using a dosing pump, with an adjustable flow rate upstream of a diffuser in the pipeline of the water to be treated.

These reagents are introduced either at concentrations just sufficient to neutralize all the chlorine in water (stoichiometric quantity), or at concentrations involving a 100% overdose for chlorine neutralization (twice the stoichiometric quantity). The quantities reported in Table 21.13 imply that the reducing agents are 100% pure. Depending on their purity, the quantity to be incorporated must be corrected.

| Reagent | Formula | Molecular weight | Stoichiometric quantitie mg·mg$^{-1}$ $Cl_2$ | Usual range mg·mg$^{-1}$ $Cl_2$ |
|---|---|---|---|---|
| **Sulfurous gas** | $SO_2$ | 64.09 | 0.903 | 1–1.2 |
| **Sodium sulfite** | $Na_2SO_3$ | 126.04 | 1.775 | 1.8–2.0 |
| **Sodium bisulfate** | $NaHSO_3$ | 104.06 | 1.465 | 1.5–1.7 |
| **Sodium metabisulfite** | $Na_2S_2O_5$ | 190.10 | 1.33 | 1.3–1.5 |
| **Sodium thiosulfate** | $Na_2S_2O_3$ | 112.12 | 0.556 | 0.6–0.9 |
| **Calcium thiosulfate** | $CaS_2O_3$ | 184 | 0.53 | 0.5–0.7 |
| **Calcium metabisulfite** | $Ca_2S_2O_5$ | 224 | 1.57 | 1.5–1.8 |

**Table 21.13.** *Dechlorination stoichiometric quantities using reducing agents*

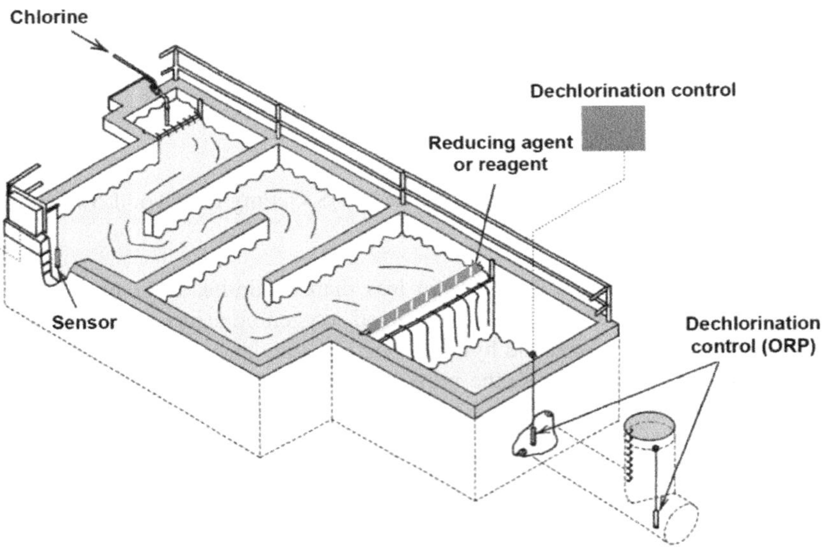


**Figure 21.16.** *Implementation of dechlorination with a reducing agent. For a color version of this figure, see www.iste.co.uk/gaid/watertreatment5.zip*

## 21.6.6. *Disinfection performances with chlorine*

The performances against bacteria, viruses and *Giardia* are summarized from Tables 21.14–21.16. However, they may differ for specific microorganisms.

### 21.6.6.1. *Bacteria*

| Bacterial log reduction | Chlorine Ct ($mg \cdot L^{-1} \cdot min^{-1}$) (pH 7) |
|---|---|
| | 20°C |
| 1 | 0.1–0.2 |
| 2 | 0.4–0.8 |
| 3 | 1.5–3 |
| 4 | 10–12 |

**Table 21.14.** *Required Ct values ($mg \cdot L^{-1} \cdot min^{-1}$) for bacteria*

### 21.6.6.2. *Viruses*

| Viral log reduction | Chlorine Ct ($mg \cdot L^{-1} \cdot min^{-1}$) (pH 6.5–8.5) | | | | |
|---|---|---|---|---|---|
| | 5°C | 10°C | 15°C | 20°C | 25°C |
| 2 | 4 | 3 | 2 | 1 | 1 |
| 3 | 6 | 4 | 3 | 2 | 1 |
| 4 | 8 | 6 | 4 | 3 | 2 |

**Table 21.15.** *Required Ct values ($mg \cdot L^{-1} \cdot min^{-1}$) for viruses*

### 21.6.6.3. *Parasites*

| Log inactivation *Giardia* | Chlorine Ct ($mg \cdot L^{-1} \cdot min^{-1}$) (pH 6.5–8.5) | | | | |
|---|---|---|---|---|---|
| | 5°C | 10°C | 15°C | 20°C | 25°C |
| **1.0-log** | 39–79 | 29–59 | 20–39 | 15–30 | 10–20 |
| **2.0-log** | 78–157 | 59–118 | 39–79 | 29–59 | 19–39 |
| **3.0-log** | 117–236 | 88–177 | 59–118 | 44–89 | 29–59 |

**Table 21.16.** *Required Ct values ($mg \cdot L^{-1} \cdot min^{-1}$) and a chlorine residual of 0.5 $mg \cdot L^{-1}$*

*Cryptosporidium* are even more resistant protozoa than *Giardia*. The Ct values to be implemented are such that chlorine is considered “ineffective” (Ct of 1,000 $mg \cdot min \cdot L^{-1}$ at pH 6 for inactivating 1 log or 90%). Chlorine is efficient against bacteria and viruses, although its action is limited for protozoa and has no effect on *Cryptosporidium* under normal processing conditions.

At this point, it is essential to draw a distinction between chlorine dose and chlorine residual. Confusion often leads to errors that can result in ineffective disinfection. The chlorine dose is the quantity of chlorine applied per liter of water to be treated.

If the water's chlorine demand is high (presence of ammonium ions, organic matter), the efficient free chlorine residual is very low or even inexistent. It is necessary to think in terms of dissolved free chlorine residual, the only one which is actually effective and which must be as stable as possible.

### 21.6.7. *Synthesis of chlorine reactions*

Table 21.17 summarizes the most common reactions observed in groundwater and surface water.

| **Applications** | **Typical doses $mg \cdot L^{-1}$** | **Optimum pH values** | **Reaction times** | **Efficiencies** | **Remarks** |
|---|---|---|---|---|---|
| **Iron** | 0.62 $mg \cdot mg^{-1}$ Fe | 7–7.5 | 3 min | Very good | – |
| **Manganese** | 0.77 $mg \cdot mg^{-1}$ Mn | 7.3–8 | >>1 h | Very slow kinetics | Begins to be effective above pH 8.5 |
| **Ammonia** | 8–10 $mg \cdot L^{-1}$ | 6–8 | 10–20 min | Very good | Break point |
| **Biological growth reduction** | 2–3 $mg \cdot L^{-1}$ | 6–8 | – | Good | By-products in the presence of organic substances |
| **Tastes and smells** | Variable depending on the type of molecule | 6–8 | Variable (20–60 min) | Variable | Efficiency depending on the type of molecules |
| **Color** | Variable | 5–6.9 | 10–30 min | Good | Formation of by-products |
| **Algae** | 5–10 $mg \cdot L^{-1}$ | – | Shock chlorination (1–2 h/d) | Good | Formation of by-products |
| **Zebra molds** | 3–8 $mg \cdot L^{-1}$ | – | Shock chlorination (1–2 h/d) | Good | Formation of by-products |

**Table 21.17.** *Most commonly applied chlorine reactions*

## 21.7. Calcium hypochlorite

The main formation reaction of $Ca(OCl)_2$ is obtained by bringing gaseous chlorine into contact with lime milk:

$$2Cl_2 + 2Ca(OH)_2 \rightarrow Ca(OCl)_2 + CaCl_2 + 2H_2O$$

In practice, a lime excess between 1 and 2% is implemented so as to trigger the reaction and to obtain a pH of 11.2–11.5 in the final solution.

Calcium hypochlorite is prepared by precipitation or crystallization from the mixture of sodium hypochlorite-calcium chloride or hypochlorous acid-lime milk.

Calcium hypochlorite $(CaOCl)_2$ is available as powder, granules or tablets (65–70% $mg{\cdot}mg^{-1}$ active chlorine). Commercial slurry concentrations are 35% potential chlorine. In its pre-hydrated form, active chlorine is found in the $Ca(OCl)_2{\cdot}2H_2O$ form.

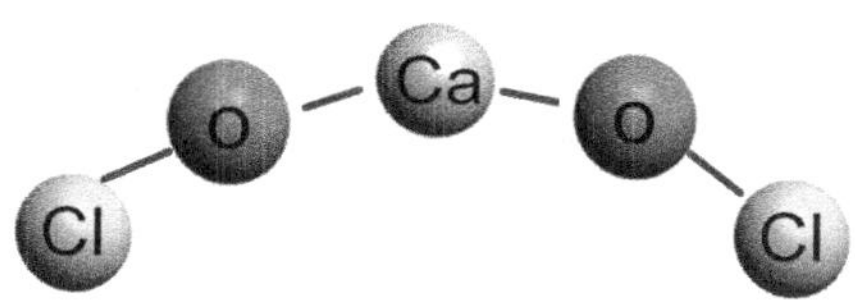


**Figure 21.17.** *Structural formula of calcium hypochlorite*

Calcium hypochlorite is stable when kept in dry places so as to limit its decomposition and can be stored for several months. However, it reacts with the humidity in the air, forming gaseous chlorine. For its use on site, it needs to be diluted in a tank before being pumped, and then mixed with the water to be treated. Due to the impurities it contains (chlorate, metals, etc.), it is recommended not to exceed a concentration in water of 60 $g{\cdot}L^{-1}$.

The percentage of chlorine in calcium hypochlorite is estimated as follows:

$$Cl_2 + 2e^- \rightarrow 2Cl^-$$

$$Ca(OCl)_2 + 2H^+ + 4e^- \rightarrow CaO + H_2O + 2Cl^-$$

The equivalent weight of chlorine is:

$$35.45 = 2\frac{35.45}{2}$$

with the two electrons from the dissociation reaction.

**Figure 21.18.** *Calcium hypochlorite*

The equivalent weight of chlorine for calcium hypochlorite is:

$$35.745 = \frac{142.98}{4}$$

with the four electrons from the dissociation reaction, the percentage of available chlorine is then:

$$\%\text{chlorine} = 100\frac{35.745}{35.45} = 99.17\%$$

## 21.8. Chlorine dioxide disinfection

Chlorine dioxide is a chlorine compound in oxidation state 4, whose formula is $ClO_2$ and has a molar mass of 67.45 g. It is a relatively strong oxidant, but four-fifths of its oxidizing potential is only usable at an acidic pH. At a pH close to neutrality, the transformation of chlorine dioxide will mainly take place in the form of chlorite, in accordance with the first reaction:

$ClO_2 + e^- \rightarrow ClO_2^-$ E: 0.95V at 25°C

$ClO_2^- + 4\ H_3O^+ + 4e^- \rightarrow Cl^- + 6H_2O$

Both reactions show that there are five electrons involved. Note that 1 mole of chlorine dioxide contains 67.5 g and is equivalent to 5 × 35.5 = 177.5 g of $Cl_2$. That is:

$$1 \text{ g } ClO_2 = 177.5/67.5 = 2.63 \text{ g of } Cl_2.$$

Chlorine dioxide is highly soluble in water. Its solubility depends on temperature and pressure (70 $g \cdot L^{-1}$ at 20°C and 1 atm). Its dissolution in water at neutral pH gives rise to a mixture of chlorous and chloric acids:

$$2ClO_2 + H_2O \rightarrow HClO_2 + HClO_3$$

In a basic medium, chlorine dioxide undergoes an irreversible dismutation into chlorite and chlorate ions. The reaction is irreversible at a pH greater than 11:

$$2ClO_2 + 2OH^- \rightarrow ClO_2^- + ClO_3^- + H_2O$$

In a neutral or an acid medium, chlorine dioxide reacts slowly, forming chloric acid:

$$6ClO_2 + 3H_2O \rightarrow 5HClO_3 + HCl$$

Chlorine dioxide is an orange, unstable gas. It explodes under the effect of heat when exposed to light or in contact with organic matter. For this reason, it is prepared in solution at the site of use.

### 21.8.1. *Chlorine dioxide preparation*

Chlorine dioxide is an unstable product that dissociates easily in the gaseous phase. Albeit explosive when present in air at levels greater than 10%, it is harmless when dissolved in water.

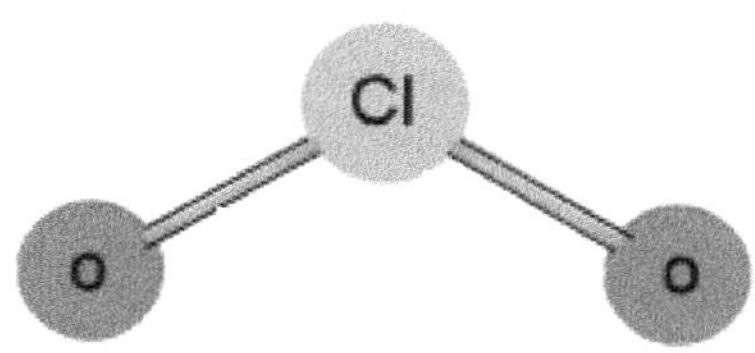


**Figure 21.19.** *Structural formula of chlorine dioxide*

In order to avoid any risk associated with transport and storage, chlorine dioxide is prepared at the place of use (in situ) and should be manipulated immediately. It is prepared by the chemical reaction between chlorine and sodium chlorite. This reaction takes place in the liquid phase and can occur in two manners:

1) due to the action of elemental chlorine on sodium chlorite:

$$2NaClO_2 + Cl_2 \rightarrow 2ClO_2 + NaCl$$

For the production of chlorine dioxide from gaseous chlorine, in modern reactors with a chlorinated water enrichment loop, to obtain 1 g of chlorine dioxide, the requirements are: 0.6 g of chlorine and 1.34 g of sodium chlorite;

2) by the action of hypochlorite in the presence of hydrochloric acid:

$$NaOCl + HCl \rightarrow Cl_2 + NaCl + H_2O$$

$$\downarrow$$

$$Cl_2 + 2NaClO_2 \rightarrow 2ClO_2 + NaCl$$

The second reaction pathway requires the use of three reagents (NaOCl, HCl and $NaClO_2$), but has the advantage of not involving gaseous chlorine, as is the case for pathway (1).

Controlling the stoichiometric reaction is very important to minimize possible by-products:

– If the chlorine concentration is insufficient, a mixture of chlorine and chlorites is obtained, which means that the reaction is incomplete. Although there are still no regulations on the chlorite content in drinking water, it is advisable to maintain a lower content than 0.3–0.4 $mg \cdot L^{-1}$. While their role is not yet clearly understood, chlorites are suspected to have a similar effect on the human body to nitrites (hemoglobin reduction, formation of carcinogenic products due to the reaction with certain amines contained in drugs, etc.). To introduce a sufficient quantity of chlorine (from gaseous chlorine) into the $ClO_2$ preparation reactor, the devices available on the market work with a chlorine enrichment loop.

– If chlorine is excessively dosed, a mixture of chlorine and chlorine dioxide is obtained, and this excess chlorine can give rise to haloforms (THM) or to sapid compounds.

– Chlorine dioxide can also be obtained from the action of a strong (hydrochloric, sulfuric) acid onto sodium chlorite, as follows:

$4HCl + 5NaClO_2 \rightarrow 4ClO_2 + 2H_2O + 5NaCl$ (in the case of hydrochloric acid)

1 $mg \cdot L^{-1}$ of $ClO_2$ equivalent to 2.63 mg/l of $Cl_2$ in oxidizing power.

This preparation method can give rise to various side reactions, including the formation of undesirable products such as chlorates. In addition, this preparation method consumes 25% $NaClO_2$ more than the reaction with chlorine, for the same quantity of chlorine dioxide.

Sodium chlorite used for preparing chlorine dioxide is usually delivered as a 25% solution (density 1.21–1.22) in canisters, containers or tankers. It must be stored in a tank made of an inert material (PVC, polyethylene, etc.).

Due to its oxidizing nature, sodium chlorite must be stored away from heat, humidity and light, in a ventilated room exclusively reserved for the storage of this product. Chlorine dioxide has a faster and more persistent disinfecting action than chlorine. Explosive at concentrations over 10% by volume in air, chlorine dioxide is preferably produced in situ, for safety against the reaction of sodium chlorite with gaseous chlorine, or in a hydrochloric acid medium.

Sodium chlorite can be delivered as a powder for the preparation of a solution at a maximum concentration of 200 $g \cdot L^{-1}$ (it is then necessary to use softened water to avoid the precipitation of carbonates), or in a concentrated solution at 300 $g \cdot L^{-1}$ and specific gravity 1.23.

It is a dangerous product to handle and concentrated solutions should be stored at a temperature over 10°C. The reagents and the reactor (generator) are set in independent rooms with direct openings to the outside.

A chlorine dioxide preparation and dosing facility mainly includes:

– a gaseous chlorine storage area;

– a concentrated sodium chlorite storage area;

– a chlorine dioxide generator;

– a chlorine dioxide transfer tank with dosing pumps.

The operation of the generator is controlled by a level control system, placed in the transfer tank. The chlorine dioxide room is a specific space equipped with natural ventilation.

The storage of chlorine cylinders is done in a room adjoining the dioxide and equipped with a security door.

After dilution in the transfer tank, the dioxide solution is absorbed by two dosing pumps (including one in standby). The dosing pumps are equipped with frequency converters to modulate the reagent rate injected.

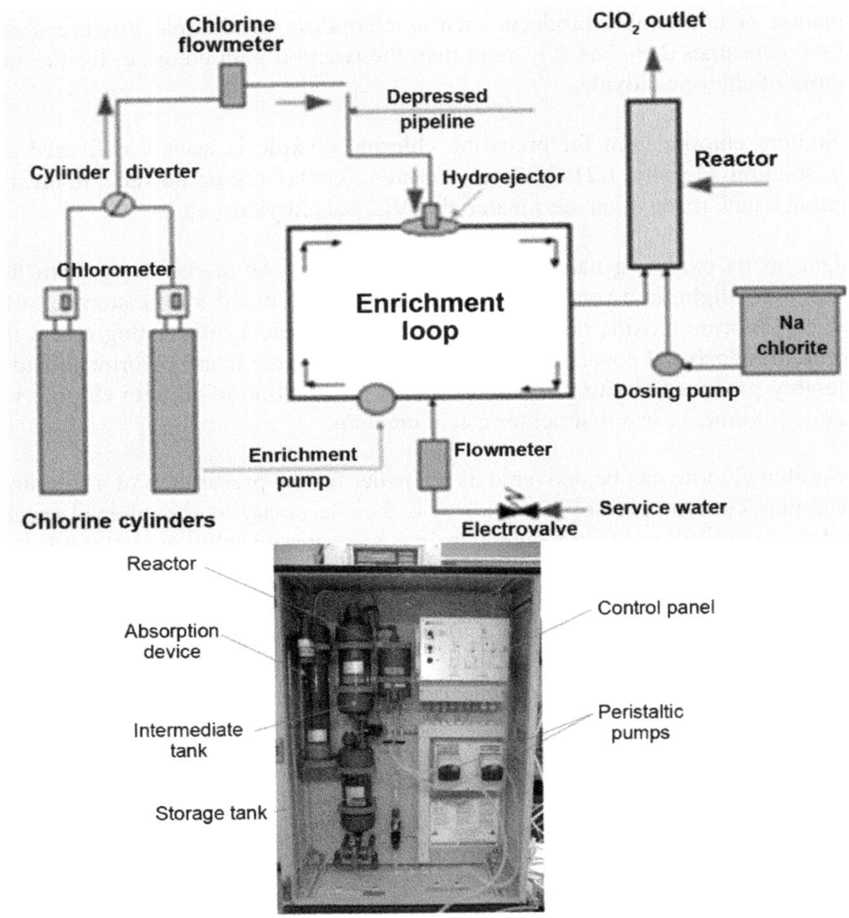


**Figure 21.20.** *Example of a chlorine dioxide preparation facility. For a color version of this figure, see www.iste.co.uk/gaid/watertreatment5.zip*

The usually applied treatment rates are between 0.3 and 0.6 g $ClO_2$ per $m^3$ of treated water. The autonomy required by the specifications, over a period of one to two months, is achieved thanks to chlorine cylinders (equipped with an automatic diverter), and calculated on the basis of an average treatment rate of 0.4 $g \cdot m^{-3}$.

As a safety measure, the chlorine dioxide production unit is equipped with a gas detector, a sensor for gaseous chlorine and a flashing light.

The instrumentation provided for this workstation generally includes a pH meter for verifying that the generator is operating correctly. Chlorine production involves the use of flow meters (for measuring the chlorine flow), capacitive sensors (for starting/stopping the generator), chlorite tank threshold levels, online analyzers (residual chlorine and regulation of the chlorine injection) and float switches (bulbs serving as indicators for the operator).

### 21.8.2. *Chlorine dioxide performances*

Chlorine dioxide has a faster and more persistent disinfecting action than chlorine.

In general, the application conditions for disinfection are the following:

– bactericidal conditions: 0.2–0.3 $mg \cdot L^{-1}$, for 5–10 min;

– virucidal conditions: 0.3–0.5 $mg \cdot L^{-1}$, for 30 min.

It has been found that, unlike chlorine (whose different forms depend on pH and have a variable disinfecting power), the effects of chlorine dioxide on bacteria, viruses and protozoa are not affected by changes in pH (in the range 6–9).

The inactivation kinetics of microorganisms by chlorine dioxide generally follow the Chick–Watson model:

$$(\text{Log } Nt/No = -k\, C^n\, t)$$

Depending on the nature of microorganisms, the value of coefficient n varies. It is equal to 2.5 for *Fluorescent pseudomonas*, 2 for *E. coli*, 0.55 for *Naegleria Gruberi* and 1 for *Giardia* and *Cryptosporidium*.

As with other disinfectants, the efficiency of chlorine dioxide depends on temperature. Tables 21.18–21.20 show the different efficiencies of chlorine dioxide against the different microorganisms.

#### 21.8.2.1. *Bacteria*

| Bacterial log reduction | $ClO_2$- Ct ($mg \cdot L^{-1} \cdot min^{-1}$) (pH 7) |
|---|---|
| | 20°C |
| 1 | 2–4 |
| 2 | 8–10 |
| 3 | 20–30 |
| 4 | 50–70 |

**Table 21.18.** *Required Ct values ($mg \cdot L^{-1} \cdot min^{-1}$) for bacteria*

#### 21.8.2.2. *Virus*

| Viral log reduction | $ClO_2^-$ Ct ($mg \cdot L^{-1} \cdot min^{-1}$) (pH 6.5–8.5) | | | | |
|---|---|---|---|---|---|
| | 5°C | 10°C | 15°C | 20°C | 25°C |
| 2 | 5.6 | 4.2 | 2.8 | 2.1 | 1.4 |
| 3 | 17.1 | 12.8 | 8.6 | 6.4 | 4.3 |
| 4 | 33.4 | 25.1 | 16.7 | 12.5 | 8.4 |

**Table 21.19.** *Required Ct values ($mg \cdot L^{-1} \cdot min^{-1}$) for viruses*

#### 21.8.2.3. *Parasites*

| Log reduction *Giardia* | $ClO_2^-$ Ct ($mg \cdot L^{-1} \cdot min^{-1}$) (pH 6–9) | | | | | |
|---|---|---|---|---|---|---|
| | ≤1°C | 5°C | 10°C | 15°C | 20°C | 25°C |
| 0.5 | 10.0 | 4.3 | 4.0 | 3.2 | 2.5 | 2.0 |
| 1.0 | 21.0 | 8.7 | 7.7 | 6.3 | 5.0 | 3.7 |
| 1.5 | 32.0 | 13.0 | 12.0 | 10.0 | 7.5 | 5.5 |
| 2.0 | 42.0 | 17.0 | 15.0 | 13.0 | 10.0 | 7.3 |
| 2.5 | 52.0 | 22.0 | 19.0 | 16.0 | 13.0 | 9.0 |
| 3.0 | 63.0 | 26.0 | 23.0 | 19.0 | 15.0 | 11.0 |

**Table 21.20.** *Ct values ($mg \cdot L^{-1} \cdot min^{-1}$) for the inactivation of Giardia cysts using chlorine dioxide at pH between 6 and 9*

The data collected in these tables show that it is possible to carry out post-disinfection with chlorine dioxide having previously defined the desired goals of disinfection (in particular its bactericidal action), while ensuring that the chlorite concentration will not exceed the value of 0.2 $mg \cdot L^{-1}$ at the consumer's tap.

### 21.8.3. *Dechlorination*

The chemical reactions for the neutralization of chlorine dioxide are the following and involve the same reducing agents used for chlorine and chlorine dioxide:

– With *sulfurous gas* ($SO_2$):

$$SO_2 + H_2O \rightarrow H_2SO_3 \text{ (sulfurous acid)}$$

$$ClO_2 + H_2SO_3 \rightarrow ClO^- + H_2SO_4$$

that is, 1.33 mg $SO_2$/mg $ClO_2$.

Since sodium bisulfite ($NaHSO_3$) and sodium sulfite ($Na_2SO_3$) are easy to implement, they are a good alternative to $SO_2$:

- *Sodium bisulfite*: $NaHSO_3 + 2ClO_2 + H_2O \rightarrow NaHSO_4 + 2ClO_2^- + 2H^+$

that is, 0.77 mg $NaHSO_3$/mg $ClO_2$.

By neutralizing chlorites:

$$4NaHSO_3 + 2ClO_2^- \rightarrow 4NaHSO_4 + 2Cl^-$$

$$5NaHSO_3 + 2ClO_2 + H_2O \rightarrow 5NaHSO_4 + 2HCl$$

and therefore, 3.85 mg $NaHSO_3$/mg $ClO_2$.

- *Sodium sulfite*: $Na_2SO_3 + 2ClO_2 + H_2O \rightarrow Na_2SO_4 + 2ClO_2^- + 2H^+$

that is, 0.93 mg $Na_2SO_3$/mg $ClO_2$.

The neutralization of chlorites requires an excess of sodium sulfite:

$$4Na_2SO_3 + 2ClO_2^- \rightarrow 4Na_2SO_4 + 2Cl^-$$

that is, $5Na_2SO_3 + 2ClO_2 + H_2O \rightarrow 5Na_2SO_4 + 2HCl$

and therefore, 4.66 mg $Na_2SO_3$/mg $ClO_2$.

– With *sodium metabisulfite* ($Na_2S_2O_5$):

$$2Na_2S_2O_5 + 2ClO_2 + 2H_2O \rightarrow Na_2SO_4 + HCl + 3NaHSO_4 + Cl^-$$

that is, 2.81 mg sodium metabisulfite ($Na_2S_2O_5$)/mg $ClO_2$.

– With *sodium thiosulfate* ($Na_2S_2O_3$): the reaction is the following:

$$5Na_2S_2O_3 + 8ClO_2 + 9H_2O \rightarrow 10NaHSO_4 + 8HCl$$

*sodium thiosulfate, chlorine dioxide, sodium bisulfate, hydrochloric acid*

that is, 1.04 mg $Na_2S_2O_3$/mg $ClO_2$.

This reaction can be accompanied by the formation of sodium chlorite:

$$5Na_2S_2O_3 + 2NaClO_2 + H_2O \rightarrow 5Na_2SO_4 + 2NaHCl$$

that is, 3.1 mg $Na_2S_2O_3$/mg chlorite.

– With *calcium thiosulfate* ($CaS_2O_3$):

$$2CaS_2O_3 + 2ClO_2 + 6H_2O \rightarrow CaSO_4 + CaCl_2 + 3H_2SO_4$$

*calcium thiosulfate, hypochlorous acid, calcium sulfate, hydrochloric acid, sulfuric acid*

that is, 0.93 mg calcium thiosulfate ($CaS_2O_3$)/mg $ClO_2$.

– With *calcium metabisulfite* ($Ca_2S_2O_5$):

$$Ca_2S_2O_5 + HOCl + H_2O \rightarrow 2CaSO_4 + SO_2 + HCl$$

$$SO_2 + HOCl + H_2O \rightarrow HCl + H_2SO_4$$

$$Ca_2S_2O_5 + 2HOCl + H_2O \rightarrow 2CaSO_4 + 2HCl + H_2SO_4$$

*calcium metabisulfite, hypochlorous acid, calcium sulfate, hydrochloric acid, sulfuric acid*

that is, 2.13 mg calcium metabisulfite ($Ca_2S_2O_5$)/mg HOCl or 1.57 mg $Ca_2S_2O_5$/mg$Cl_2$.

| Reagents | Formulae | Molecular weight | Stoichiometric quantities mg·mg$^{-1}$ $ClO_2$ | Usually injected quantities mg·mg$^{-1}$ $ClO_2$ |
|---|---|---|---|---|
| **Sulfurous gas** | $SO_2$ | 64.09 | 1.33 | 1.4–1.6 |
| **Sodium sulfite** | $Na_2SO_3$ | 126.04 | 1.04–4.66* | 1.2–5.0 |
| **Sodium bisulfate** | $NaHSO_3$ | 104.06 | 0.77–3.85* | 0.9–4.0 |
| **Sodium metabisulfite** | $Na_2S_2O_5$ | 190.10 | 2.81 | 2.8–3.1 |
| **Sodium thiosulfate** | $Na_2S_2O_3$ | 112.12 | 1.04 | 1.1–1.3 |
| **Calcium thiosulfate** | $Ca_2S_2O_3$ | 126 | 0.93 | 1.1–1.3 |

*Excess reagent so as to neutralize chlorites.

**Table 21.21.** *Reagent quantity for chlorine dioxide dechlorination*

#### 21.8.3.1. *Recommendations to limit the formation of chlorites and chlorates*

The recommendations to limit the formation of chlorites $ClO_2^-$ and $ClO_3^-$ are:

– controlling the generator setting in the event of a chlorine lack or excess;

– avoid injecting chlorine dioxide upstream of ozonation;

– neutralizing residual ozone upstream of a chlorine dioxide injection;

– optimizing and limiting disinfection treatment rates during the cold water periods;

– whenever possible, avoid the mixture of oxidants such as for example, network rechlorination after a disinfection with chlorine dioxide.

### 21.8.4. *The advantages of dioxide over chlorine*

Unlike chlorine, dioxide does not form additional compounds with organic matter (chlorophenols, THMs). In terms of disinfection, chlorine dioxide remains active across a wide pH spectrum, whereas chlorine disinfection is favored at pH values below 7–7.5.

Ammonia does not react with chlorine dioxide (therefore, the notion of breakpoint does not apply). The inactivation kinetics of $ClO_2$ on microorganisms is faster than those of chlorine.

Dioxide is more effective than chlorine on spore-forming bacteria and parasites. The residual power of chlorine dioxide is higher than that of chlorine.

### 21.8.5. *The special case of the use of chlorine dioxide at a station equipped with ozone*

When water disinfection is done with ozone, there is an increasing demand to inject a small quantity of chlorine dioxide before the water leaves the station so as to maintain the bacteriostatic power of distributed water in the network. There is an incompatibility between $ClO_2$ and $O_3$. If these reagents are simultaneously present in the water, they mutually destroy one another, to such extent that there is no residual ozone or residual $ClO_2$ left in the water.

The reactions are as follows:

$$ClO_2 + O_3 + H_2O \rightarrow 2ClO_3^- + O_2 + 2H^+$$

Before the introduction of $ClO_2$, it is imperative to make sure that there is no residual ozone left. In practice, the dioxide injection point must be placed as far downstream as possible (often at the discharge or suction recovery pumps) in order to use the treated water tank for the removal of ozone. A priori, it is impossible to estimate the time lapse required for the reduction of ozone, because this mainly depends on the quality of water and its temperature. In any case, a time interval of at least 20 min is recommended between injections.

### 21.8.6. *Advantages and drawbacks of using chlorine dioxide*

| Advantages | Drawbacks | Advantages compared to chlorine |
|---|---|---|
| – High germicidal power independent of pH within the range of 6–10<br>– Does not generate any taste<br>– Removal of Fe and Mn complexed with organic matter<br>– Limitation of algae development in the facilities<br>– Improved flocculation<br>– No formation of bromates<br>– No formation of trihalomethanes<br>– No formation of chlorophenols<br>– Persistence of its disinfecting power in the network twice than chlorine | – Incompatibility with ozone: the dioxide injection must be applied into water no longer containing ozone<br>– No action on ammonia<br>– Possible presence of chlorites in the network<br>– Formation of chlorites in the presence of organic matter | – No formation of chloramines or organic derivatives → lower dosage<br>– Its germicidal power even persists at high pH values<br>– Shorter reaction time (15–20 min)<br>– Residual quickly disappears<br>More effective against viruses |

**Table 21.22.** *Advantages and drawbacks of using chlorine dioxide*

## 21.9. Chloramination

### 21.9.1. *Principle*

Ammonia easily dissolves into water and the form in which it is found is highly dependent on pH and temperature. The proportion of the non-ionized form of ammonia increases when the pH exceeds 7.5. Equal concentrations of $NH_3$ and $NH_4^+$ exist at a pH around 9.3 and 25°C.

When chlorine (or hypochlorite) is introduced into water containing ammonia, chloramines are formed, which amount to a fraction of "combined chlorine" (the remainder being constituted by organic chloramines). The organic nitrogen compounds obtained are monochloramine ($NH_2Cl$), dichloramine ($NHCl_2$) and trichloramine ($NCl_3$).

The proportions of mono-, di- and trichloramine formed depend on a number of factors such as pH, temperature, alkalinity, the chlorine dose introduced and the $Cl_2$/N-$NH_4$ proportion.

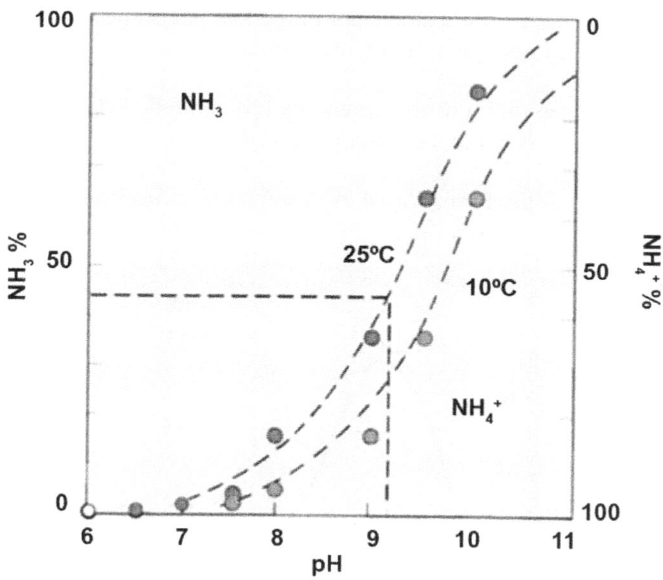


**Figure 21.21.** *Ionized and non-ionized forms of ammonia. For a color version of this figure, see www.iste.co.uk/gaid/watertreatment5.zip*

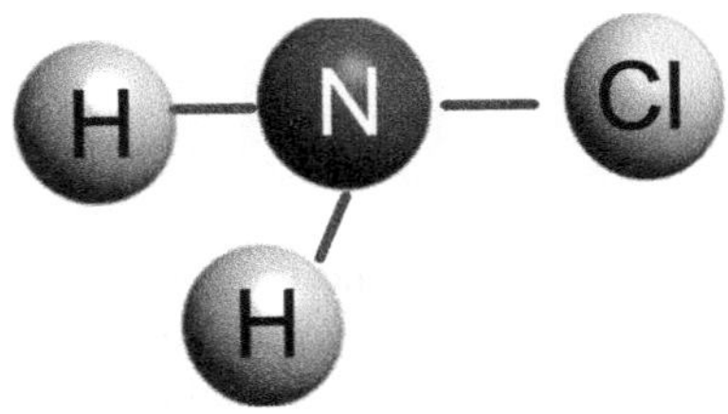


**Figure 21.22.** *Structural formula of monochloramine*

Monochloramine ($NH_2Cl$) is normally the desired chloramine species due to its biocidal properties and relative stability.

The optimum conditions for the formation of monochloramine are pH 8.3, 25°C and a weight ratio from 4:1 to 5:1 $Cl_2$/N-$NH_4$. Normally, at pH conditions of 6.5–9.0 and a temperature between 10 and 25°C, monochloramine is formed within seconds.

Chlorine and chloramines react with organic compounds forming organic chloramines, which have little disinfectant properties, consume the disinfectant and interfere with the measurement of inorganic chloramines.

A minimum concentration of free ammonia (no combined chloramines) must be sought (<0.1 mg·$L^{-1}$ N-$NH_4$) in the water produced.

Figure 21.23 shows the formation of chloramines as a function of pH.

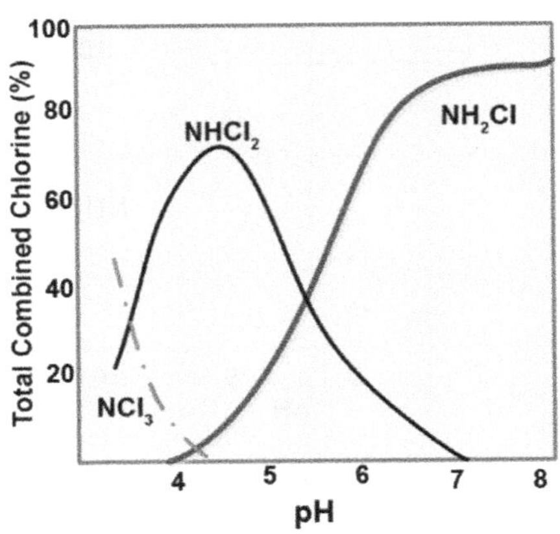


**Figure 21.23.** *Formation of chloramines as a function of pH*

Figure 21.23 shows that monochloramine is mainly formed when the $Cl_2$/N ratio applied is lower than 5:1, expressed as a mass ratio.

Gaseous chlorine as well as sodium and calcium hypochlorite can be used to prepare chloramines, since it is the hypochlorite ion ($OCl^-$) which acts as the essential reagent.

The germicidal power of monochloramine – although lower than that of chlorine and chlorine dioxide – has been extensively proven.

| **Disinfectants** | **E(V) at 25°C** |
|---|---|
| Hypochlorous acid (HOCl) | 1.49 |
| Hypochlorite ($OCl^-$) | 0.90 |
| Chlorine dioxide ($ClO_2$) | 1.71 |
| Monochloramine ($NH_2Cl$) | 0.75 |
| Dichloramine ($NHCl_2$) | 0.79 |

**Table 21.23.** *The oxidation–reduction potential of chloramines compared to other disinfectants*

The major advantage of chloramines is that they grant disinfection without generating halogenated by-products with organic matter (THMs) and have an unparalleled residual power (higher than that of chlorine).

The main drawbacks of using monochloramine are:

– flavors in the treated water, due to dichloramine traces;

– the progressive appearance of nitrites within the network (bacterial nitrification);

– controlling the stoichiometric dosages between chlorine (whose excess is at the origin of THMs formation) and ammonia (whose excess is at the origin of the increase in the network's biofilm).

### 21.9.2. *Implementation*

Chloramination comes down to introducing both ammonia (or ammoniacal water) and chlorine (gaseous chlorine or hypochlorite) in the water to be treated, while controlling the pH reaction conditions and stoichiometry. A Cl/N-$NH_4$ ratio between 3 and 5 (mg $Cl \cdot mg^{-1}$ N-$NH_4$) is generally recommended, as well as a pH greater than or equal to 7, so as to promote the formation of monochloramine, whose germicidal power is known.

In general, the injection points are based on the addition of ammonia ($NH_3$) followed by chlorine. When chlorine is added in the first place, there may be taste and odor problems due to the reaction of chlorine with organic compounds (phenolics, toxins, etc.). The usual $NH_2Cl$ dosages are between 2.0 and 4.0 $mg{\cdot}L^{-1}$. The residual tolerated in the distribution networks is 0.5 $mg{\cdot}L^{-1}$.

Since the germicidal efficacy of monochloramine is 200 times lower than that of free chlorine, very long contact times (60–120 min) are required to attain adequate disinfection performances.

When ammonia lingers (<0.5 $mg{\cdot}L^{-1}$) in filtered water, two chloramination solutions can be implemented:

– applying a breakpoint in order to remove all the residual ammonia and then implementing chloramination with controlled $Cl_2/N$ ratios;

– applying chloramination acknowledging the ammonia present. Nevertheless, given the fact that this value strongly varies throughout the year, it is important to keep it under control with an ammoniometer in order to avoid chlorine deficits.

### 21.9.3. *Performances of monochloramine*

The inactivation of *E. coli* by monochloramine shows that it reacts swiftly with four amino acids: cysteine, cystine, methionine and tryptophan. The inactivation mechanism involves the inhibition of proteins or protein-mediated processes, such as respiration.

As with free chlorine, the virus inactivation mechanism depends on factors such as type of virus and disinfectant concentration.

#### 21.9.3.1. *Bacteria*

| Bacterial log reduction | $NH_2Cl$ Ct ($mg{\cdot}L^{-1}{\cdot}min^{-1}$) (pH 7) |
|---|---|
| | 20°C |
| 1 | 4–6 |
| 2 | 12–20 |
| 3 | 30–75 |
| 4 | 200–250 |

**Table 21.24.** *Required Ct values ($mg{\cdot}L^{-1}{\cdot}min^{-1}$) for bacteria*

#### 21.9.3.2. *Virus*

| Viral log reduction | $NH_2Cl$ Ct ($mg \cdot L^{-1} \cdot min^{-1}$) (pH 6.5–8.5) | | | | |
|---|---|---|---|---|---|
| | 5°C | 10°C | 15°C | 20°C | 25°C |
| 2 | 857 | 643 | 428 | 321 | 214 |
| 3 | 1,423 | 1,067 | 712 | 534 | 356 |
| 4 | 1,988 | 1,491 | 994 | 746 | 497 |

**Table 21.25.** *Required Ct values ($mg \cdot L^{-1} \cdot min^{-1}$) for viruses*

It takes between 2 and 8 h of reaction time with 1 $mg \cdot L^{-1}$ of residual $NH_2$ Cl to achieve the contact.

A 2-log inactivation for polioviruses and coxsackieviruses B1 and B5 can be obtained with chloramine based on a Ct of 643 $mg \cdot L^{-1} \cdot mn^{-1}$ at 15°C. Contrary to this, it only takes 0.25 $mg \cdot L^{-1}$ of free chlorine and a 15-min reaction time with chlorine to achieve the same inactivation.

#### 21.9.3.3. *Parasites*

Chloramines are less powerful disinfectants against viruses and protozoa. It is difficult to meet performance goals for protozoans such as *Cryptosporidium* and *Giardia* because unduly extensive Cts are required.

| *Giardia* log inactivation | $NH_2Cl$ Ct ($mg \cdot L^{-1} \cdot min^{-1}$) (pH 6.5–8.5) | | | | |
|---|---|---|---|---|---|
| | 5°C | 10°C | 15°C | 20°C | 25°C |
| 0.5 | 365 | 310 | 250 | 185 | 125 |
| 1 | 735 | 615 | 500 | 370 | 250 |
| 1.5 | 1,100 | 930 | 750 | 550 | 375 |
| 2 | 1,470 | 1,230 | 1,000 | 735 | 500 |
| 2.5 | 1,830 | 1,540 | 1,250 | 915 | 625 |
| 3 | 2,200 | 1,850 | 1,500 | 1,100 | 750 |

**Table 21.26.** *Ct ($mg \cdot L^{-1} \cdot min^{-1}$) for the inactivation of* Giardia *with monochloramine*

### 21.9.4. *Dechloramination*

Dechloramination can be achieved using $SO_2$, sodium bisulfite ($NaHSO_3$) or sodium metabisulfite ($Na_2S_2O_5$).

With *sulfurous gas* $SO_2$:

$$SO_2 + H_2O \rightarrow H_2SO_3 \text{ (sulfurous acid)}$$

$$NH_2Cl + H_2SO_3 + H_2O \rightarrow NH_4Cl + H_2SO_4$$

$$NHCl_2 + 2\ H_2SO_3 + 2\ H_2O \rightarrow NH_4Cl + HCl + 2\ H_2SO_4$$

$$NCl_3 + 3\ H_2SO_3 + 3\ H_2O \rightarrow NH_4Cl + 2\ HCl + 3\ H_2SO_4$$

These reactions take place within seconds. For each mg of neutralized hypochlorite, 2.8 $mg \cdot L^{-1}$ in alkalinity ($CaCO_3$) are consumed. But, as the additional $SO_2$ concentrations are weak (2–5 $mg \cdot L^{-1}$), the pH variation is minor. Dechloramination leads to the formation of ammonium chloride, which is ultimately found in the produced water. The choice of chloramination as a disinfection solution must be carefully considered, particularly when no chloramine residual is desired in the water leaving the station.

Since sodium bisulfite ($NaHSO_3$) and sodium metabisulfite ($Na_2S_2O_5$) are easy to implement, they are a good alternative to $SO_2$:

$$NaHSO_3 + NH_2Cl + H_2O \rightarrow NaHSO_4 + NH_4Cl$$

$$Na_2S_2O_5 + NH_2Cl + 3H_2O \rightarrow 2NaHSO_4 + NH_4Cl$$

Other reagents can be used, such as sodium sulfite ($Na_2SO_3$) and sodium thiosulfate ($Na_2S_2O_3$). On the other hand, their use requires rigorous controls because an overdose leads to a reduction in the concentration of dissolved oxygen. This is the reason why they are rarely used in drinking water stations.

It is little used in Europe, mainly due to the permissible limit of ammonia in distributed water (guide level [GD] = 0.05 $mg \cdot L^{-1}$ and maximum acceptable concentration [MAC] = 0.5 $mg \cdot L^{-1}$, according to EEC standards). In Europe, ammonia is considered an undesirable product, interfering with chlorine disinfection. Current practice favors the removal of ammonia upstream of disinfection by biological means if the conditions permit it. This solution reduces water's chlorine demand and consequently, the chlorine levels injected in the final chlorination. The formation of THMs is strictly limited.

### 21.9.5. *Advantages and drawbacks of chloramination*

The advantages and drawbacks of chloramination are enumerated in Table 21.27.

| Advantages | Drawbacks |
|---|---|
| – Germicidal power independent of pH within the range of 6–9<br>– Longer persistence than chlorine and chlorine dioxide<br>– Twice the persistence of its disinfecting power in the network over chlorine<br>– Chloramines are not as reactive with organics as is free chlorine in the formation of THMs<br>– The monochloramine residual is more stable and lasts longer than free chlorine or chlorine dioxide, offering better protection against bacterial regrowth in distribution networks<br>– Biofilm control tends to reduce biofilm-induced corrosion<br>– Chloramines do not react with organic compounds<br>– No formation of bromates, trihalomethanes and chlorophenols | – Taste and odor problems in case of overdose<br>– Nitrite problems in case of excess ammonia Lower inactivation against viruses and protozoa than chlorine, ozone and chlorine dioxide<br>– Cannot be applied in all cases, depending on the microorganisms to be removed and the performances required<br>– Low $Fe^{2+}$ and $Mn^{2+}$ oxidation<br>– Less effective as a disinfectant at high pH than at low pH values<br>– Residual dichloramines can be problematic<br>– excess ammonia in the distribution system can lead to nitrification problems, especially in the network areas with low disinfectant residuals<br>– Different germicidal quality and organoleptic qualities depending on the size of networks |

**Table 21.27.** *Advantages and drawbacks of chloramination*

## 21.10. Proportion of chlorine in chlorine disinfectants

Table 21.28 shows the proportion of available chlorine in different chlorine disinfectants. This approach makes it possible to compare their efficiency.

| Disinfectants | Molecular weight | Chlorine equivalents | % of chlorine present | % of available chlorine |
|---|---|---|---|---|
| $Cl_2$ | 71 | 1 | 100 | 100 |
| $Cl_2O$ | 87 | 2 | 81.7 | 163.4 |
| $ClO_2$- | 67.5 | 5 | 52.5 | 260 |
| $CaOCl_2$ | 127 | 1 | 56 | 56 |
| $Ca(OCl)_2$ | 143 | 2 | 49.6 | 99.2 |
| HOCl | 52.5 | 2 | 67.7 | 135.4 |
| $NaClO_2$ | 90.5 | 4 | 39.2 | 157 |
| NaOCl | 74.5 | 2 | 47.7 | 95.4 |
| $NHCl_2$ | 86 | 2 | 82.5 | 165 |
| $NH_2Cl$ | 51.5 | 2 | 69 | 138 |

**Table 21.28.** *Proportion of chlorine in chlorine disinfectants*

– Chlorine present:

$$\% \text{ of } Cl_2 \text{ present} = 100\ \frac{\text{Chlorine weight in the disinfectant}}{\text{Disinfectant's molecular weight}}$$

– Available chlorine: Chlorine's oxidation power is based on the value of the disinfectant's chlorine valence, which is then reduced to valence –1:

$$HOCl + H^{+} + 2e^{-} \rightarrow Cl^{-} + H_2O$$

Two electrons are at stake. Then available chlorine is then:

$$\% \text{ of available } Cl_2 = (\text{Chlorine equivalent})\ (\% \text{ of } Cl_2 \text{ present})$$

With HOCl, the percentage of chlorine present is:

$$100 \times 35.5/52.5 = 67.7\%$$

With $ClO_2$, the percentage of chlorine present is:

$$100 \times 35.5/67.5 = 52.5\%$$

## 21.11. Disinfection with ozone

Ozone is extensively used for the disinfection of potable water. It is actually the best disinfectant among the oxidizing reagents available. Its efficiency gain compared to chlorinated compounds is especially remarkable at the level of viruses and parasite cysts such as *Giardia* and *Cryptosporidium*.

Ozone can be used at different points at the water treatment station for different purposes. Ozone is used primarily as a pre-treatment oxidant. As well, it is used as a pre-disinfectant for inter-ozonation and as a disinfectant for post-ozonation. In post-disinfection, ozone is becoming less popular due to the formation of bioassimilable dissolved organic carbon (BDOC), which can cause bacterial regrowth in the distribution network.

### 21.11.1. *General remarks on ozone*

With the chemical formula $O_3$, ozone is a naturally existing gas in the upper atmosphere, where it is formed from oxygen under the action of ultraviolet radiation. Its presence is signaled by a characteristic smell, which can sometimes be detected at lower altitudes in stormy weather (due to the action of atmospheric electrical discharges on the oxygen in the air).

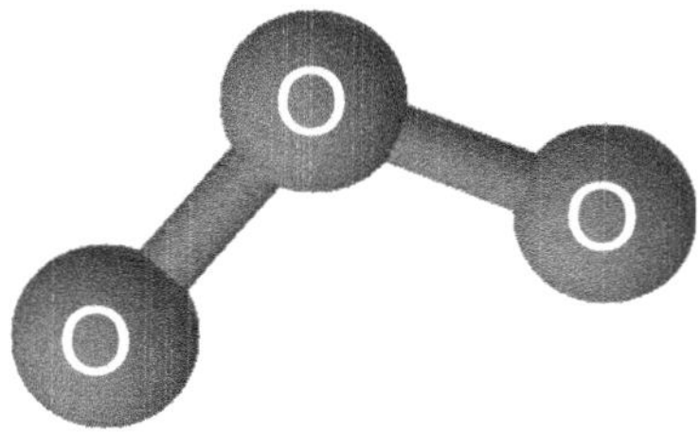


**Figure 21.24.** *Structural formula of ozone*

Ozone is used in water treatment due to its strongly oxidizing properties. It is relatively unstable in an aqueous solution: under the usual pH conditions found in water treatment (pH 6–8), it remains stable for around 10–20 min.

The disappearance of ozone in water is explained by a natural phenomenon of radical self-decomposition initiated by $OH^-$ ions, and consumption by oxidant-consuming substances present in the water (organic matter, oxidizable mineral matter).

Water's ozone demand is the quantity of ozone to be introduced to detect a non-zero residual after a 6-min contact time.

Due to its instability in water, ozone does not provide a lasting residual. In the majority of cases, disinfection by ozonation must be complemented by a secondary reagent to relay the action of ozone. Usually only a small quantity of chlorine or chlorine dioxide suffices to ensure a bacteriostatic effect in the distribution network. However, there are cases in which ozone is used on its own: with water of exceptional quality, small networks and/or systems in good condition, etc.

In addition to its exceptional disinfecting power (destruction of pathogenic germs 20–30 times faster than chlorine), ozone is used because of its many other properties. To mention only a few:

– the removal of certain pesticides;

– the destruction of detergents in very high proportions;

– the complete removal of phenols;

– the destruction of algae in large proportions;

– the reduction of tastes and odors;

– the removal of organic-origin color;

– the significant reduction of chloroform-extractable matter (which constitute a comprehensive analytical indicator of organic micropollutants).

Conversely, ozone has no perceptible effect on ammonia (very slow oxidation kinetics).

### 21.11.2. *Ozone production*

In the field of water treatment, ozone is prepared by passing a big arc discharge through an oxygen current. The source of oxygen used is generally air or oxygen-enriched air. In some cases, pure oxygen is used. The ozone thus produced is always diluted in the carrier gas used for its preparation: the ozone concentrations obtained from air or oxygen are 10–20 g per m$^3$ of ozone-enriched air or 40–60 g per m$^3$ ozone-enriched oxygen.

An ozone facility includes four sections:

– air treatment;

– ozone generation;

– ozone transfer into the water to be treated;

– recovery and treatment of ozone vents.

#### 21.11.2.1. *Air treatment*

The air used for the production of ozone must be perfectly dust-free and extremely dry (dew point lower than –50°C) to grant production efficiency and to safeguard the materials used in the ozonizers. During the formation of ozone, air ionization forms nitrogen oxides, including nitric anhydride which is transformed into a highly corrosive acid in the presence of humidity. Furthermore, as the decomposition of ozone is accelerated at high temperature, to improve production efficiency, it is essential to work with cold air.

Regardless of the technology chosen for air treatment, either high-pressure (6–7 bar) or low-pressure (LP) applications (1.5–2 bar), the devices include:

– an overpressure or compression stage after the filtration of ambient air;

– a drying stage comprising:

- a first condensation stage,

- a second drying stage onto an adsorbent.

The choice between high-pressure and LP applications results from an optimization of investment and operating costs. In practice, high-pressure systems are implemented at facilities handling low flow rates, whereas LP systems tend to be used at facilities handling high flow rates.

### 21.11.2.2. *Ozone generation*

As ozone is unstable, it has to be directly produced at the site of use. An industrial ozonizer is an assembly of elementary generators composed of two electrodes:

– a central stainless steel electrode;

– a ground electrode made of a stainless steel tube cooled by water circulation.

The space between these two electrodes is occupied by two dielectrics: one with a high permittivity thin glass and the other with air or oxygen.

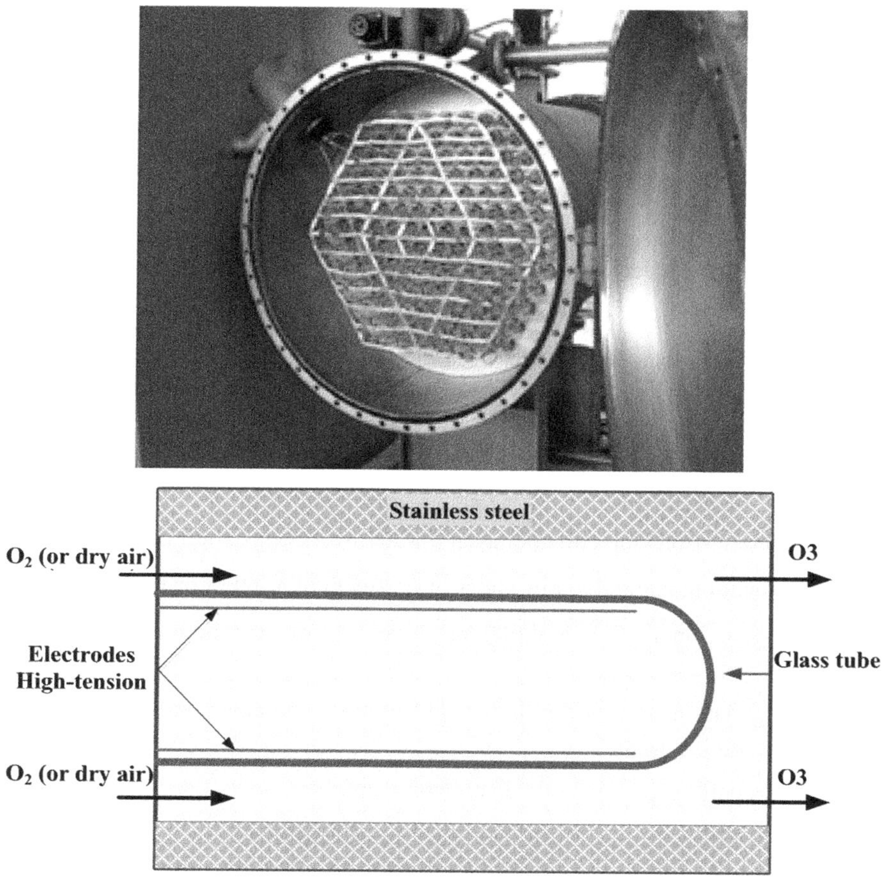


**Figure 21.25.** *Diagram of an elementary tubular ozonizer. For a color version of this figure, see www.iste.co.uk/gaid/watertreatment5.zip*

When these two electrodes are de-energized (for a voltage greater than a threshold value), the gas ionizes and becomes an electrical conductor: the appearance of luminous discharges can be observed, accompanying the formation of ozone. The production of ozone (expressed in kg of ozone/h) can be adjusted by regulating the applied voltage.

Most of the energy is dispersed in the form of heat which is carried away by the cooling water circulating around the ground electrode. The cooling water is generally drawn from the station's service water circuit and returned to the treated water tank. While the content of this water is somewhat high in chlorides, it is necessary to provide a closed cooling water circuit to prevent any attack on the stainless steel.

#### 21.11.2.3. *Ozone transfer into the water to be treated*

In general, the ozone transfer into the water is obtained by means of different processes: the use of self-priming turbines, hydro-ejectors or emulsifiers, the diffusion of ozone-enriched gas by porous media.

In the case of (post-ozonation) disinfection, given the fact that water is clarified and free from suspended solids, it is the technique with porous diffusers that is generally used.

In the case of pre-ozonation, taking into account the risks of diffuser clogging, the ozone is generally transferred by means of a turbine or an emulsifier.

### 21.11.3. *Ozone demand*

The different ozone injection points at a water treatment station include:

– pre-ozonation: applied to raw water or upstream of a settler or flotation unit;

– inter-ozonation I: after a settler or flotation unit, and before the first-stage filtration;

– inter-ozonation II: after the first-stage filtration, combined with a second stage activated carbon filtration;

– post-ozonation upstream of a disinfection tank.

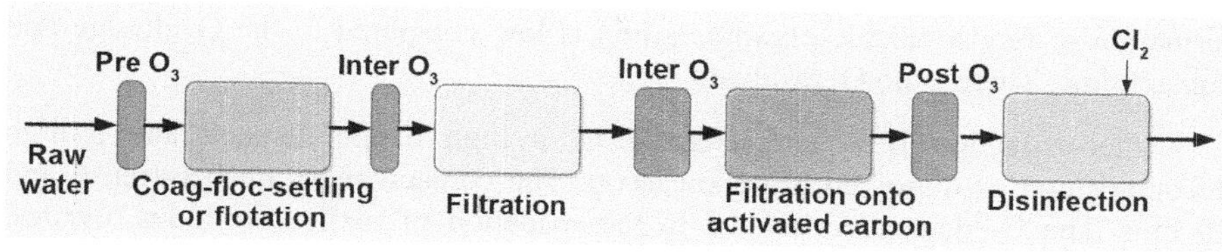


**Figure 21.26.** *Different $O_3$ injection points. For a color version of this figure, see www.iste.co.uk/gaid/watertreatment5.zip*

Depending on the water quality and injection point, the $O_3$ dosages and contact times are variable. The $O_3$ demand depends on the quality of the water to be treated. Table 21.29 lists the different injection sites and consequences observed after dosing $O_3$. The advantage of placing the ozone before the activated carbon filtration is that the biodegradable organic matter produced during ozonation can be removed by the biological activity present on the material.

– Pre-ozonation is intended for the oxidation of iron and manganese which generate insoluble oxides later removed by settling or filtration. It promotes the lysis of algae and reduces part of the color. It prompts a better flocculation. The ozone demand is rapid and takes place under contact times shorter than 3 min. The most general case is 2 min. There is no $O_3$ residual.

| Water to be treated | $O_3$ injection point | Remarks |
|---|---|---|
| Raw water turbidity<br>$O_3$ dosage: 0.5–1 mg·L$^{-1}$ | Pre-ozonation | Low $O_3$ demand<br>No production of by-products<br>Low DOC decomposition |
| Turbidity: 1-5 NTU<br>$O_3$ dosage: 0.5–1.5 mg·L$^{-1}$ | After settling<br>or flotation<br>Inter-ozonation I | Low $O_3$ demand<br>Low DOC decomposition |
| Turbidity: < 1 NTU<br>$O_3$ dosage: 1–3 mg·L$^{-1}$ | After the first-stage filtration<br>Inter-ozonation II | High $O_3$ demand<br>Formation of by-products<br>High DOC decomposition |
| Turbidity: < 1 NTU<br>$O_3$ dosage: 1–2 mg·L$^{-1}$ | After the first-stage filtration<br>Post-ozonation (disinfection) | High $O_3$ demand<br>Formation of by-products<br>High DOC decomposition |

**Table 21.29.** *$O_3$ demand depending on the different injection points*

– Inter-ozonation I is characterized by a low ozone demand and a low decomposition of dissolved organic carbon. In general, the contact times are shorter than 3 min. The $O_3$ demand is related to the oxidation of residual iron, manganese, color and the partial oxidation of toxins. Pre-disinfection can take place. The formation of biodegradable organic carbon is low compared to the $O_3$ dosage and contact time. There is no $O_3$ residual.

– Inter-ozonation II is characterized by a high ozone demand and a high decomposition of dissolved organic carbon. The contact times are between 6 and 10 min. The $O_3$ demand is related to the oxidation of toxins, bromides, nitrites, ammonia (pH dependent) and dissolved organic carbon. The use of ozone generates BDOC leading to biological growth, which can also increase the corrosion rates in the distribution systems, unless removed by biologically active filtration. An addition of hydrogen peroxide promotes the formation of OH• and accelerates the reaction kinetics. Pre-disinfection takes place. The $O_3$/DOC ratio is between 0.5 and 1.2. The formation of by-products (bromates and organics) takes places at this stage. A residual of 0.4 $mg \cdot L^{-1}$ is often required at the outlet of $O_3$ tanks to ensure that the reactions were efficient.

– Post-ozonation acts as a first disinfection stage. It is followed by an addition of chlorine in view of protecting the network. A neutralization of the $O_3$ residual is obtained by the injection of chlorine. This is because ozone reacts with other oxidants, such as chlorine, chlorine dioxide and monochloramine. The organic substances still present before post-ozonation can form biodegradable organic carbon, which contributes to the formation of a biofilm in the network. The formation of bromates is also favored. For all these reasons, this solution is becoming less widespread, especially if the design of this stage is planned for the inactivation of parasites such as *Cryptosporidium* and *Giardia*, which requires high dosages of $O_3$.

However, the use of ozone in post-ozonation reduces the subsequent chlorine or chlorine dioxide demand, maintaining a residual essentially composed of stable chlorine at a far lower level.

### 21.11.4. *The implementation of ozonation*

#### 21.11.4.1. *Ozonation in tanks*

Ozonation tanks are civil engineering works 4–6 m high and 10–15 m long on average, designed depending on the maximum ozone rate to be applied. These tanks are separated into several compartments where the water and the ozone circulate in co-current or countercurrent, improving the diffusion of the gas in the liquid.

The configuration of the tanks is generally the following (Figure 21.27):

– fist countercurrent tank, designed for a 2–4 min contact time: this is the oxidation compartment, receiving 60% of the overall injected ozone rate;

– second co-current or countercurrent tank (if there is an intermediate baffle), designed for a 4–6 min contact time in which 40% of the rate is injected;

– in some cases, the two tanks in which the ozone is injected are followed by a third contact tank (2 min).

The most widespread configuration groups three tanks with a contact time of 2 min each.

If the facility is equipped with porous media, the ozone-enriched air is injected into the first two compartments, generally an air flow rate of 50–70% into the first compartment, and 30–50% into the second one.

The ozone residual in water is destroyed by injecting sodium bisulfite or sodium disulfite (sodium hydrogen sulfite [$NaHSO_3$]). This is delivered as a commercial solution at 310 $g \cdot L^{-1}$ sulfur dioxide $SO_2$ (i.e. 500 $g \cdot L^{-1}$ of $NaHSO_3$, d = 1.32).

The application conditions for disinfection are the following: bactericidal conditions: 0.1–0.2 $mg \cdot L^{-1}$, for 1–2 min and virucidal conditions; 0.3–0.5 $mg \cdot L^{-1}$, for 4 min.

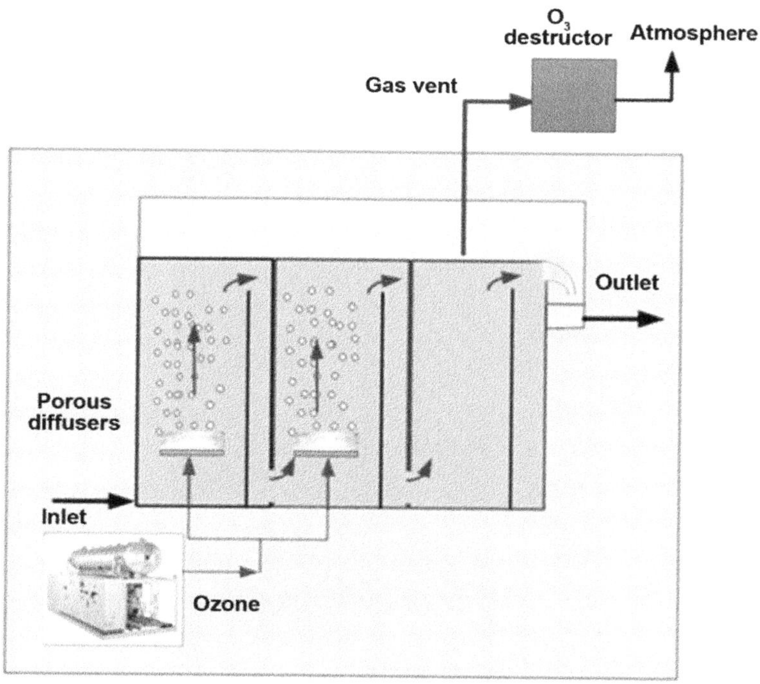


**Figure 21.27.** *Implementation of ozonation in reaction tanks. For a color version of this figure, see www.iste.co.uk/gaid/watertreatment5.zip*

#### 21.11.4.2. *In-line injection*

Ozone is injected into a water stream under negative pressure, which is generated in a venturi section, drawing ozone into the water stream. Usually, the full flow is pumped at a higher pressure to increase the vacuum available for ozone injection.

**Figure 21.28.** *Ozone diffusion in water using a porous diffuser. For a color version of this figure, see www.iste.co.uk/gaid/watertreatment5.zip*

Once the ozone is injected in a parallel line, the ozone-water mixture may be subjected to strong turbulence in order to obtain a good ozone dispersion in the water. Figure 21.29 illustrates the in-line and sidestream ozone injection systems. The Ct is obtained in a contact cell placed downstream of the injection system. The volume of this tank depends on the treatment goals, depending on the residual $O_3$ concentration and contact time, which are intimately related.

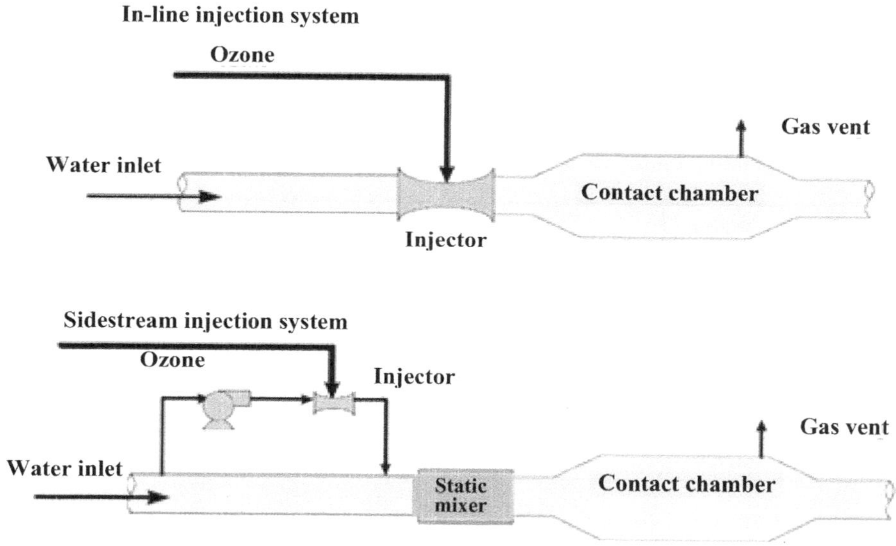


**Figure 21.29.** *In-line ozone injection modes*

### 21.11.4.3. *Static mixer injection*

Static mixers are well suited for high-performance ozone transfer in the water. A diffusion of ozone by static mixer makes it possible to better harness the Ct. The entire water flow is brought into contact with ozone. The ozone transfer into water is then very fast, taking place in less than a second. The fluid considered is single-phase, which grants a laminar flow in the structures.

Disinfection is then carried out in a concrete contact tank, 4–6 m deep.

The parameters considered are:

– the water flow rate, inducing either an in-line injection or a side jet;

– the gas flow depends on the concentration of ozone-enriched gas and the required ozone dose;

– the available space for setting up the static mixer (static mixer length);

– the water height available at the static mixer outlet (degassing column).

The optimization of all these parameters leads to transfer efficiencies over 90%.

Veolia has many stations where this system is applied both in pre-ozonation and inter-ozonation. Table 21.30 shows the operating parameters for some disinfection station references, located in the United Kingdom.

The injected dose of ozone is between 2 and 2.5 mg·L$^{-1}$ and the ozone gas concentration is between 80 and 100 g·Nm$^{-3}$. The water temperature is between 15 and 20°C, depending on the case.

| Site | Water flow $m^3 \cdot h^{-1}$ | Static mixer diameter m | Number of static mixers | Head losses Mbar | Transfer efficiencies % |
|---|---|---|---|---|---|
| **Ardleigh** | 792 | 500 | 9 | 186 | 92 |
| **Walton** | 1,160 | 500 | 4 | 180 | 98 |
| **Clay lane** | 3,541 | 900 | 4 | 137 | 89 |
| **Egham** | 2,920 | 700 | 4 | 196 | 97 |

**Table 21.30.** *A few (Veolia) references for the in-line injection of $O_3$ in the UK*

a) b)

**Figure 21.30.** *(a) Static mixer and (b) reaction chamber (Veolia site). For a color version of this figure, see www.iste.co.uk/gaid/watertreatment5.zip*

### 21.11.4.4. *In-line injection using a venturi*

Figure 21.30 illustrates the operating principle of the venturi tube mixer. The venturi tube is named after the Venturi effect, which states that if the flow section decreases, then the pressure also decreases. This is easily explained by Bernoulli's principle and the continuity equation:

– Bernoulli's equation: if the head losses between points 1 and 2 are negligible:

$$\rho\frac{v1^2}{2} + P1 = \rho\frac{v2^2}{2} + P2$$

where:

– ρ: water density;

– *v*i: flow velocity at point i;

– Pi: pressure at point i;

– continuity equation:

$$v_1S_1 = v_2S_2$$

where Si is the conduit section at point i.

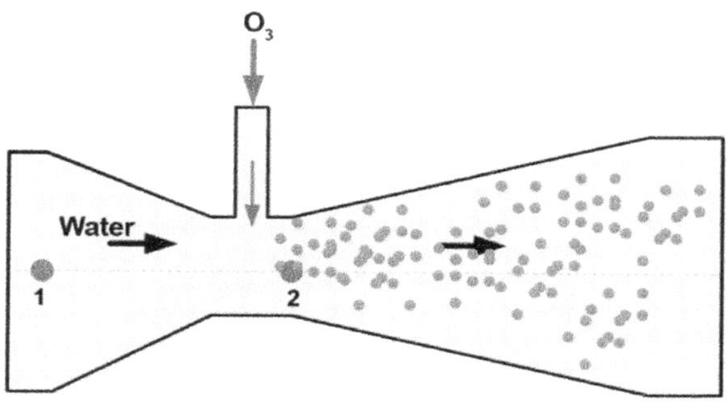


**Figure 21.31.** *Schematic diagram of an in-line injection of $O_3$ with a venturi. For a color version of this figure, see www.iste.co.uk/gaid/watertreatment5.zip*

The pressure at the narrowest point of the venturi (point 2) is lower than the pressure at the venturi's inlet (point 1) by:

$$\Delta P = P2 - P1 = \rho\frac{v1^2}{2}[\left(\frac{S1}{S2}\right)^2 - 1]$$

We can design the venturi (in particular the $S_1/S_2$ ratio) to obtain a pressure lower than atmospheric pressure at point 2. This results in a suction effect on a gas inlet, which is placed at the narrowest point of the venturi. Ozone is suctioned in and mixed with the water.

The solutions with the static mixer and the venturi are compact solutions. All of the injected ozone is entrained downstream of the mixer. The transfer of ozone from the gaseous phase to the liquid phase takes place downstream of the mixer in a reaction chamber designed to match the contact time adapted to the treatment goals.

#### 21.11.4.5. *Recovery and treatment of ozone vents*

A fraction of the ozone injected into the water is released in the gas phase in the upper section of the ozonation tanks. Depending on the case, and depending on the constraints within the station's immediate environment, this ozone-enriched air can be:

– directly into the atmosphere using high vents: this solution may not be completely satisfactory if there is still a smell of ozone in the vicinity of the facility;

– discharged into the atmosphere after dilution with fresh air (10 volumes of fresh air per volume of ozone-enriched air), ejecting this mixture at a velocity encouraging further dilution into the atmosphere;

– recycled in pre-ozonation, with or without additional fresh ozone;

– destroyed by heat at a temperature close to 350°C for a few seconds, or by passage across a palladium or manganese catalytic bed.

**Figure 21.32.** *Ozone facilities. For a color version of this figure, see www.iste.co.uk/gaid/watertreatment5.zip*

### 21.11.5. *Performances of ozone disinfection*

#### 21.11.5.1. *Bacteria*

| Bacterial log reduction | Ozone Ct ($mg \cdot L^{-1} \cdot min^{-1}$) (pH 7) |
|---|---|
| | 20°C |
| 2 | 3–4 |
| 3 | 6–8 |
| 4 | – |

**Table 21.31.** *Required Ct values ($mg \cdot L^{-1} \cdot min^{-1}$) at 20°C*

#### 21.11.5.2. *Virus*

| Viral log reduction | Ozone Ct ($mg \cdot L^{-1} \cdot min^{-1}$) (pH 6.5–8.0) | | | | |
|---|---|---|---|---|---|
| | 5°C | 10°C | 15°C | 20°C | 25°C |
| 2 | 0.6 | 0.5 | 0.3 | 0.25 | 0.15 |
| 3 | 0.9 | 0.8 | 0.5 | 0.40 | 0.25 |
| 4 | 1.2 | 1.0 | 0.6 | 0.50 | 0.30 |

**Table 21.32.** *Required Ct values ($mg \cdot L^{-1} \cdot min^{-1}$) at 20°C*

| Ozone required $mg \cdot L^{-1}$ | | |
|---|---|---|
| **Virus** | **2 log inactivation** | **3 log inactivation** |
| **Coxsackie B3** | 0.095 | 0.6 |
| **Polio 3** | 0.082 | 0.22 |
| **Polio 1** | 0.042 | 0.095 |
| **Echo 1** | 0.044 | 0.086 |
| **Coxsackie B5** | 0.053 | 0.076 |
| **Polio 2** | 0.039 | 0.052 |

**Table 21.33.** *Ozone doses for the inactivation of some viruses*

Table 21.33 shows the doses of $O_3$ required to inactivate 2 and 3 logs of viruses at 25°C, pH 7.0 and a contact time of 10 min.

#### 21.11.5.3. *Parasites*

The Ct values required for the reduction of *Giardia* parasites with ozone are summarized in Table 21.34.

| ***Giardia* log inactivation** | **Ozone Ct ($mg \cdot L^{-1} \cdot min^{-1}$) (pH 6.5–8.0)** | | | | | |
|---|---|---|---|---|---|---|
| | **0.5°C** | **5°C** | **10°C** | **15°C** | **20°C** | **25°C** |
| **0.5** | 0.8 | 0.5 | 0.4 | 0.4 | 0.3 | 0.2 |
| **1** | 1.7 | 1.0 | 0.8 | 0.7 | 0.5 | 0.3 |
| **1.5** | 2.3 | 1.5 | 1.3 | 1.0 | 0.8 | 0.5 |
| **2** | 3.3 | 2.0 | 1.7 | 1.3 | 1.0 | 0.7 |
| **2.5** | 3.7 | 2.5 | 2.1 | 1.7 | 1.2 | 0.8 |
| **3** | 4.5 | 3.0 | 2.5 | 2.0 | 1.5 | 1.0 |

**Table 21.34.** *Influence of temperature on Ct value*

*Cryptosporidium* cysts are more difficult to remove. In relation to ozonation, Table 21.35 groups some of these values for systems measuring ozone residuals exclusively at the tank's outlet.

In the case of ozone, parameters C and t have an equivalent weight (doubling the value of one and reducing the other by a factor of two results in the same degree of inactivation). The Ct generally implemented is 1.6 $mg \cdot min^{-1} \cdot L^{-1}$ (that is to say, 0.4 $mg \cdot L^{-1}$ maintained for 4 min). For this value, with water at 20°C, the degree of inactivation is 3.0 log for *Giardia* cysts. In the case of *Cryptosporidium*, the same Ct results in less than 0.5 log at 20°C.

| **(Log) reduction *Cryptosporidium*** | **Ozone Ct ($mg \cdot L^{-1} \cdot min^{-1}$) (pH 6.5–8.0)** | | | |
|---|---|---|---|---|
| | **1°C** | **5°C** | **10°C** | **20°C** |
| 0.5 | 12.8 | 8.7 | 5.4 | 2.1 |
| 1 | 25.6 | 17.5 | 10.9 | 4.2 |
| 2 | 51.2 | 35.0 | 21.7 | 8.4 |

**Table 21.35.** *Ct ($mg \cdot L^{-1} \cdot min^{-1}$) for the inactivation of* Cryptosporidium *cysts by ozone*

The ozone concentration considered is the residual at the outlet of the ozonation tank.

### 21.11.6. *De-ozonation*

The chemical reactions for the neutralization of ozone are the following and involve the same reducing agents used for chlorine and chlorine dioxide:

– With *sulfurous gas* ($SO_2$):

$$SO_2 + H_2O \rightarrow H_2SO_3 \text{ (sulfurous acid)}$$

$$O_3 + H_2SO_3 \rightarrow O_2 + H_2SO_4$$

that is, 1.33 mg $SO_2$/mg $O_3$.

Since *sodium bisulfite* ($NaHSO_3$) and *sodium sulfite* ($Na_2SO_3$) are easy to implement, they are a good alternative to $SO_2$:

$$NaHSO_3 + O_3 \rightarrow NaHSO_4 + O_2$$

that is, 2.16 mg $NaHSO_3$/mg $O_3$

$$Na_2SO_3 + O_3 \rightarrow Na_2SO_4 + O_2$$

that is, 2.65 mg $Na_2SO_3$/mg $O_3$.

– With *sodium thiosulfate* ($Na_2S_2O_3$): the reaction is:

$$Na_2S_2O_3 + 2O_3 + H_2O \rightarrow Na_2SO_4 + H_2SO_4 + O_2$$

that is, 1.64 mg $Na_2S_2O_3$/mg $O_3$.

– With an excess of thiosulfate, then:

$$3Na_2S_2O_3 + 4O_3 + 3H_2O \rightarrow 3Na_2SO_4 + 3H_2SO_4$$

that is, 2.46 mg $Na_2S_2O_3$/mg $O_3$.

– With *calcium thiosulfate* ($CaS_2O_3$), the reaction is:

$$CaS_2O_3 + 2O_3 + H_2O \rightarrow CaSO_4 + H_2SO_4 + O_2$$

that is, 1.58 mg $CaS_2O_3$/mg $O_3$.

– With an excess of thiosulfate, then:

$$3CaS_2O_3 + 4O_3 + 3H_2O \rightarrow 3CaSO_4 + 3H_2SO_4$$

that is, 2.37 mg $Na_2S_2O_3$/mg $O_3$.

– With *sodium metabisulfite* ($Na_2S_2O_5$): the reaction is:

$$Na_2S_2O_5 + 2\ O_3 \rightarrow Na_2S_2O_6 + O_2$$

that is, 1.98 mg $Na_2S_2O_5$/mg $O_3$.

| Disinfectants | Reagent | Molecular weight | Stoichiometric quantities $mg \cdot mg^{-1}\ O_3$ | Usually injected quantities $mg \cdot mg^{-1}\ O_3$ |
|---|---|---|---|---|
| **Sulfurous gas** | $SO_2$ | 64.09 | 1.33 | 1.3–1.5 |
| **Sodium sulfite** | $Na_2SO_3$ | 126.04 | 2.65 | 2.6–2.8 |
| **Sodium bisulfite** | $NaHSO_3$ | 104.06 | 2.16 | 2.2–2.4 |
| **Sodium metabisulfite** | $Na_2S_2O_5$ | 190.10 | 1.98 | 2–2.2 |
| **Sodium thiosulfate** | $Na_2S_2O_3$ | 126 | 1.64–2.46 | 2.6 |
| **Calcium thiosulfate** | $CaS_2O_3$ | 152 | 1.58–2.37 | 2.5 |

**Table 21.36.** *Reagent quantities for de-ozonation*

## 21.12. Criteria for choosing a chemical disinfection technique

### 21.12.1. *Practical implementation of chemical disinfection*

Given the diversity of microorganisms present in the water to be treated, the disinfection conditions adopted must enable the effective treatment of all germ populations. This must also be valid if the water to be disinfected temporarily contains turbidity, or a certain content of organic matter which could interfere with disinfection.

In practice, to take these considerations into account, specific Ct conditions are recommended for each disinfectant. In this case, t corresponds to the average contact time in the disinfection tank, and C to the residual oxidant measured at the tank's outlet.

– Chlorine:

- bactericidal conditions: maintaining a free chlorine residual between 0.1 and 0.2 $mg \cdot L^{-1}$ for 10–15 min (at pH < 8);

- virucidal conditions: 0.3–0.5 $mg \cdot L^{-1}$, for 30–45 min (at pH < 8).

– Chlorine dioxide:

- bactericidal conditions: 0.1–0.2 $mg \cdot L^{-1}$, for 5–10 min;
- virucidal conditions: 0.3–0.5 $mg \cdot L^{-1}$, for 30 min.

– Ozone:

- bactericidal conditions: 0.1 to 0.2 $mg \cdot L^{-1}$, for 1–2 min;
- virucidal conditions: 0.3–0.5 $mg \cdot L^{-1}$, for 4 min.

### 21.12.2. *Comparative efficiency of the main techniques*

Tables 21.37 and 21.38 combine the processing criteria usually considered when choosing a disinfection technique.

| Disinfectants | Biocidal efficiencies | Remainders |
|---|---|---|
| **Chlorine HClO**<br>**$ClO^-$** | Good<br>Low | Good |
| **Chlorine dioxide** | Very good | Good |
| **Ozone** | Excellent | Low |
| **Monochloramine** | Low | Very good |

**Table 21.37.** *Comparative biocidal efficiency and persistence of various chemical disinfectants*

| | Actions on the compounds contained in the water to be treated | | | |
|---|---|---|---|---|
| | **Chlorine** | **Chlorine dioxide** | **Ozone** | **UV** |
| **$Fe^{2+}$** | ++ | ++ | ++ | + |
| **$Mn^{2+}$** | + | ++ | ++ | + |
| **$NH_4^+$** | +++ | – | – | – |
| **Tastes and smells** | +<br>•• | ++ | +++ | + |
| **Color** | ++ | ++ | +++ | + |
| **Haloform precursors** | +++<br>•• | – | + | – |
| **Increases in biodegradability** | + | ++ | +++ | ++ |

**Table 21.38.** *Actions on various compounds present in water*

COMMENT ON TABLE 21.38.–

– *Action*:

+++ *Highly efficient;*

++ *Efficient;*

+ *Poor efficiency or slow action;*

– *No action.*

– *By-products*:

•• *Highly undesirable, even harmful;*

• *Undesirable.*

– *Synthesis: roles and application conditions for $Cl_2$, $ClO_2$, $O_3$ oxidants.*

| Oxidants | Treatment rate | Injection points | Remarks |
|---|---|---|---|
| **Ozone** | Residual of 0.4 $mg \cdot L^{-1}$ after 4 minutes (Ct = 1.6 $mg \cdot L^{-1} \cdot min^{-1}$) | At the end of the line in an $O_3$ tower especially designed for this purpose | Residual effect very weak |
| **Chlorine** | Sufficient to ensure a residual of 0.3–0.5 $mg \cdot L^{-1}$ free chlorine after 30 to 45 minutes | At the end of the process, at the inlet of a treated water tank | The contact time is very important |
| **Germ development blocking in distribution water** | Residual from 0.1 to 0.3 $mg \cdot L^{-1}$ free chlorine depending on the network's nature and length | Just before the introduction in the network after disinfection | Higher rates can be used provided that this does not cause any problem (flavors, haloforms) |
| **Chlorine dioxide** | Residual from 0.3 to 0.5 $mg \cdot L^{-1}$ of $ClO_2$ after 30 min contact time | At the end of the process, at the inlet of a treated water tank | The contact time is important |
| **Germ development blocking in distribution water** | Residual from 0.1 to 0.3 $mg \cdot L^{-1}$ of $ClO_2$ depending on the nature and length of the level | Just before the introduction in the network after disinfection | Higher rates are rarely required |

**Table 21.39.** *Roles and application conditions of various chemical oxidants*

Due to the formation of chlorites, chlorine dioxide is little used, although it reduces the formation of THMs and limits the “chlorine taste” of water. Due to the risk of bromate formation, the use of ozone is less widespread.

## 21.13. Another chemical disinfectant used: bromine ($Br_2$)

This reagent is widely used for the disinfection of swimming pools. The action principle of bromine is analogous to that of chlorine. However, it has some advantages over the latter:

– bromamines have a bactericidal activity equal to that of chlorine and are much less stable than chloramines. The impact on the environment is weaker and the dosage required, less important;

– at equal pH, the fraction of active hypobromous acid is more important than the fraction of hypochlorous acid in chlorine, whose efficiency is greatly reduced above 7.5. The efficiency against viruses is better than that of chlorine. The action is faster;

– however, tests have revealed the formation of haloform-type by-products ($CHClBr_2$, $CH_2Br_2$).

Preparation is done in situ from sodium bromide (NaBr) and chlorine:

$$NaBr + HClO \rightarrow HOBr + NaCl$$

For 1 kg of $Br_2$, 0.656 kg of NaBr and 0.453 kg of $Cl_2$ are required.

As bromamines disappear rapidly, debromination is not essential.

The cost of bromine is higher than that of chlorine. The best solution is to use it in combination with chlorine (injection of chlorine to satisfy the water’s immediate chlorine demand, and then injecting a much lower quantity of bromine).

## 21.14. Disinfection by ultraviolet radiation

### 21.14.1. *General remarks on UV radiation*

In the spectrum of electromagnetic waves, UV radiation lies between the X-rays and the violet of the visible spectrum, in the wavelength range between 100 and 400 nm (Figure 21.33).

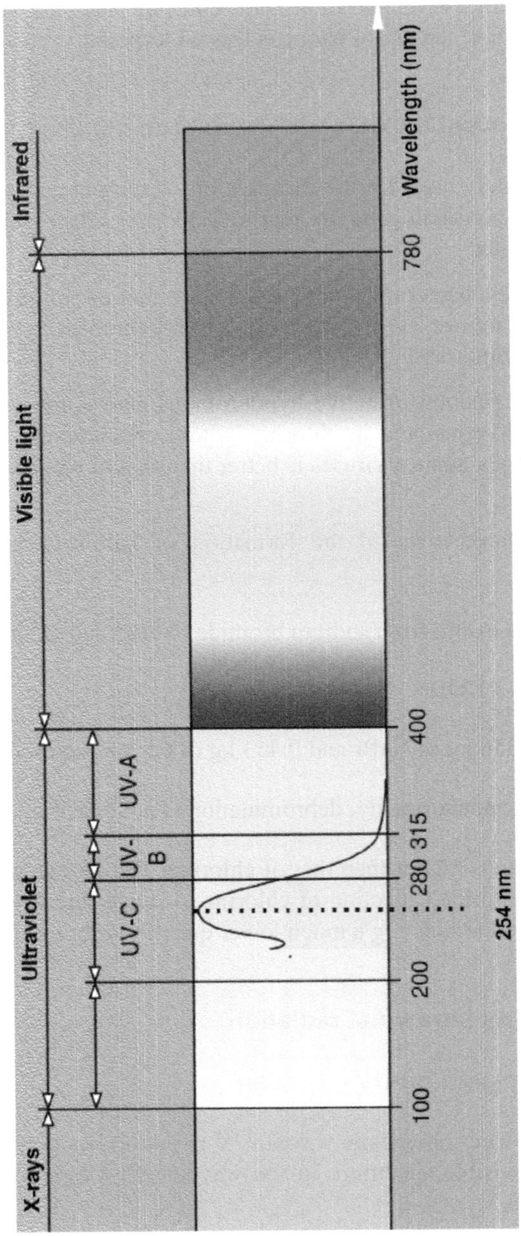


**Figure 21.33.** *Ultraviolet and visible light spectrum. For a color version of this figure, see www.iste.co.uk/gaid/watertreatment5.zip*

The UV spectrum is divided into four bands:

– UVA: 400–315 nm. Skin pigmentation;

– UVB: 315–280 nm. Production of antirachitic vitamin D;

– UVC: 280–200 nm. This region of the UV spectrum is used for its microbicidal power, with maximum germicidal action at d = 253.7 nm (265 nm according to some authors);

– UV: <200 nm. Radiation from this part of the UV spectrum induces the production of ozone from oxygen, absorbed by most substances, including air.

UV-C (wavelength between 200 and 280 nm) are the most germicidal rays, the optimal germicidal region being located at 254 nm. The exposure of bacteria and viruses to UV radiation with a wavelength close to 254 nm leads to the deterioration of their cellular material (DNA and RNA). The evolution of knowledge shows that other wavelengths can provoke more general cellular transformations. The treatment therefore involves exposing the water to be treated to a UV radiation source.

## 21.14.2. *Inactivation mechanisms*

Two UV effects are currently known: bactericidal and bacteriostatic. The bactericidal effect relates to a direct bursting of the cell membrane subjected to UV irradiation. The bacteriostatic effect concerns the modification of the genetic capital of irradiated microorganisms. UV creates thymine dimers in the DNA chain, which results in the dissociation of unsaturated double bonds. The cell can then neither reproduce nor create the proteins required for its survival. It thus becomes inactivated and no longer has any infectivity potential. Microorganisms have sensitivity to UV strongly dependent on the absorption curve of their genetic material. Nevertheless, all the results obtained so far confirm the similarity between the DNA absorption curve and spectral sensitivity for wavelengths between 230 and 280 nm.

Compared to chemical disinfection processes, the technique involving ultraviolet radiation – whose bactericidal properties have long been known – has the advantage of not inducing toxicity, and not generating sapid by-products (as in the case of chlorine). However, it has no residual action.

Disinfection is only efficient provided that a sufficient UV dose is applied. The UV radiation dose is expressed as an energy flux, in $mW \cdot s \cdot cm^{-2}$ units (milliwatt seconds per square centimeter), which is the product of the intensity emitted by the lamp and the water's retention time in the reactor. The minimum dose required for disinfection depends on several factors (including the sensitivity of microorganisms), but is generally considered to be comprised between 16 and 40 $mW \cdot s \cdot cm^{-2}$.

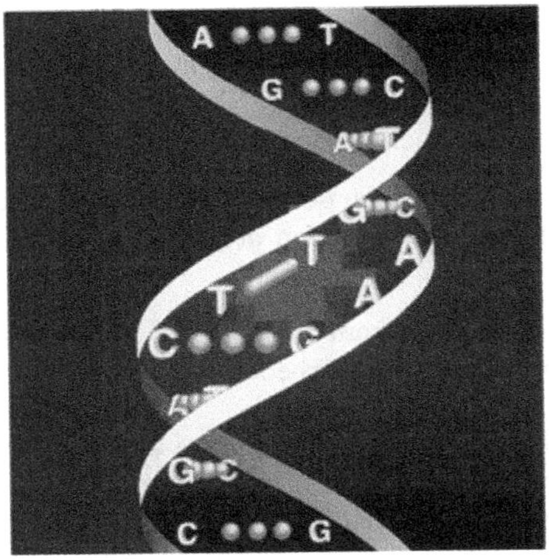


**Figure 21.34.** *Deterioration of cellular material using UV. For a color version of this figure, see www.iste.co.uk/gaid/watertreatment5.zip*

### 21.14.3. *Lethal dose and inactivation kinetics*

The absorption of radiation follows the Lambert–Beer law:

$$I = (P/S)(T/100)^{-d}$$

where:

– P: UV power emitted by the lamp (W);

– S: surface at distance d from the lamp = $2\pi(d\text{-}r)L$ where r is the lamp's radius and L, its length;

– d: water depth (cm);

– I: intensity received at distance d from the lamp.

The radiation emitted by UV lamps is partly absorbed by the water, to such extent that microorganisms only receive part of the radiation. The percent transmission across 10 mm of an effluent (T%) is the preponderant parameter for design. It is easily measured using a spectrometer at 254 nm and a quartz-cuvette that is 1 cm thick.

Thus, considering only the absorption term $(T/100)^{-d}$ (neglecting the P/S dispersion), for a 3 cm water layer, the UV dose emitted at the surface of the lamp is divided by 3 for a UV transmission of 65%, and by 15 for a UV transmission of 40%.

The water sheet must be thin to obtain the best UV reduction performance. UV transmission has a decisive influence on the design.

The lethal exposure dose is calculated by:

$$D = \frac{Pt}{S} e^{-kx}$$

where:

– P: germicidal power of the UV source (in W);

– S: UV-emitting surface (in $cm^2$);

– t: exposure time of a volume element (in s);

– k: UV absorption coefficient of the water to be treated (in $cm^{-1}$). It varies from 2 to 10 $m^{-1}$ (0.02 to 0.1 $cm^{-1}$) for drinking water;

– x: water sheet thickness (in cm);

– D: inactivation dose ($mJ \cdot cm^{-2}$).

The quantity of energy received by a microorganism subjected to UV radiation is equal to:

$$D = I \times t$$

where:

– I: power received per unit area, depending on the power of the lamp used, its geometry, the trajectory followed by the UV radiation, and the UV absorption coefficient in the water to be treated ($mW \cdot cm^{-2}$);

– t: exposure time in seconds;

– D: energy received per unit area or exposure dose expressed in $mW \cdot s \cdot cm^{-2}$ (or in $mJ \cdot cm^{-2}$).

In short, the UV dose ($mJ \cdot cm^{-2}$) is the total radiant energy (all germicidal wavelengths) passing from all directions through an infinitesimally small sphere of cross-sectional area dA ($cm^2$), divided by dA.

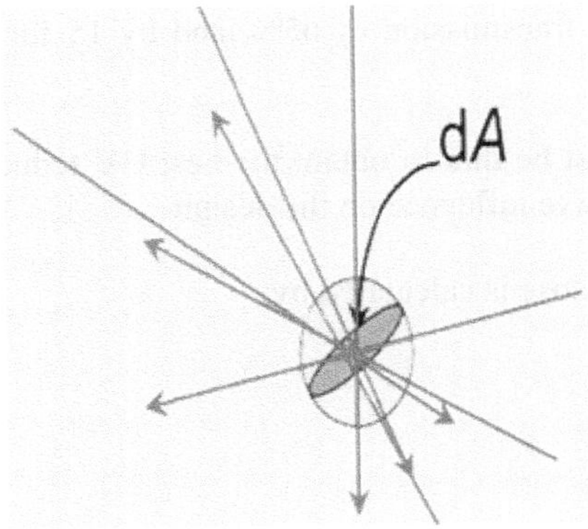


**Figure 21.35.** *Radiation on a spherical particle*

The law for the inactivation of microorganisms is:

$$\text{Log}\frac{\text{No (initial concentration)}}{\text{N (final concentration)}} = \text{kD}$$

The reduction in log units is directly proportional to the dose. The inactivation coefficient k specific to each type of microorganism reflects the specific sensitivity of various germs to the UV.

If turbulence is good, the probability that the germs will pass at the maximum distance from the lamps (e.g. along the edges) is low. The minimum intensity is replaced by the average intensity, which is calculated by integrating the intensity received at each point and emitted by the various lamps along the reactor's volume. Design is usually calculated according to the lowest UV transmission and peak flow.

#### 21.14.3.1. *Equivalent reduction dose (ERD)*

The ERD was defined a few years ago when drinking water requirements encouraged manufacturers and builders to maintain more specific guarantees concerning UV disinfection. The ERD dose is based on the use of biodosimeters which serve as a reference. A biodosimeter can be a microorganism such as bacteriophage MS2. It can also be *E. coli*, *Bacillus* spores or any other microorganism known to be sensitive to UV.

Thus, a concentrated solution of the biodosimeter microorganism is injected upstream of the UV reactor, and after homogenization, a sample is collected upstream and downstream of the reactor. It is then possible to deduce the reduction of the biodosimeter considered. This reduction corresponds to a certain UV dose calculated under conventional conditions at the laboratory (measured in a collimator). This dose is called the ERD dose.

This dose depends on the conditions of the experiment, such as water's UV transmission, water temperature, the presence of suspended solids or other products in the water, which may influence the dose. But, above all, it depends on the biodosimeter considered and makes it possible to calculate more accurate doses in relation to the microorganisms tested.

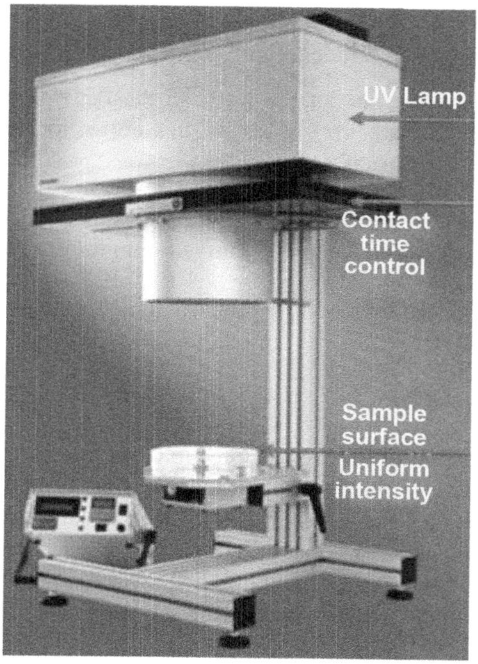


**Figure 21.36.** *Collimator*

This device (the collimated beam) is used at bench scale to irradiate samples containing known microorganisms. It contains a UV lamp that focuses the UV beam in a perpendicular direction to the sample on a Petri dish. The intensity of the LP lamp is 0.1959 mW·cm$^{-2}$ and the wavelength is 253.7 nm. The UV dose is deduced by multiplying the average intensity by the irradiation time. This system irradiates a suspension of known microorganisms, with constant stirring, so that the UV rays are almost perpendicular to the sample. In this way, the intensity received is considered to be constant at all points on the sample's surface.

If water absorbs UV at the wavelength(s) of interest, then it is necessary to account for the decrease in irradiance arising from absorption as the beam passes through the water.

The water factor is defined as follows:

$$\text{Water factor} = Io\frac{(1-10^{-a\,l})}{a\,l\,Ln(10)}$$

where:

– $a$: decadic absorption coefficient ($cm^{-1}$) or absorbance for a 1 cm path length;

– $l$: vertical path length (cm) of the water in the Petri dish. For polychromatic light (e.g. for a medium pressure UV lamp), this correction must be made over a narrow band (usually 1–5 nm) of wavelengths.

The formula is derived from integrating the Beer–Lambert law over the sample depth and holds true only for a completely mixed sample.

Several doses are tested by modifying the irradiation time.

### 21.14.4. *Implementation*

The light sources used in UV disinfection are mercury vapor lamps. The particularity of mercury is that is has a resonance line at 253.7 nm located in the maximum disinfection efficiency band. In this type of lamp, the electric discharge between the electrodes causes the excitation of the mercury atoms, which, on falling back into their ground state, emit radiation. The relative intensities of the different rays emitted depend on the mercury pressure in the lamp.

#### 21.14.4.1. *LP lamps*

Ultraviolet rays are produced by LP or medium-pressure (MP) mercury lamps, which have a lower germicidal efficiency than LP lamps. “Low-pressure” lamps mainly emit at 254 nm. LP mercury lamps, which generate 85% of their energy at a wavelength of 254 nm, are the most commonly used. Their wavelength is in the optimal germicidal range, between 250 and 265 nm. The design, construction and operation of these lamps is similar to that of fluorescent tubes, except that they are made of UV-transparent quartz, rather than phosphor-coated glass. The optimum operating temperature of the lamp is around 40°C, so the lamp is normally separated from the water by a “sleeve” to prevent cooling by the water.

The intensity of UV radiation emitted decreases with lamp age. Typical lamp life is about 14,000 h, after which the output is reduced by over 30%, requiring a lamp replacement.

Given the limited power of these lamps, a large number must be used to ensure a correct disinfection (approximately 500 lamps for 1,000 $m^3 \cdot h^{-1}$). The lamps are arranged in skids (modules comprising a given number of lamps) in closed reactors where the water circulates.

The main advantage of LP is its good efficiency and its mono-chromatic spectrum, targeting the germicidal wavelength at 254 nm and to avoiding the formation of by-products at UVC disinfection doses.

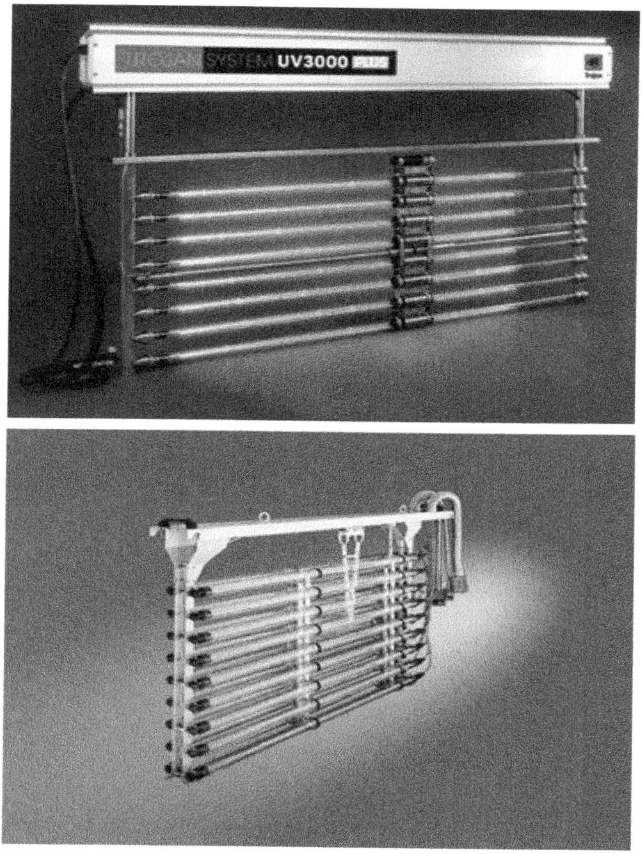


**Figure 21.37.** *UV lamps from Trojan and Wedeco suppliers. For a color version of this figure, see www.iste.co.uk/gaid/watertreatment5.zip*

#### 21.14.4.2. *Medium-pressure lamps*

These lamps have a broader emission spectrum, resulting in a lower UV efficiency (around 6–15%). Their unit power is far higher than LP lamps (>1,000 W). Therefore, the number of lamps to be implemented is much lower (a few lamps will suffice).

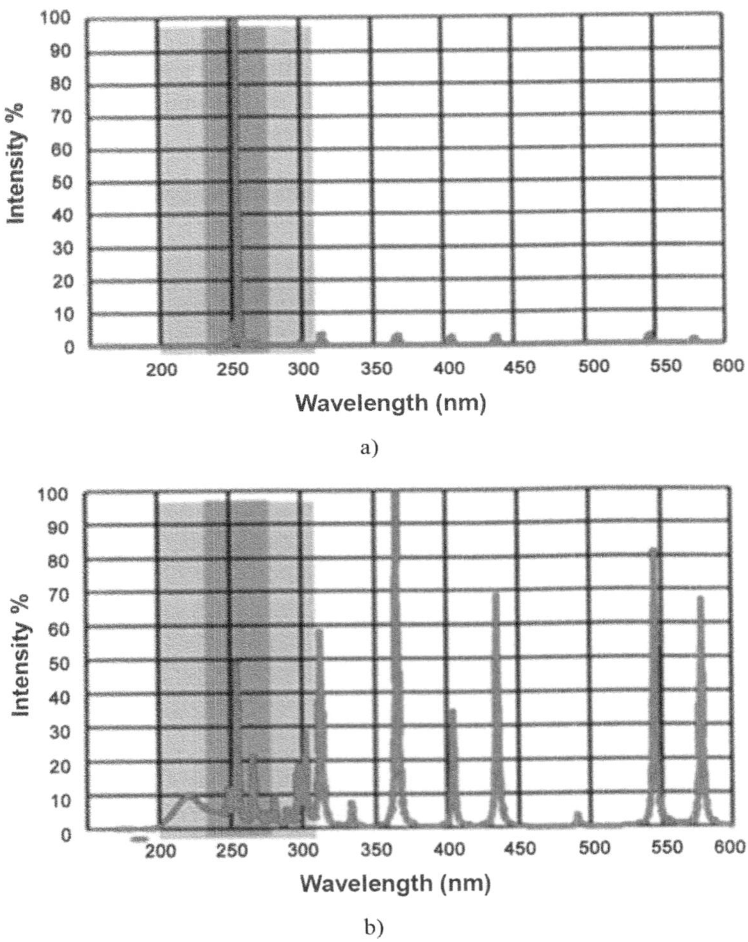


**Figure 21.38.** *(a) Low- and (b) medium-pressure lamp wavelengths. For a color version of this figure, see www.iste.co.uk/gaid/watertreatment5.zip*

The advantage of the MP lamp is its power, treating high flow rates with a minimum of lamps. However, due to their polychromatic spectrum, medium-pressure (MP) lamps have the disadvantage of forming unwanted by-products.

| Parameters | UV lamps | | |
|---|---|---|---|
| | **Low-pressure low intensity** | **Low-pressure high intensity** | **Medium-pressure high intensity** |
| **Consumed power:**<br>**– W**<br>**– KW** | 70–100 | 200–500<br>1–1.2 | 2–5 |
| **Efficiency, %** | 30–40 | 25–35 | 12–18 |
| **Temperature, °C** | 35–45 | 90–150 | 500–800 |
| **Pressure, mm Hg** | 0.007 | 0.001–0.01 | – |
| **Lamp length, m** | 0.75–1.50 | Variable | Variable |
| **Lamp diameter, mm** | 15–20 | Variable | Variable |

**Table 21.40.** *Characteristics of UV lamps*

### 21.14.4.3. *Operating principle*

Disinfection takes place in a treatment chamber where the water is irradiated by one or more UV generators. Each generator is isolated from contact with the water by a quartz sheath, which is permeable to UV rays. Disinfection is obtained by the circulation of a thin annular water sheet around these sheaths. The hydraulic regime of the water flow in the chamber must be turbulent and piston-like in order to ensure the treatment's homogeneity (same exposure time for all germs).

The facilities are equipped with a radiometer to monitor UV emission performances and detect any malfunctions.

The efficiency of disinfection is related to the UV radiation dose applied, itself defined by the product of the UV intensity and the exposure time of germs to radiation.

The UV reactor is set up in-line on the filtered water pipeline. Cleaning can be carried out automatically with the equipment already installed on the reactor manually with a cleaning kit. This kit includes a container with a cleaning reagent and a recirculation pump set up on a mobile cart.

During cleaning, the reactor inlet/outlet valves are closed, while the cleaning kit is connected to the cleaning taps. The cleaning product is recirculated for about twenty minutes. This operation will be implemented approximately every 6 months.

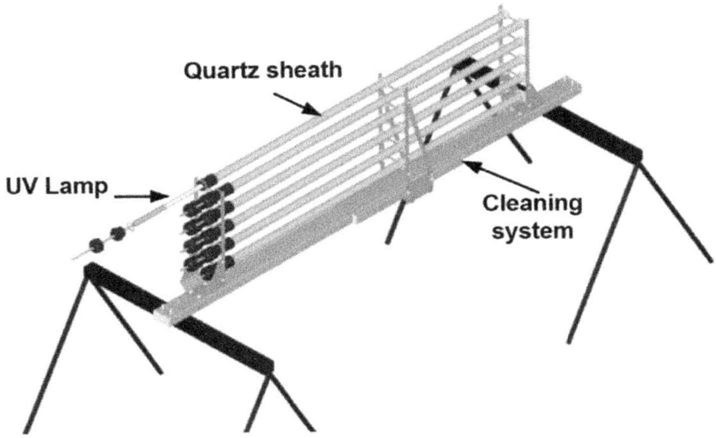


**Figure 21.39.** *UV lamp system*

In the event an intervention should be required, the reactor can be bypassed. In that case, chlorine should ensure the disinfection goals.

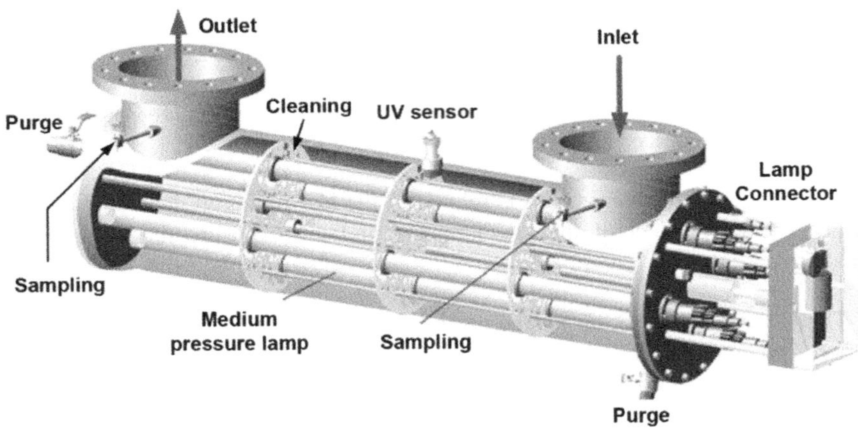


**Figure 21.40.** *Closed reactor used for drinking water facilities*

The minimum dose required for disinfection depends on several factors, including microorganism sensitivity, and is applied at 40 $mJ \cdot cm^{-2}$ (400 $J \cdot m^{-2}$). Depending on the performances expected and the type of microorganisms, this dose can vary between 40 and 70 $mJ \cdot cm^{-2}$.

Several new treatment technologies have been developed for the inactivation of *Cryptosporidium*. These include UV or pulsed white light systems, as well as combined filtration–irradiation or adsorption–irradiation techniques, which increase UV exposure. Pulsed UV and pulsed white light devices generate high-intensity, short-duration pulses of radiation and are reported to achieve a more successful inactivation of oocysts than conventional UV systems.

### 21.14.5. *UV disinfection design parameters*

The design of a UV device first requires measuring the transmissivity of the water to be treated through a 1 cm water sheet, for example. In the absence of water availability, the estimate of UV transmittance is calculated in relation to treatment stations and equivalent qualities of produced water. For the reactor's design, it is the lowest transmissivity of the water to be treated that is taken into account.

The second parameter to consider is the flow to be treated in the reactor. It is the maximum flow which can pass through the UV reactor that is taken into account.

#### 21.14.5.1. *UV dose ($mJ \cdot cm^{-2}$ or $J \cdot m^{-2}$)*

As previously mentioned, it is the product between the intensity ($W \cdot m^{-2}$) and the irradiation time. It can be expressed either in $mJ \cdot cm^{-2}$ or in $J \cdot m^{-2}$, knowing that $1\ mJ \cdot cm^{-2} = 10\ J \cdot m^{-2}$.

The intensity depends on the type of lamp, its power, age, quality of the quartz sheath and water quality (transmittance).

#### 21.14.5.2. *Electrical power consumed*

The calculation of the electrical power consumed in $Wh \cdot m^{-3}$ is based on the following principle.

The relationship between the UV dose and the power supply per $m^3$ of water to be treated (UV radiation) is defined by the following formula:

$$\mathrm{PUV}\ (\mathrm{Wh} \cdot \mathrm{m}^{-3}) = \frac{\mathrm{DoseUV}\ \mathrm{K_{Fe}}}{\mathrm{R}\ \mathrm{K_{lamp}}}$$

where:

– $K_{Fe}$: coefficient related to the dissolved iron content in the water to be treated. For an iron concentration lower than 0.5 mg·L$^{-1}$, $K_{Fe} = 1$;

– $K_{lamp}$: rate of electrical power remaining at the end of lamp life (after 12,000 hours). It is equal to 85% (15% power loss).

R is a Dose Joules·m$^{-2}$/Dose Wh·m$^{-3}$ ratio

$$R = Y \text{ Exp } (2.728\ T)$$

where:

– T: transmittance (in %);

– Y: variable coefficient between 4 and 6 depending on the type of UV reactor.

#### 21.14.5.3. *Calculation of the number of lamps*

The total number of lamps is equal to the total electrical power to be applied, divided by the unit power of a lamp.

#### 21.14.5.4. *Calculation of installed power (kWh)*

$$\text{PUV(kW)} = \text{Kt} \times \text{Number of lamps}/1{,}000$$

Calculation of the power consumed per m$^3$ of water (kW·m$^{-3}$·d$^{-1}$):

$$\text{PUV}(\text{kW} \cdot \text{m}^{-3} \cdot \text{d}^{-1}) = \text{PUV radiation applied} \times 1{,}000/\text{Flow}$$

The water flow is expressed in m$^3$ per day.

### 21.14.6. *Factors affecting the efficiency of a UV treatment*

The conditions for a proper disinfection depend on several parameters.

#### 21.14.6.1. *Water transmittance*

The quality of water subjected to UV radiation is fundamental. The UV treatment facility is designed according to the UV transmittance of the water to be treated. The lower the UV transmittance, the larger the size power of the UV treatment facility, and the higher its costs (installation, operation). Transmittance is inversely proportional to turbidity, suspended solids, color (humic acids), dissolved residual iron and manganese.

The presence of these compounds can create a "UV demand" analogous to a chlorine demand. The method used to quantify this interference is the transmittance measurement, which represents the amount of light available to irradiate a microorganism at a given distance.

#### 21.14.6.2. *Turbidity and suspended solids*

In addition to attenuating the penetration of UV rays into the water, turbidity also has a protective effect from the action of these rays for microorganisms.

Suspended solids (SS) act on two levels: they directly absorb radiation and they reduce the available UV dose. They harbor certain microorganisms (adsorbed onto their surface or trapped inside) in the effluent to be disinfected (barrier effect), which then become protected against the action of UV rays.

The turbidity of the water to be disinfected must therefore be as low as possible. Needless to say, this is also necessary in relation to the parasitic risk.

#### 21.14.6.3. *Deposit formation on the sheaths*

The UV lamps are protected from contact with water by a quartz sheath. Its transparency is a crucial element for disinfection efficiency. However, several phenomena cause its opacification:

– a slimy film due to fat or residual iron oxidation contained in the effluent;

– scaling (more intense phenomenon with MP lamp technology because the operating temperature is higher, which leads to an increased risk of mineral precipitation);

– algal development;

– condensation inside the quartz sheaths.

The compounds liable to form deposits that impair radiation efficiency are mainly:

– calcium and magnesium, in the case of scaling water (the risks of scaling are accentuated by the heating effect taking place nearby the lamps);

– iron and manganese;

– suspended solids and miscellaneous debris;

– residual iron or aluminum salts used in coagulation.

A longitudinal water flow along the lamps is likely to promote the gradual precipitation of matter on the walls, which can be more critical than the transverse installation mode. Lamp deposits and fouling reduce UV radiation scattering, therefore requiring periodic mechanical and chemical cleaning.

#### 21.14.6.4. *The influence of residual iron concentration*

The residual iron concentration in water strongly affects UV transmittance. This is because iron compounds absorb UV by reducing UV transmittance. Figure 21.41 illustrates the decrease in UV transmittance of distilled water with an increasing dissolved iron concentration.

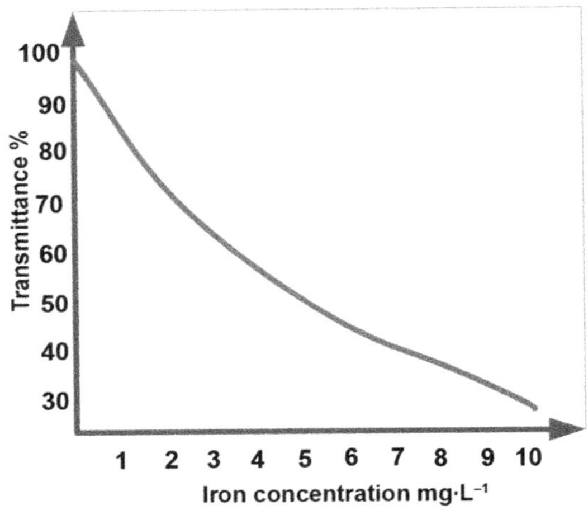


**Figure 21.41.** *Influence of iron concentration on UV transmittance*

The accumulation of iron on the particles present in the water tends to increase the fouling of quartz sheaths, limiting the diffusion of UV radiation. The impact of this fouling on disinfection is directly proportional to the quantity of particles deposited on the sheaths.

However, residual iron concentrations in drinking water are limited to 50 $\mu g \cdot L^{-1}$ and strongly limit the impact of this parameter on transmittance. However, care should be taken to respect this threshold and avoid a decrease in disinfection efficiency.

#### 21.14.6.5. *UV emitted intensity*

It is a function of:

– the number of lamps in operation;

– the voltage applied (almost proportional variation of intensity in relation to voltage);

– the fouling condition of sheaths, which stresses the importance of regular cleaning;

– lamp degradation: the light intensity emitted by the lamps decreases by 35% at the end of their life. Due to internal irradiation, the quartz of lamps undergoes a loss by solarization, a phenomenon which constitutes one of the causes of aging.

#### 21.14.6.6. *Contact time*

The contact time must be sufficient and in accordance with the suppliers' specifications.

#### 21.14.6.7. *Lamp aging*

Lamps wear out over time under the combined action of various factors:

– evaporation of the electron-emitting electrodes (due to ignition without preheating);

– deposit of metal oxides on the internal wall of lamps (if there is no protective coating);

– surface attack of the quartz of the lamps' internal wall by the particularly reactive plasma, because it is highly charged with radicals or dissociated ionic species (the same happens if there is no coating);

– damage to the quartz crystal lattice under the effect of UV photon shocks (depending on the quality of the quartz).

The quartz sheaths also wear out, but to a lesser extent, taking into account the possible mechanical stresses due to the cleaning system and the vibrations created by the passage of water, and the damage to the quartz crystal lattice under the impact of UV photons.

#### 21.14.6.8. *Lamp lifespan*

Lamp suppliers have developed internal lamp protection coatings for extending lamp lifespan, while limiting the harmful effects of their aging during the operation periods. LP lamps currently have lifespans easily ranging from 8,760 to 16,000 h.

At the end of the guaranteed life, the radiation emitted by the lamp, in the range of germicidal wavelengths defined by the manufacturer, must be of intensity equal to or greater than 80% of the radiation intensity measured for a new lamp. Lamp lifespan can be assessed from the information provided by the intensity sensors: a lamp is considered to be at the end of its life when, at maximum electrical power applied to the lamp, the intensity detected by the sensors is lower than or equal to 80% of the intensity provided by a new lamp.

#### 21.14.6.9. *Quartz sheath and ballasts*

Quartz sheaths and ballasts are guaranteed for a period of 10 years, *prorata temporis*.

#### 21.14.6.10. *Lamp operation management*

Whenever possible, it is advisable to avoid switching lamps ON/OFF for the following reasons: when switching on, there is a latency time (approximately 2 min) during which the lamp has not reached its steady state, and during which the water is not disinfected.

The number of ON/OFF operations of a lamp is limited. Manufacturers generally recommend not exceeding four operations per day because each ON/OFF reduces the lamp life by approximately 10 h.

### 21.14.7. *UV radiation performances*

As chemical disinfectants, UV irradiation does not kill microorganisms but inactivates them by suppressing their ability to multiply and infect the host. While the infectivity decrease estimates are known for viruses and bacteria, they are much more difficult to ascertain for parasites such as *Cryptosporidium* and *Giardia*. In recent years, cell culture infectivity analysis methods have emerged as an alternative to different analytical techniques in vivo and in vitro. Among the techniques for assessing the infectivity and disinfection efficiency for *Cryptosporidium parvum* oocysts, a cell culture method has been developed which reproduces the environment where the microorganism would thrive in vivo as much as possible.

The sensitivity of *Cryptosporidium* between 276 and 287 nm represents approximately 80% of the one observed between 250 and 261 nm. The sensitivity of viruses and *Bacillus subtilis* spores is equivalent to wavelengths 254 and 280 nm. Finally, irradiation at 280 nm reveals an efficiency very close to the one observed at 254 nm for various coliforms and viruses. During disinfection with LP lamps, it is easy to extrapolate the static inactivation at 254 nm, into the equivalent reduction dose obtained on a reactor. On the other hand, in the case of MP lamps, it is necessary to relate the efficiency of each wavelength to that of the 254 nm wavelength.

Figure 21.42 shows that a dose lower than 30 $mJ \cdot cm^{-2}$ inactivates up to 4 log of *Cryptosporidium parvum* parasites. Table 21.41 presents some data relating to the dose–responses of ultraviolet (UV) radiation for various microorganisms.

| | Lamp type | UV dose (mJ·cm$^{-2}$) | | | |
|---|---|---|---|---|---|
| **Microorganism** | | 1 log | 2 log | 3 log | 4 log |
| | | **Bacteria** | | | |
| **Coliforms** | LP | <3 | 5 | 10 | 18 |
| ***Escherichia coli*** | LP | <3 | 5 | 8 | 15 |
| ***Aeromonas hydrophila*** | LP | 1.1 | 2.6 | 3.9 | 5.9 |
| ***Campylobacter jejuni*** | LP | 1.6 | 3.4 | 4 | 4.6 |
| ***Citrobacter diversus*** | LP | 5 | 7 | 9 | 11.5 |
| ***Klebsiella pneumoniae*** | LP | 12 | 15 | 17.5 | 29 |
| ***Legionella pneumophila*** | LP | 1.6 | 3.2 | 4.8 | 6.4 |
| ***Legionella pneumophila*** | MP | 1.9 | 3.8 | 5.8 | 7.7 |
| ***Salmonella spp.*** | LP | <2 | 2 | 3.5 | 7 |
| ***Salmonella typhimurium*** | – | 2 | 3.5 | 5 | 9 |
| ***Shigella dysenteriae*** | LP | 0.5 | 1.2 | 2 | 3 |
| ***Shigella sonei*** | LP | 3.2 | 4.9 | 6.5 | 8.2 |
| ***Staphylococcus aureus*** | – | 3.9 | 5.4 | 6.5 | 10.4 |
| ***Streptococcus faecalis*** | – | 6.6 | 8.8 | 9.9 | 11.2 |
| ***Vibrio cholerae*** | LP | 0.8 | 1.4 | 2.2 | 2.9 |
| ***Yersinia enterocolitica*** | LP | 1.7 | 2.8 | 3.7 | 4.6 |
| ***Yersinia ruckeri*** | LP | 1 | 2 | 3 | 5 |
| ***Spore Bacillus subtilis*** | LP | 24 | 35 | 47 | 79 |
| | | **Parasites** | | | |
| ***Cryptosporidium parvum*** | LP | <3 | 5.8 | 12 | 22 |
| ***Giardia lamblia*** | LP | <3 | 5.2 | 11 | 22 |
| | | **Viruses** | | | |
| **MS2 (Phage) on *E. Coli*** | LP | 15 | 32 | 60 | 85 |
| **Canine calicivirus** | LP | 7 | 15 | 22 | 30 |
| **Adenovirus type 15** | LP | 40 | 80 | 122 | 165 |
| **Adenovirus type 40** | LP | 55 | 105 | 155 | 186 |
| **Poliovirus 1** | LP | 7 | 17 | 28 | 37 |
| **Coxsackievirus B3** | LP | 8 | 16 | 24.5 | 32.5 |
| **Reovirus Type 1** | LP | - | - | 16 | 36 |

| | Lamp type | UV dose (mJ·cm$^{-2}$) | | | |
|---|---|---|---|---|---|
| **Rotavirus SA-11** | LP | 7 | 13 | 20 | 35 |
| **Hepatitis A** | – | 5.5 | 9.8 | 15 | 25 |
| **Echovirus I** | LP | 8 | 16.5 | 25 | 33 |
| **Echovirus II** | LP | 7 | 14 | 20.5 | 28 |

**Table 21.41.** *UV dose (mJ·cm$^{-2}$) required for various microorganisms by log removal*

In most cases, data were generated from LP lamps where the lamp intensity can be measured and multiplied by the exposure time, and thus estimate the dose applied.

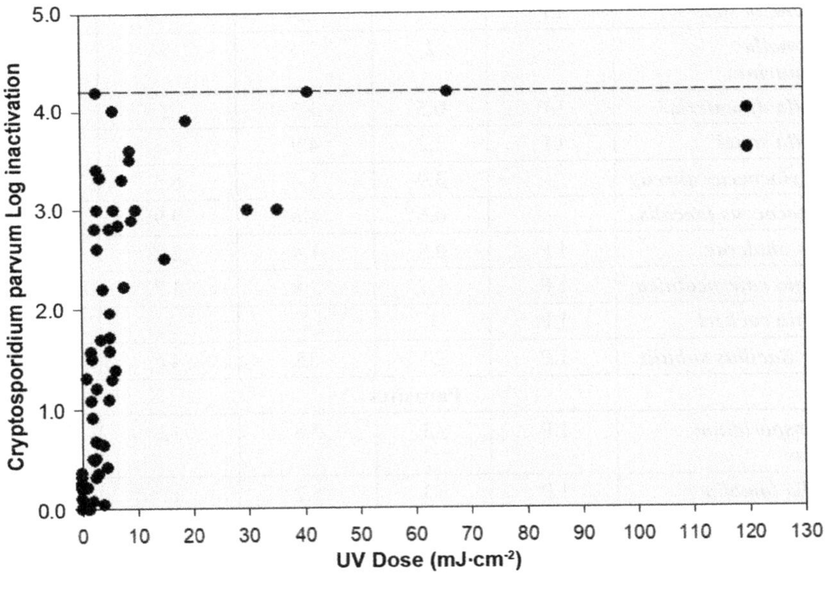


**Figure 21.42.** *Inactivation of Cryptosporidium parvum*

Adenoviruses show a strong UV resistance: a UV dose of 165–186 mJ·cm$^{-2}$ is required to inactivate 4 log with an LP lamp.

Figures 21.43 and 21.44 show the inactivation doses for MS2 and for various microorganisms present in water.

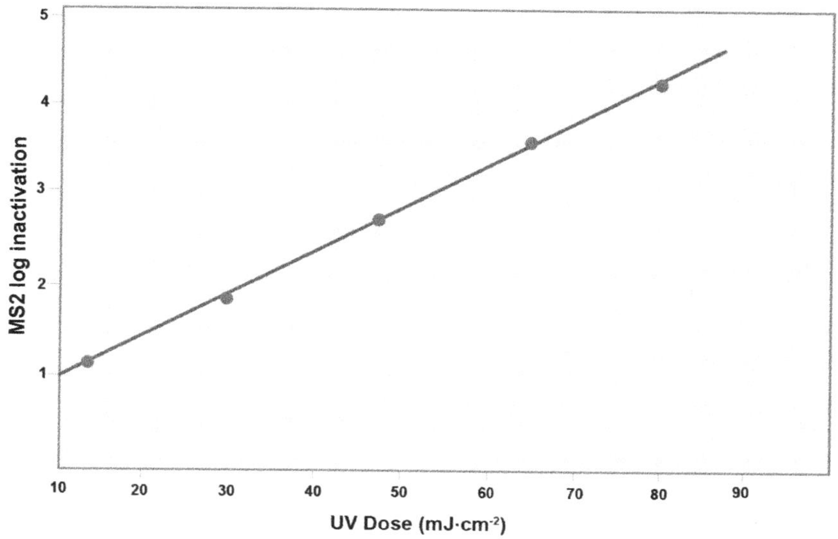


**Figure 21.43.** *MS2 log inactivation (on E. coli)*

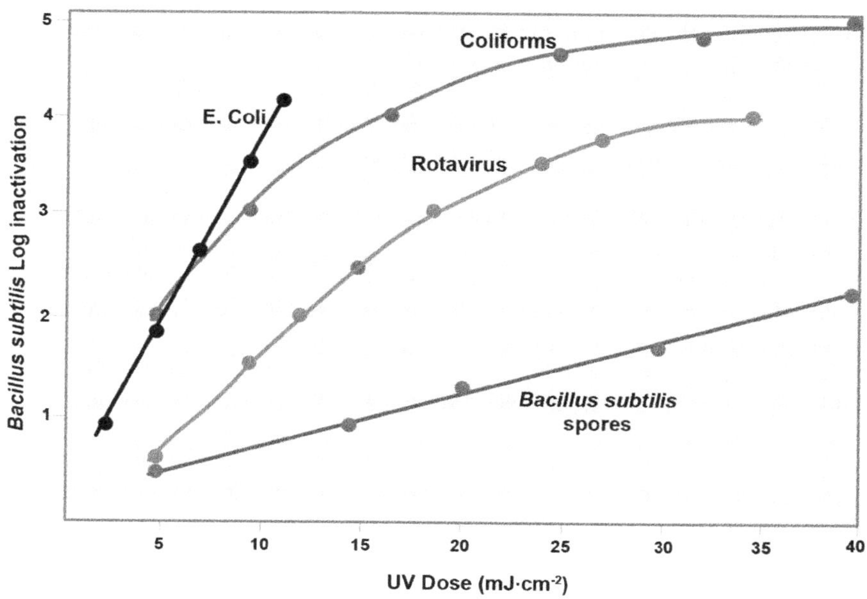


**Figure 21.44.** *Log inactivation of various microorganisms*

### 21.14.8. *Photoreactivation*

The controversial argument relating to the use of UV as a disinfection treatment is that some microorganisms irradiated with UV have the ability to repair the damage caused by UV to their genetic material when they reach the distribution networks. This happens when a second disinfectant is not added for network protection (chlorine, chlorine dioxide).

Some microorganisms (bacteria or parasites) have the ability to repair themselves after UV irradiation.

Repair can occur by two mechanisms:

– Dark repair of nucleic acids can occur through a multi-enzymatic process dependent on the available metabolic energy. This does not involve any external factor. It is referred to as "dark" repair.

– Repair during irradiation with visible light, known as "photo" repair. Photoreactivation refers to the enzymatic repair corresponding to the monomerization of the pyrimidine base dimers (thymine and cytosine) formed under the effect of UV. A part of the two DNA strands, unless modified, can serve as a copy for the synthesis of a complementary strand, leading to the reconstitution of the original DNA. The dimerization of pyrimidine bases induced during UV irradiation can be reversible, thanks to an enzyme whose synthesis is activated at wavelengths of 300–500 nm, under the action of sunlight. In this case, sunlight is indispensable, although metabolic energy is not needed. Throughout their evolution, the bacteria capable of photoreactivation have developed a cell defense mechanism to fight against the damage caused by the sun's ultraviolet rays.

Numerous studies have shown that, despite DNA repair after irradiation with visible light, there was no infectivity recovery for *Cryptosporidium* irradiated by a LP lamp. This phenomenon suggests that the damage caused by UV rays is not limited to pyrimidical bases, but is also found on other molecules. The use of other wavelengths (MP lamps) is an additional asset for triggering reactions on other families of molecules. However, these mechanisms have mainly been implemented for the UV disinfection of treated wastewater, later discharged into a receiving environment.

In this sense, many microorganisms capable of photoreactivation-repair to the damage caused to their DNA by UV rays can be enumerated: total coliforms, fecal coliforms, *Escherichia coli*, *Streptococcus faecalis*, *Streptomyces*, *Saccharomyces*, *Aerobacter*, *Micrococcus*, *Erwinia*, *Proteus*, *Penicillium*, etc.

Veolia's experience with numerous drinking water stations – where no second disinfectant is added – has revealed no photoreactivation in the networks.

### 21.14.9. *Advantages and drawbacks of UV disinfection*

| Advantages | Drawbacks |
|---|---|
| – Significant reduction of *Cryptosporidium* and *Giardia* parasites, far higher than the one obtained with chemical disinfectants, in particular chlorine dioxide and chlorine, unless preceded by ozonation. According to recent studies, it might be possible to inactivate 3 to 4 log (99.9–99.99%) of *Cryptosporidium* under normal operating conditions.<br>– Regarding the efficiency on *Giardia,* 4 log removal of infectivity for doses between 10 and 100 $J{\cdot}m^{-2}$. However, the results often depend on the infectivity testing method.<br>– Good bactericidal and virucidal efficiency.<br>– Physical process only modifying the chemical composition of water a little, and not accompanied by the formation of undesirable compounds found in chemical disinfection, or problems of tastes and odors.<br>– No need for on-site storage of reagents, avoiding risks for the operating staff and safety-related issues.<br>– Efficiency is not dependent on pH (as is the case for chlorine).<br>– Very short contact time required (a few seconds), directly ensured in the irradiation chamber.<br>– Easily inserted into an existing water pipeline.<br>– Possibility of continuous UV emission control.<br>– Reduction of residual color.<br>– Ease of implementation and operation. | – No residual power liable to preserve the good quality of treated water in the distribution network, and no possibility of directly assessing the effectiveness of the treatment by measuring a residual, as in the case of a chemical oxidant.<br>– Formation of nitrites from nitrates, but this is very moderate at the UV doses usually applied.<br>– Efficiency dependent on maintenance.<br>– Efficiency influenced by certain water quality parameters which must be taken into account during the project's design and during operation.<br>– Treatment more expensive than chlorine, the cost is directly proportional to the size of the facility.<br>– Restricted field of application if it is the only disinfection treatment (in the absence of residual effect), but procedure to be taken into consideration for the treatment of drinking water in view of its advantages. |

**Table 21.42.** *Advantages and drawbacks of UV disinfection*

### 21.14.10. *Conclusions on UV disinfection*

The UV treatment is justified compared to chemical disinfectants as a first disinfection stage because of the many advantages it presents:

– no addition of chemical reagent;

– the water has no taste or odor problems;

– no resistance of microorganisms against UV rays is currently known;

– UV rays are not corrosive;

– especially very efficient against parasites such as *Cryptosporidium* and *Giardia* and against various bacteria such as Legionella, in particular. Especially knowing that chemical disinfectants are only effective against parasites at high concentrations, something difficult to envisage in the treatment of drinking water due to consequences they entail: taste problems, side reactions, etc. UV technology (ERD 40 $mJ \cdot cm^{-2}$) becomes a plausible alternative and is part of the "multi-barrier treatment" concept, which combines optimal clarification and a possible refining stage, on the one hand, with the addition of a second oxidant intended for network protection, on the other.

Secondary disinfection (chlorine or derivatives) contributes to the protection of distribution networks. In this case, the disinfectant demand should be lower than the one observed during conventional single-stage disinfection.

In addition, the regulations regarding bromates (10 $\mu g \cdot L^{-1}$) encourages operators to reduce the ozone concentration applied in inter-ozonation and post-ozonation to limit bromate production. This decrease results in a decrease in the disinfectant effect of this stage. They compensate for this decrease by increasing the final doses of chlorine disinfectant or its derivatives. The consequence could be an evolution of the by-product concentrations. The alternative solution in recent years is to integrate a first stage of UV treatment compensating for the ozone deficit, but without affecting the secondary chemical disinfectant demand.

Especially since UV radiation has no effect on the various physicochemical parameters (nitrates, nitrites, nitrosamines, ammonium, DOC) and physicochemical characteristics of water, at the operating dose of 40 $mJ \cdot cm^{-2}$.

## 21.15. Comparative criteria between the various chemical disinfectants

The chosen method for disinfecting potable water depends on a variety of factors:

– the disinfectant's efficiency against waterborne pathogens (bacteria, viruses, protozoa and helminths);

– the simplicity and reliability with which the process can be monitored and controlled;

– its ability to produce a residual offering additional protection against possible contamination;

– the esthetic quality of treated water;

– the nature and concentration of the by-products formed. The propensity of various disinfection methods to produce health-related by-products and the possibility of removing or avoiding these undesirable by-products are important factors to take into account when making final decisions.

Economic factors also play a role in the final decision.

| **Criteria** | **Bleach** | **Ozone** | **Gaseous chlorine** | **Chlorine dioxide** | **UV** |
|---|---|---|---|---|---|
| **Active disinfectant** | $HClO^-$ | $O_3$ | $HClO^-$ | $ClO_2$- | $\lambda = 254$ nm |
| **Persistence** | Good | Nil | Good | Good | Nil |
| **Taste/odor** | Characteristic | Low | Characteristic | Low | Nil |
| **Formation of by-products** | THMs | Aldehydes and bromates | THMs | Chlorites and chlorates | |

**Table 21.43.** *Comparative criteria between the various chemical disinfectants*

The main mechanisms that have been brought forward to account for the action of disinfectants are as follows: damage to the cell membrane, alteration of cell permeability, alteration of the protoplasm's colloidal nature, DNA and/or RNA alteration and enzyme activity inhibition.

Table 21.44 shows the disinfection mechanisms for chlorine, ozone and UV.

Due to the formation of chlorites, chlorine dioxide is little used, although it reduces the formation of THMs and limits the water's "chlorine taste". Due to the risk of bromate formation, the use of ozone is less widespread. UV disinfection is being increasingly used and is always followed by an injection of chlorine or $ClO_2$ to thoroughly complete a difficult disinfection, for the oxidation of $NH_4$ or nitrites and for network protection.

| Chlorine | Ozone | UV |
|---|---|---|
| Oxidation | Direct oxidation and destruction of the cell membrane with rejection of cell constituents outside the cell | Photochemical damage to DNA and RNA inside the cells |
| Reactions with available chlorine | Free radical reactions following $O_3$ decomposition | As DNA and RNA carry genetic information for reproduction, the damage caused inactivates microorganisms |
| Protein precipitation | Damage to nucleic acid constituents (purines and pyrimidines) | Nucleic acids are the most important energy absorbers, between 240 and 280 nm |
| Change in cell membrane permeability | Breaking of C-N bonds | – |
| Hydrolysis | | |

**Table 21.44.** *Disinfection mechanisms for chlorine, ozone and UV*

## 21.16. References

AWWARF, Compagnie Générale des Eaux (1991). *Ozone in Water Treatment: Application and Engineering.* Lewis Publishers, Boca Raton, FL.

Benoufella, F. and Gaid, K. (1986). Extraction et identification par chromatographie en phase gazeuse des acides aminés présents dans les eaux destinées à la consommation. *Trib. du CEBEDEAU*, 515, 39–46.

Benoufella, F. and Gaid, K. (1987). Decomposition of amino acids in dilute solution under the effect of active chlorine. *International Conference on Reservoir Limnology and Water Quality*, Prague, Czech Republic.

Benoufella F., Gaid, K., Laplanche., A. (1990). Etude de la chloration d'acides aminés aliphatiques. Produits réactionnels et mécanismes. *J. Français d'Hydrologie*, 21, 1, 41–48.

Bishop, M.M., Morgan, J.M., Cornwell, B., Jamison, D.K. (1993). Improving the disinfection detention time of a water plant clearwell. *J. Am. Water Works Assoc.*, 85(3), 68–75.

Black, A.P. and Hannah, S.A. (1965). Measurement of low turbidities. *J. Am. Water Works Assoc.*, 57, 901–918.

Bolton, J.R. and Karl, G.L. (2003). Standardization of methods for fluence (UV dose) determination in bench-scale UV experiments. *J. Environ. Eng.*, 129(3), 209–215.

Boudaud, N., Machinal, C., David, F., Freval-Le Bourdonnec, A., Jossent, J., Bakanga, F., Arnal, C., Jaffrezic, M.P., Oberti, S., Gantzer, C. (2012). Removal of MS2, Qb and GA bacteriophages during drinking water treatment at pilot scale. *Water Res.*, 205, 1–14.

Bukhari, Z., Hargy, T.M., Bolton, J.R., Dussert, B., Clancy, J.L. (1999). Medium pressure UV light for oocyst inactivation. *J. Am. Water Works Assoc.*, 91, 86–94.

Clark, R.M. (1990). Modeling Inactivation of Giardia Lamblia. *J. Environ. Eng.*, 116(5), 837–853.

Courcier, J.P., Welte, B., Decerle, D., Jedor, B., Thibert, S. (2014). Limiter la formation des sous-produits de désinfection : le cas des bromates et des trihalométhanes dans les eaux destinées à la consommation humaine. *Techniques et sciences municipales*, 6(109), 71–83.

Craik, S.A., Weldon, D., Finch, G.R., Bolton, J.R., Belosevic, M. (2001). Inactivation of *Crytosporidium parvum* oocysts using medium and low-pressure ultraviolet radiation. *Water Res.*, 35, 1387–1398.

Dorn, J.M. (1974). A comparative study of disinfection on viruses and bacteria by monochloramine. Master's Thesis, University of Cincinnati, Cincinnati, OH.

Fair, G.M., Morris, J.C., Chang, S.L., Weil, I., Burden, R.P. (1948). The behavior of chlorine as a water disinfectant. *J. Am. Water Works Assoc.*, 40(10), 1051–1061.

Finch, G.R., Haas, C.N., Oppenheimer, J.A., Gordon, G., Trussell, R.R. (2001). Design criteria for inactivation of cryptosporidium by ozone in drinking water, ozone: Science and engineering. *J. Int. Ozone Assoc.*, 23(4), 259–284.

Feliers, C. and Laplace, C. (2001). Évaluation de la dose dans un réacteur vertical en utilisant la bio-dosimétrie, l'actinométrie et le calcul CFD. In *2nd World Water Congress*, Berlin, Germany.

Gaid, K. (2002). Les nouveaux paramètres liés à la désinfection, répercussions sur les installations d'eau potable. *AGHTM Conference*, 16 October, Toulouse, France.

Gaid, K. (2005). Sécurité sanitaire de la qualité en eau potable : l'expérience de Veolia Water. *Journées municipales Eau*, 10 May, Omnium Traitement et Valorisation – Veolia. Saint Maurice, France.

Gaid, K., Benoufella F., Laplanche., A. (1989). Chloration de quelques acides aminés présents dans les eaux à potabiliser. Cas de la valine et de l'alanine. *Prés. aux journées sur la désinfection des eaux*, 28–29 September, Lille, France.

Gaid, K., Cleret, D., Prados, M. (2003). Comment concevoir et exploiter les traitements pour limiter les sous-produits de désinfection. AGHTM congress, 14 June, Casablanca. *Techniques et Sciences Municipales*, 5(99), 18–28.

Gaid, K., Girodet, P., Bord, G., Féliers, C. (2005). La désinfection aux ultraviolets : une alternative de choix. *L'eau, l'industrie et les nuisances*, 282, 45–52.

Grayman, W.M., Deininger, R.A., Green, A., Boulos, P.F., Bowcock, R.W., Godwin, C. (1996). Water quality and mixing models for tanks and reservoirs. *J. Am. Water Works Assoc.*, 88(7), 60–73.

Haas, C.N. and Engelbrecht, R.S. (1980). Chlorine dynamics during inactivation of coliforms, acid-fast bacteria and yeasts. *Water Res.*, 14, 1744–1755.

Hart, F.L., Allen, R., DiAlesio, J., Dzialo, J. (1975). Modifications improve chlorine contact chamber performance, Part 1.*Water Sewage Works*, 122(9), 73–75.

Helmi, K., Menard-Szczebara, F., Lenes, D., Jacob, P., Jossent, J., Barbot, C., Delabre, K., Arnal, C. (2010). Adenovirus, MS2 and PhiX174 interactions with drinking water biofilms developed on PVC, cement and cast iron. *Water Sci. Technol.*, 61(12), 3198–3207.

Hibler, C., Hancock, C.M., Perger, L.M., Wegrzn, J.G., Swabby, K.D. (1987). Inactivation of Giardia cysts with chlorine at 0.5°C to 5.0°C. *American Water Works Association*, Denver, CO, 39 (Technical Research Series).

Hileman, B., Long, J.R., Kirschner, E.M. (1994). Chlorine industry running flat out despite persistent health fears. *Chem. Eng. News*, 21, 12–26.

Hill, A.G. and Rice, R.G. (1982). Historical background, properties and applications. In *Handbook of Ozone Technology and Applications*, volume 1, Rice, R.G., Netzer, A. (eds). Ann Arbor Science Publishers, Ann Arbor, MI.

Hoff, J.C. and Geldreich, E.E. (1981). Comparison of the biocidal efficiency of alternative disinfectants. *J. Am. Water Works Assoc.*, 73(1), 58–70.

Hoyer, O. (2000a). The status of UV technology in Europe, IUVA News. *Int. Ultraviolet Assoc.*, 2(1), 22–27.

Hoyer, O. (2000b). UV disinfection of drinking water supply, IUVA News. *Int. Ultraviolet Assoc.*, 2(5), 22–25.

Jacangelo, J., Darby, J.L., Loge, F., Tchobanoglous, G., Heath, M., Swaim, P. (1995). Comparison of UV irradiation to chlorination: Guidance for achieving optimal UV performance. *Water Environment Research Foundation*, 91-WWD-1.

Li, H., Finch, G.R., Smith, D.W., Belosevic, M. (2001). Chlorine dioxide inactivation of cryptosporidium parvum in oxidant demand-free phosphate buffer. *J. Environ. Eng.*, 127(7), 594–603.

Loeb, B.L. (2002). Ozone: Science & engineering, the first 23 years. *J. Int. Ozone Assoc.*, 24(6), 399–412.

Mackey, E.D., Cushing, R.S., Crozes, G.F. (2001). Practical aspects of UV disinfection. *AWWA Research Foundation and American Water Works Association*, Denver, CO.

Malley, J.P. (2000). Engineering of UV disinfection systems for drinking waters. *UV News, Int. UltraViolet Assoc.*, 2(3), 8–12.

Morris, J.C. (1971). Chlorination and disinfection – State of the art. *J. Am. Water Works Assoc.*, 63(8/12), 769–774.

Morowitz, H.J. (1950). Absorption effects in volume irradiation dosimetry. *Science*, 111, 229–230.

Norton, C.D. and LeChevallier, M.W. (1997). Chloramination: Its effect on distribution system water quality. *J. Am. Water Works Assoc.*, 89(7), 66.

Olivieri, V.P., Dennis, W.H., Snead, M.C., Richfield, D.T., Kruse, C.W. (1980). Reaction of chlorine and chloramines with nucleic acids under disinfection conditions. In *Water Chlorination: Environmental Impact and Health Effects*, vol. 3, Jolley, R.J. (ed.). Ann Arbor Science Publishers, Inc., Ann Arbor, MI.

Prados, M. (2003). Pratique de la désinfection. Direction Technique Veolia. Collection Doc-Tech no. 12. Paris, France.

Qualls, R.G. and Johnson, J.D. (1983). Bioassay and measurement of dose in ultraviolet disinfection. *Appl. Environ. Microbiol.*, 45, 872–885.

Qualls, R.G., Flynn, M.P., Donald, J. (1983). The role of suspended particles in ultraviolet disinfection. *Journal of Water Pollution Control Federation*, 55(10), 1280–1285.

Rice, R. and Gomez-Taylor, M. (1986). Occurrence of by-products of strong oxidants reacting with drinking water contaminants – Scope of the problem. *Environ. Health Perspect.*, 69, 31.

Santurette, N. (1996). Micro et macro-organismes dans les eaux d'alimentation : risques sanitaires, moyens de lutte". Générale des Eaux, Direction Technique. Collection Doc Tech no. 8. Paris, France.

Sharp, A. and Avery, P. (1993). Model solutions to chlorine contact. *Water Services*, 97 (1173), 30–36.

Shibata, K., Benson, A.A., Calvin, M. (1954). The absorption spectra of suspensions of living microorganisms. *Biochem. Biophys. Acta*, 15, 461–476.

Shin, G.A., Linden, K.G., Sobsey, M.D. (2005). Low pressure ultraviolet inactivation of pathogenic enteric viruses and bacteriophages. *J. Environ. Eng. Sci.*, 4, 7–11.

Sohn, J., Amy, G., Boireau, A., Dourdin, G., Gatel, D., Cavard, J. (1998). In plant *versus* distribution system formation of chlorination by product. *WQTC*, 6, 9–12.

Smith, D.B., Clark, R.M., Pierce, B.K., Regli, S. (1995). An empirical model for interpolating C.t values for chlorine inactivation of *Giardia Lamblia*. *Aqua J. Water Supply Res. Technol.*, 44(5), 203–211.

de Traversay, C., Bonnard, R., Adrien, C., Luck, F. (1999) Static mixer as a disinfectant process? *Proc. IOA Conf.*, 2(16), 251–261, 22–26.

de Traversay, C., Bonnard, R., Adrien, C., Luck, F. (2000). Static mixer: A reactor for the ozone process. *International Specialised Symposium on Fundamental and Engineering Concepts for Ozone Reactor Design*, International Ozone Ass., Toulouse, France, 155–162.

Vázquez Bravo, B., Vonder Haar, T., Mariñas, B.J. (2011). Inactivation of adenovirus, coxsackievirus and norovirus with polychromatic medium pressure ultraviolet light. Report, Department of Civil and Environmental Engineering.

Wright, H.B. (2000). Dose requirements for UV disinfection, UV news. *Int. UltraViolet Assoc.*, 2(3), 28–34.

# 22

# Disinfection By-products

## 22.1. General aspects

The main purpose of disinfection is to prevent the spread of waterborne diseases through the inactivation of microbial pathogens. Partly due to its low cost, chlorine is the most popular chemical disinfectant in the production of drinking water. One of its other advantages is its stability. As a result, residual chlorine is maintained in the distribution network, which prevents bacterial revival or development. In addition to its disinfectant activity, chlorine reacts with the organic and inorganic molecules present in the water.

Natural organic matter (NOM) is common in most streams, rivers and lakes. Emerging sources of NOM include algal organic matter (AOM) and wastewater effluent organic matter (EfOM). This organic matter is often a precursor to by-product formation. Even properly designed and well-operated drinking water plants are not all able to remove all organic matter from surface water. As a result, the chlorine that is used to treat for disease-causing microorganisms reacts with the remaining organic matter and forms disinfection by-products (DBPs) in the water produced and the distribution system.

Dissolved organic carbon (DOC) present in water can form a variety of DBPs during the final chlorination process in water treatment. Some of these by-products are known to be potentially carcinogenic and are being regulated.

The reactions with organic molecules can give rise to DBPs, many of which are harmful or potentially harmful to human health. In the future, it is highly likely that regulations on DBPs will become more stringent and include additional by-products, as the health risk for these substances becomes increasingly known.

One of the main classes of these by-products is haloacetic acids (HAAs), which comprise the sum of nine species: monochloroacetic acid (MCA), dichloroacetic acid (DCA) and trichloroacetic acid (TCA); monobromoacetic acid (MBA), dibromoacetic acid (DBA) and tribromoacetic acid; monochloro-, bromodichloro- and dibromochloroacetic acid. However, only the first five of the nine HAA species mentioned above are regulated by the sanitary authorities in some countries and the control of these species should be considered in the management of water-treatment facilities. Conventionally, the terms haloacetic acids formation potential (HAAFP) is used to reflect the potential of the water sample to form HAA DBPs.

The risk posed by DBPs must be weighed against the hazards resulting from microbial contamination, due to an incomplete disinfection. Various solutions are available to minimize the formation of DBPs. They involve modifying ordinary practices, changing chlorine injection points or even reducing precursors upstream of treatment stations. These last points are particularly relevant. In total, more than 600 DBPs caused by various oxidants such as chlorine, chloramine, ozone and chlorine dioxide have been identified in finished drinking water or laboratory tests.

However, the sum of halogenated DBPs usually only accounts for about 45–60% of the total organic halogen (TOX) compounds. This amounts to saying that approximately 40–55% of the by-products analyzed are not properly identified. Among all the known DBPs, trihalomethanes (THMs) and HAAs are considered the dominant groups.

Total THMs and the sum of five HAAs are regulated in the United States at 80 $\mu g \cdot L^{-1}$ and 60 $\mu g \cdot L^{-1}$, respectively, whereas the EU has a THM limit set at 100 $\mu g \cdot L^{-1}$.

## 22.2. Reaction by-products

The composition of the DBPs formed is affected by the type of disinfectant used, the disinfection conditions and the nature of the precursors present in the water.

This is because NOM is a complex mixture of many organic substances, which vary both over time and depending on the geographical area. The main NOM organic families are humic substances (HS), carboxylic acids, amino acids, proteins and carbohydrates. Actually, there are at least 12 important functional groups in NOM: carboxylic acid, enolic hydrogen, phenolic hydrogen, quinine, alcoholic hydroxyl, ether, ketone, aldehyde, ester, lactone, amide and amine.

Most NOM has an autochthonous or allochthonous origin (naturally present in the land catchment area). Terrestrial NOM is generally derived from lignin, characterized by its strong aromatic structure, whereas substances originating from algae, microorganisms, etc., tend to have weak aromatic structures and tend to be loaded with nitrogen compounds. Terrestrial NOM is therefore humic, non-polar and hydrophobic in nature. On the other hand, microbial NOM is non-humic, polar and has a hydrophilic character. Therefore, the characteristics of catchment basins affect both the fractional and chemical composition of NOM. Given the fact that the NOM classification rarely refers to a specific chemical identity, there is uncertainty about the identity of the DBP reactive precursors in drinking water.

The reaction of chlorine with the organic compounds present in water can lead to the formation of organohalogen products. Total organohalogens (TOX) include THMs and HAAs.

THMs and HAAs are formed through complex reaction mechanisms between chlorine and NOM. In general, high molecular weight aromatic substances such as humic acids are considered remarkably reactive in relation to this oxidant. The main THMs are as follows: chloroform, bromoform, bromodichloromethane (BDCM), dibromochloromethane (DBCM) and chlorodibromomethane (CDBM).

The nine most common HAAs are as follows: MCA, DCA, TCA, MBA, and DBA, bromochloroacetic acid, bromodichloroacetic acid, chlorodibromoacetic acid and tribromoacetic acid.

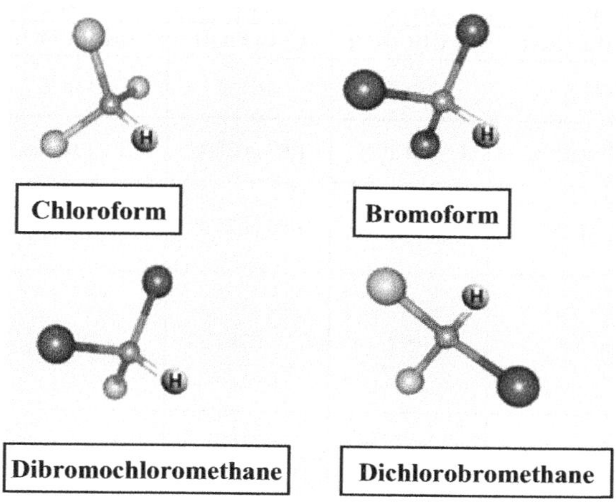


**Figure 22.1.** *Trihalomethanes (THMs). For a color version of this figure, see www.iste.co.uk/gaid/watertreatment5.zip*

Often it is the first five HAAs in the list, called the HAA5, which are taken into consideration. The six compounds deemed as probably or possibly carcinogenic to humans are as follows: BDCM, bromate, bromoform, chloroform, DCA and CDBM.

| Properties | Chloroform | Bromoform | Dichlorobromomethane | Dibromochloromethane |
|---|---|---|---|---|
| **Formulae** | $CHCl_3$ | $CHBr_3$ | $CHCl_2Br$ | $CHBr_2Cl$ |
| **Molecular weight** | 119.1 | 252.7 | 163.8 | 208.3 |
| **Density (g·cm$^{-3}$)** | 1.49 | 2.89 | 1.98 | 2.45 |
| **Water solubility (g·mL$^{-1}$)** | 7,500 at 25°C | 3,190 at 30°C | 3,320 at 30°C | 1,050 at 30°C |
| **Octanol-water partition coefficients (log KOW)** | 1.49 | 2.38 | 2.10 | 2.24 |
| **Henry's constant at 20°C (Pas·m$^3$·mol$^{-1}$)** | 440 | 54 | 160 | 86 |

**Table 22.1.** *Characteristics of trihalomethanes*

| Properties | Monochloroacetic acid MCA | Dichloroacetic acid DCA | Trichloroacetic acid TCA | Monobromoacetic acid MBA | Dibromoacetic acid DBA |
|---|---|---|---|---|---|
| **Formulae** | $ClCH_2COOH$ | $Cl_2CHCOOH$ | $Cl_3CCOOH$ | $BrCH_2COOH$ | $Br_2CHCOOH$ |
| **Molecular weight** | 94.5 | 128.9 | 163.3 | 138.9 | 217.8 |
| **Density (g·cm$^{-3}$)** | 1.40 at 25°C | 1.56 at 20°C | 1.62 at 25°C | 1.93 at 20°C | MD |
| **Water solubility (g·mL$^{-1}$)** | 1.09 at 25°C | Miscible | 1.50 at 25°C | 1.75 at 25°C | 2.11 at 25°C |
| **Octanol–water partition coefficients (Log KOW)** | 0.34 | 0.92 | 1.70 | 0.41 | 1.22 |
| **Dissociation constants (pKa) at 25°C** | 2.87 | 1.26 | 0.66 | 2.69 | N/A |

**Table 22.2**. *Characteristics of the five main HAAs*

In addition to THMs and HAAs, DBPs amount to other families (Table 22.3) represented by nitrosamines, acetonitriles, haloaldehydes, haloketones, halogenated acetamide and halogenated nitromethane.

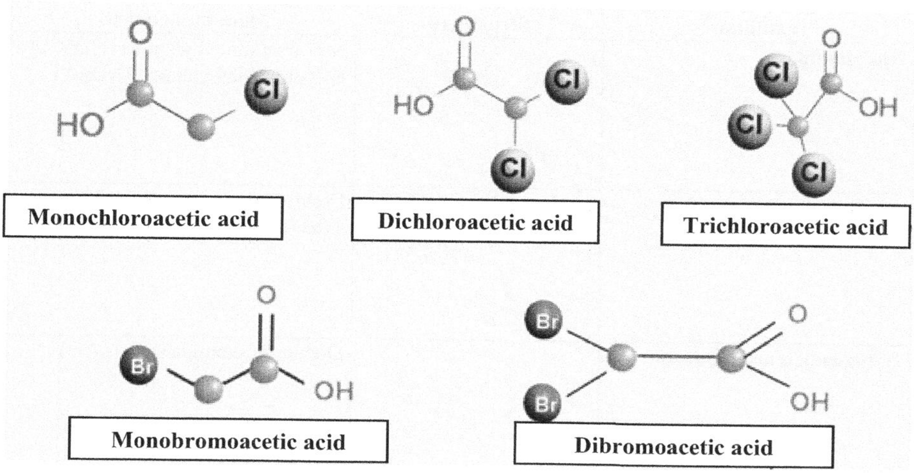


**Figure 22.2.** *Haloacetic acids (HAA5). For a color version of this figure, see www.iste.co.uk/gaid/watertreatment5.zip*

| Families | Structures | Main by-products |
|---|---|---|
| **Trihalomethanes** | 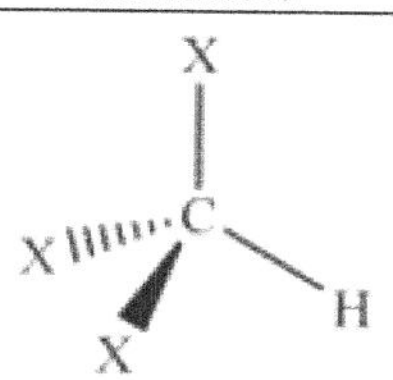 | Chloroform ($CHCl_3$)<br>Bromoform ($CHBr_3$) |
| **Halogenated acetic acids** | 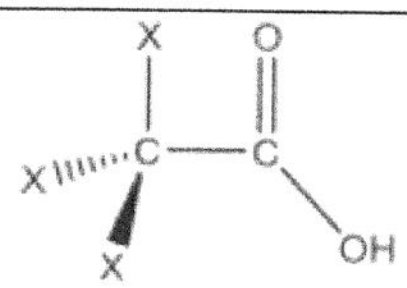 | Monochloroacetic acid ($ClCH_2COOH$)<br>Dichloroacetic acid ($Cl_2CHCOOH$)<br>Trichloroacetic acid ($Cl_3CCOOH$)<br>Monobromoacetic acid ($BrCH_2COOH$)<br>Dibromoacetic acid ($Br_2CHCOOH$)<br>Tribromoacetic acid ($Br_3CCOOH$)<br>Bromochloroacetic acid ($BrClCHCOOH$)<br>Dibromochloroacetic acid ($Br_2ClCCOOH$)<br>Bromodichloroacetic acid ($Cl_2BrCCOOH$) |

| Families | Structures | Main by-products |
|---|---|---|
| **Nitrosamines** | 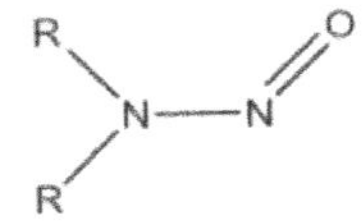 | N-Nitrosodimethylamine (NDMA)<br>N-Nitrosodiethylamine (NDEA) |
| **Acetonitriles** | 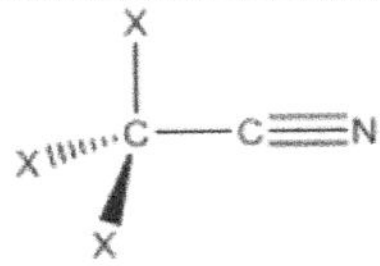 | Dichloroacetonitrile (DCAN)<br>Trichloroacetonitrile (TCAN) |
| **Halogenated acetamides** | 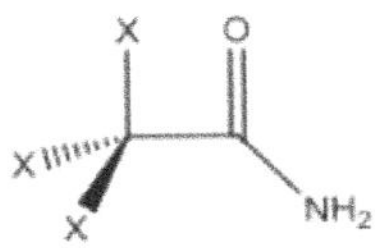 | Dichloroacetamide (DCAcAm) |
| **Halogenated aldehydes** | 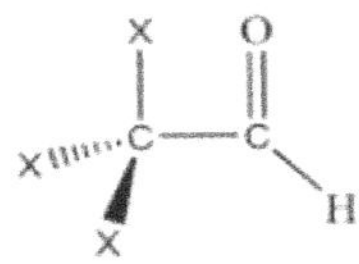 | Dichloroacetaldehyde (DCA)<br>Trichloroacetaldehyde (TCA) |
| **Halogenated ketone** | 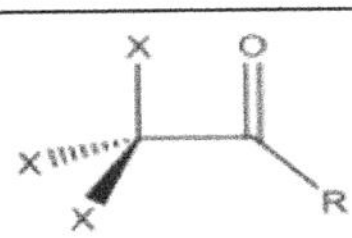 | 1,1,1-Trichloropropanone (TCP) |
| **Halogenated nitromethanes (HNMs)** | 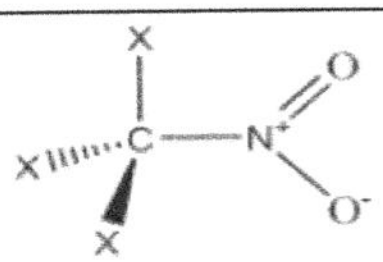 | Trichloronitromethane (TCNMM)<br>Chloropicrin |

**Table 22.3.** *Main by-products identified during disinfection*

## 22.3. Formation and evolution of chlorination by-products

The generic term DBPs corresponds to all the by-products formed at the level of water treatment stations, regardless of the oxidant's nature and injection point.

The same oxidizing agent can be used at the different junctures of a treatment chain in order to oxidize or disinfect water before distribution:

– The oxidant (ozone, chlorine dioxide or potassium permanganate) can be added upstream of the treatment by way of a procedure known as chemical pre-oxidation. The purpose is to oxidize certain undesirable compounds and thus facilitate their ultimate removal (iron, manganese, arsenic, color, algae, etc.) at the settling or flotation stages.

– The oxidant (ozone, chlorine dioxide, chlorine or potassium permanganate) can be added in the middle of the treatment, in which case the procedure is called chemical inter-oxidation (inter-chlorination, inter-ozonation, etc.). The aim is to control the development of various microorganisms or to oxidize certain residual compounds, in view of facilitating their removal at the filtration stage (iron, manganese, arsenic, color, algae, etc.).

– The oxidant (chlorine, chlorine dioxide or chloramines, depending on the country) is added downstream of the treatment by way of a procedure called post-chlorination and post-chloramination. The intention is to inactivate pathogens and to prevent their reactivation in the water network.

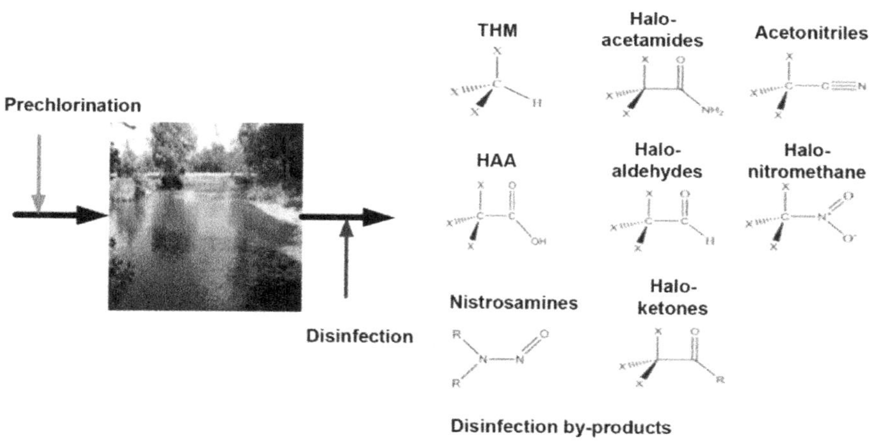


**Figure 22.3.** *DBP formation at a water treatment station. For a color version of this figure, see www.iste.co.uk/gaid/watertreatment5.zip*

Although ozone has often been used in post-ozonation, this solution is becoming less popular due to the formation of bromates and biodegradable organic carbon as well as other organic by-products.

Under these conditions, it seems clear that the practice of prechlorination, that is, the application of a high chlorine rate on water heavily loaded with organic matter, by far increases the risk of formation of chlorine oxidation by-products, especially during the hot periods. The solution adopted at present is to limit the formation of chlorination by-products (CBPs) by removing organic matter (clarification, for example, using $FeCl_3$ in an acid medium for coagulation) and ammonia (biological removal) as early as possible. Thus, chlorination is only maintained as a final disinfection treatment.

With ammonia, and for doses below the breakpoint, chlorine reacts by forming mono-, di- and trichloramines. These last two substances impart a disagreeable taste and a have certain mutagenic power, evidenced by some studies.

Phenols give rise to chlorophenols, with characteristic unpleasant tastes and odors.

## 22.4. Kinetics and formation mechanisms

### 22.4.1. *Formation kinetics*

NOM is a source of numerous precursors for the formation of CBPs. Humic substances (HS) are characterized by their aromatic rings. They represent the most reactive fraction of NOM and form CBPs such as THMs.

The formation of THMs is characterized by a rapid proliferation during the first hours, subsequently decreasing in a continuous manner, until slowly reaching the characteristic concentration of the total formation potential for THMs (Figure 22.4).

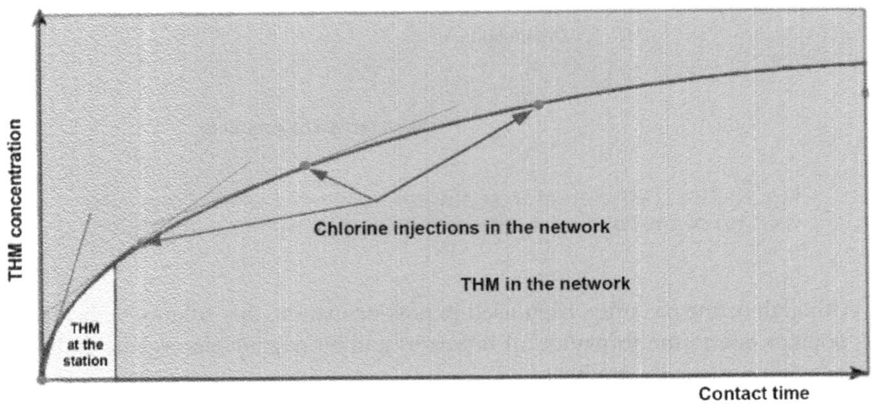


**Figure 22.4.** *THM formation at the treatment station and in the distribution network. For a color version of this figure, see www.iste.co.uk/gaid/watertreatment5.zip*

This two-speed curve is explained by the fact that NOM contains a variety of substances, each with their own reaction kinetics. The first part of the curve corresponds to the reactions taking place in the disinfection tanks, for which the contact times oscillate between 1 and 3 h. At first, chlorine reacts with the aromatic active group, encouraging the rapid formation of DBP. Reactions are very fast due to the interactions between aromatic sites and chlorine. THM concentrations can vary by 20–30% depending on the organic matter concentration and the various factors already discussed. The most reactive and productive THM precursors are mainly resorcinol-type aromatic structures, as well as other compounds from the β-diketone or β-ketoacid family.

On many sites, for 15°C temperature, a pH of 7 and a DOC concentration varying between 1 and 3 mg·L$^{-1}$, and an injected chlorine content of 1.5 mg·L$^{-1}$, the THM concentration oscillates between 30 and 50 μg·L$^{-1}$ (Figure 22.5).

The second part of the curve implies a slowing down of the kinetics because the reactions are slower, although they will continue their action in the distribution networks. As the reaction time progresses, the residual oxidant and reactive precursor decrease, slowing down the DBP formation rate. This is typical of less reactive precursors requiring longer contact times, leading to a lower formation of THMs. Although less reactive, these precursors (laden with abundant NOM) account for 70% of the total THMs formed. In particular, this fraction of precursors includes phenol-type compounds.

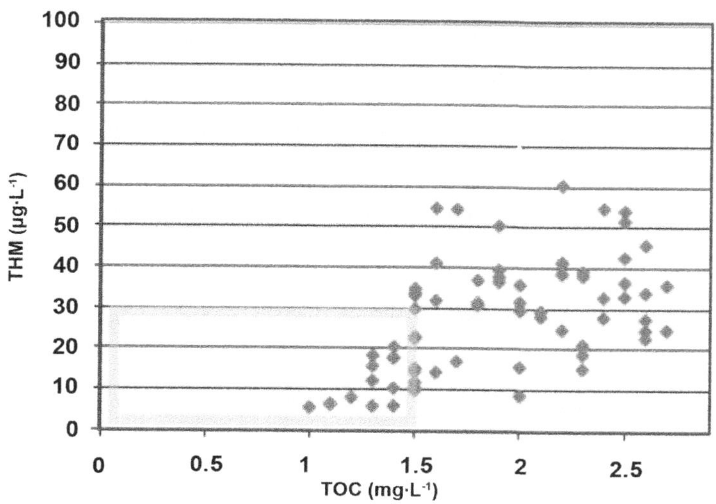


**Figure 22.5.** *THM leaving stations versus TOC in treated water. For a color version of this figure, see www.iste.co.uk/gaid/watertreatment5.zip*

Figure 22.6 shows the results of laboratory tests (24 h reaction time) on raw water with concentration of organic matter (TOC) of 7.5 mg·L$^{-1}$ and a pH of 7.4. The temperature was set at 18°C. Chlorine (gaseous chlorine) concentrations varied between 0.5 and 5 mg·L$^{-1}$.

In general, DBPs continue to form as long as the residual oxidant and the residual precursor are present. The formation rate varies depending on the properties of organic matter.

Chloroform formation kinetics are the fastest and the most critical. More than 70 µg·L$^{-1}$ are obtained after 24 h contact time. Dichlorobromomethane formation kinetics are a little slower: 50 µg·L$^{-1}$ are obtained after 24 h with a chlorine concentration of 5 mg·L$^{-1}$. Bromoform and DBCM formation kinetics are the slowest and lowest in terms of the concentrations obtained. The total sum of THMs obtained after 24 h for a concentration of 1 mg·L$^{-1}$ is 51 µg·L$^{-1}$, and 177 µg·L$^{-1}$ for a dosage of 5 mg·L$^{-1}$.

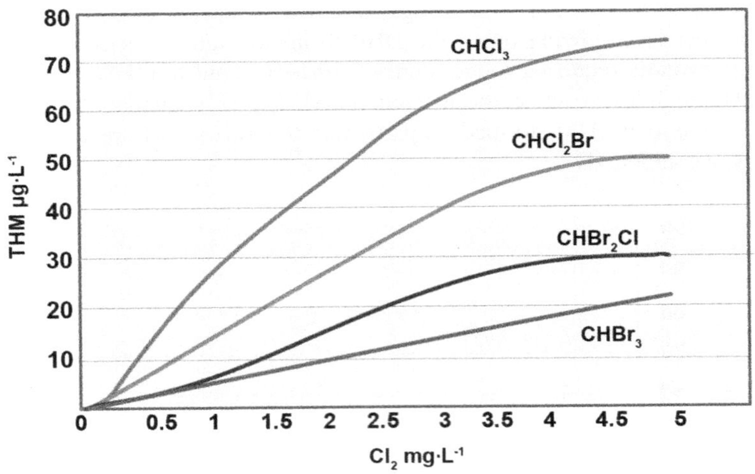


**Figure 22.6.** *THM formation as a function of the chlorine dose*

### 22.4.2. *Mechanisms*

The aromatic character of humic matter is known to be responsible for the high reactivity of these compounds with oxidants such as chlorine.

Three kinds of reactions can be described between hypochlorous acid and organic compounds:

– oxidation reactions;

– addition reactions to unsaturated bonds;

– electrophilic substitution reactions at nucleophilic locations.

The reaction begins with the rapid chlorination of the ortho carbon atom, activated by hydroxyl substitution. An intermediate form is then promptly halogenated or protonated, depending on the reaction conditions.

In the case of chlorination, there is the formation of hypochlorous acid (HOCl) and the hypochlorite ion ($OCl^-$), which in turn reacts in the presence of a bromide ion, oxidizing it into hypobromous acid (HOBr) and the hypobromite ion ($OBr^-$). Hypochlorous acid and hypobromous acid then react with organic matter (NOM) to form different DBPs, including THMs and HAAs, in particular.

THMs and HAAs are the two main groups of chlorinated DBPs present in drinking water, where they are generally found in the highest concentrations. These two groups can serve as indicators of the presence of any chlorinated DBP present in drinking water supplies.

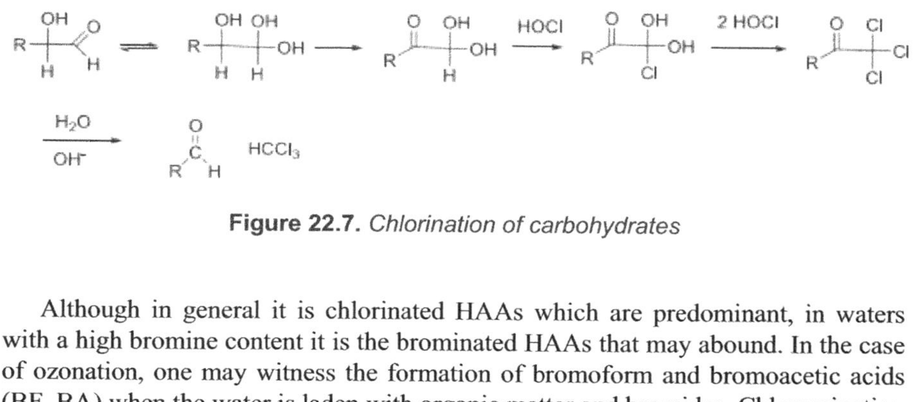

**Figure 22.7.** *Chlorination of carbohydrates*

Although in general it is chlorinated HAAs which are predominant, in waters with a high bromine content it is the brominated HAAs that may abound. In the case of ozonation, one may witness the formation of bromoform and bromoacetic acids (BF, BA) when the water is laden with organic matter and bromides. Chloramination also leads to the formation of HAAs, if chloramines are produced due to the reaction between chlorine and ammonia. Compared to chlorine, chloramines and ozone produce THMs and HAAs in negligible amounts. Chlorine dioxide can also produce HAAs.

HAA formation can be significant when drinking water is chlorinated under slightly acidic pH conditions. The formation of THMs increases with pH, whereas that of HAAs decreases, with hydrolysis probably playing an important role.

Even though the formation of HAAs and THMs depends on different pH values, there seems to be a strong correlation between their formation when the treatment conditions are relatively uniform and when the water has a low bromide content.

Longer contact times and higher water temperatures contribute to the formation of HAAs. When the water temperature is higher, the reactions are faster and the chlorine demand is stronger. In this case and in the presence of high concentrations of NOM with aromatic content (humic acids), the formation of CBPs in general, and of HAAs and THMs in particular, is favored. A predominance of trichoroacetic acid is noticed (in comparison with monochloroacetic and DCA).

In the presence of bromides, the chlorination process promotes the formation of brominated CBPs (brominated and chlorobrominated HAAs), as per the water's physicochemical properties. Bromide concentrations in surface water and groundwater can fluctuate seasonally and result from the intrusion of saltwater.

Among the multiple DBPs, chloroform is the compound that is most commonly formed for THMs, while DCA and TCA are frequent for HAAs. At the same time, haloacetonitriles (HANs) and haloketones are the other two major families of DBPs.

Limiting the formation of DBPs and CBPs necessarily involves reducing organic matter and optimizing chemical disinfection processes.

### 22.4.3. *Chlorination of HS*

Although chlorine is dosed as gas or sodium hypochlorite, it is hypochlorous acid (HOCl), which is the main reactive form for treating water. Since hypochlorous acid is an electrophilic reagent, it tends to react with the electron-rich compounds found in NOM. Oxidation, addition and electrophilic substitution reactions are all possible. Normally, only electrophilic addition is significant in reactions with organic matter. Amines and aromatic groups are all highly reactive with chlorine. Hypochlorous acid reacts rapidly with amines, producing organic and inorganic chloramines. For less reactive compounds such as alkenes, reactions with chlorine may be too slow to have a notorious impact during the disinfection period, though they may be detected in the distribution network.

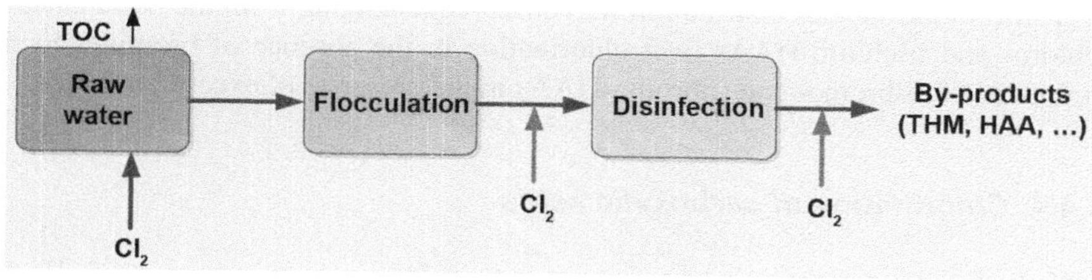


**Figure 22.8.** *Chlorination of humic substances and formation of DBPs*

HS or hydrophobic NOM include carboxylic acids, phenolic and alcoholic hydroxyls and ketones as major functional groups. The average value of $pK_a$ of HS is 4.2. Humic and fulvic acids have a high lignin content, which means that aromatic groups are strongly present. Fulvic acids generally have a molar mass (MM) between 500 and 2,000 Da, whereas humic acids are more voluminous and colloidal, with a MM between 2,000 and 100,000 Da. HS generally constitute about 30–50% of the DOC of a mid-sized river, and up to 75% of a water catchment area.

The HAA species distribution was also found to be clearly influenced by the nature of the oxidant. Dihalogenated HAAs followed by trihalogenated HAAs were generally the dominant HAA species during the $ClO_2$ treatment. For fresh water, a significant amount of mixed bromochloro-HAAs was observed after chlorination. This was due to the coexistence of HOBr ($OBr^-$) and HOCl ($OCl^-$) as reactive oxidative species, because of the presence of an appreciable quantity of bromide in freshwater samples. DCAA was dominant and contributed approximately 50% of total HAAs, followed by DBAA, BCAA, trichloroacetic acid (TCAA) and BDCAA.

In a general manner, though, one may deduce from the results that due to their relative stability, di-HAAs are more easily formed than tri-HAAs, with a significantly lower quantity of mono-HAA, for all kinds of water sources. Regardless of the type of water source, di-HAAs and tri-HAAs are regularly the predominant species. A higher incorporation of bromides, followed by HAAs, in THMs was noticed compared to fresh water treated by chlorination:

$ClO^- + Br^- + H_2O \rightarrow HOBr$ (in the presence of bromides)

HOCl (or $OCl^-$)/HOBr (or $OBr^-$) + NOM $\rightarrow$ Cl (or Br) – THMs

A shift in the species distribution was observed, changing from the dominance of dichloro- and trichloro-HAAs over chlorination in the absence of bromides to the dominance of di-bromo- and tribromo-HAAs in the presence of excess bromides.

### 22.4.4. *Chlorination of carboxylic acids*

Carboxylic functional groups are present on HS at a frequency of 5–10 per molecule. Under the pH conditions of natural waterways (pH 6–8), these groups are anionic (that is to say, dissociated). Carboxylic acids can represent up to 80% of the organic carbon in water. They are associated with aromatic, aliphatic and polymeric groups.

This indicates that reactive DBP precursors either form high quantities of THMs or HAAs, but not of both.

Since hydrophobic NOM is the fraction best removed by coagulation, this stage reduces the potential for the formation of TCAA and THMs. In waters where the hydrophilic NOM is high, it is more difficult to anticipate the type of DBPs produced. As it is the main residual constituent after coagulation–flocculation, additional treatment may be required to reduce the formation of DBPs by removing this residual precursor. Among the carboxylic acids, the β-dicarbonyl species are presumed to be the first reactive precursor group.

Nitrogen-rich NOM is present in waters affected by algal developments and sewage. NOM oxidation leads to the formation of nitrogenous DBPs including N-nitrosodimethylamine (NDMA) and dichloroacetonitrile (DCAN). Some amino acids (tryptophan and aspartic acid, in particular) are also involved in the formation of nitrogenous DBPs.

### 22.4.5. *Factors influencing the formation of DBPs*

The concentration and speciation of DBPs depends on various parameters such as water quality, NOM concentration, residual chlorine concentration, reaction time, pH and bromide concentration.

There are contradictory results regarding the identity of THM and HAA precursors in natural waters. For example, some have concluded that HAA precursors have a higher aromatic content than THM precursors. Conversely, other authors suggest that the hydrophilic fraction produces a higher proportion of HAAs in relation to THMs, compared to the hydrophobic fraction.

| Raw water qualities | Disinfection conditions with chlorine |
|---|---|
| TOC concentrations | Type of disinfectant |
| Bromide concentration | Disinfectant dosage |
| pH (↑pH, ↑THM) | Contact time |
| pH (neutral pH, THM, HAA) | Preliminary removal of TOC |
| pH (↓pH ↑HAA) | Addition of disinfectants upstream of the treatment |
| Temperature | Disinfectant combination |
| Alkalinity | Residual chlorine concentration |
| Season of the year | Retention time in the network |
| Distribution network length | |

**Table 22.4.** *Parameters influencing the formation of THMs and HAAs by chlorine disinfection*

It has been observed that waters producing high levels of THMs have a propensity to generate TCAA, and that DCAA precursors are globally less hydrophobic than TCAA ones. It is believed that the production of DCAA and TCAA is the result of different mechanisms: TCAA is not simply formed by the direct attack of chlorine onto DCAA, but a large number of intermediate reactions take place before the final formation of TCAA.

The high DCAA concentrations are related to the presence of diketones and aldehydes already at play and those created by oxidation.

In general, the presence of bromides ($Br^-$) increases the levels of halogenated DBPs. This point is crucial because not only do brominated DBPs have a higher molar mass than chlorinated DBPs, but they are often also more toxic.

Although DBPs have traditionally been associated with NOM as the source of precursors, new research has shown that pharmaceuticals, personal care products and agricultural chemicals may be transformed into by-products during the disinfection/oxidation process. In other research papers, the widely used antimicrobial agent triclosan was shown to react with chlorine to form chloroform and other chlorinated organic compounds. Bisphenol A, which is a building block of several important polymers and polymer additives, is an endocrine disruptor.

| Chemical groups | THM | HAA | N-DBP |
|---|---|---|---|
| Humic substances | Raw water | Raw water | Possible impact with halogenated nitromethanes (HNMs) |
| Carbohydrates | Raised to pH 8 | Minor impact | No impact |
| Amino acids | Low impact | Important for aspartic acid, histidine, asparagine and tryptophan | High impact |
| Proteins | Important during an algal bloom | No impact | Unlikely |
| Carboxylic acids | ß-Dicarbonyl acids | ß-Dicarbonyl acids | Minor impact |

**Table 22.5.** *Impact of the chemical constituents of water on the formation of halogenated THMs, HAAs and nitromethane*

#### 22.4.5.1. *The effect of temperature*

As with any chemical reaction, the formation rate of disinfection by-products is temperature dependent, in accordance with Arrhenius' law:

$$k = A \exp(-Ea/R \times T)$$

where:

– k: velocity coefficient (or velocity constant);

– A: frequency factor (takes into account collision frequency and steric hindrance);

– Ea: activation energy of the reaction ($J \cdot mol^{-1}$);

– R: ideal gas constant ($R = 8.314\ J \cdot mol^{-1} \cdot K^{-1}$);

– T: temperature (K).

This law shows that the formation of DBPs depends not only on the activation energy, in other words, the energy barrier to be crossed for the reaction to occur (reactions with low Ea are the most rapid), but also on temperature. An increase in temperature produces an increment in the reaction rate.

The effect of temperature seems to fade with the reaction's duration. Despite having an influence on the THM reaction kinetics, it does not modify the final equilibrium state.

As a result, the formation of by-products is higher in summer than in winter.

### 22.4.5.2. *The effect of contact time*

Contact time is one of the most important factors in THM formation. It designates the time lapse during which the disinfectant remains in contact with the precursors. An increase in the reaction time with chlorine promotes the formation of CBPs (THMs and HAAs).

As THMs are typical hydrolysis products and end-products of chlorination, the formation of THMs is increased by a rise in the reaction time.

### 22.4.5.3. *The effect of pH*

pH is an important parameter in the formation of DBPs and alkaline media to efficiently catalyze the reaction. The effect of pH on DBP formation is complex and may stimulate the formation of some products over others. An increase in pH is generally accompanied by an increment in the concentrations of the by-products formed.

The speciation of chlorine is affected by pH. At low pH, the dominant chlorine species is hypochlorous acid (HClO). At high pH, the dominant species is hypochlorite ($ClO^-$). In general, a low pH (<6.5) favors the formation of THMs as well as HAAs, trihalogenated acetaldehydes, trihalopropanones and cyanohalides.

A pH over 7.3 results in the formation of a higher concentration of THMs, but a lower concentration of halogenated HAAs and CBPs.

As the acidic pH favors the removal of organic substances during the clarification stage, the final THM and HAA concentrations are often quite close.

In general, pH effects occur as a function of the acidic or basic conditions and have an influence on reaction rates. For example, the formation of THMs from carbohydrates is higher at pH 8 than at pH 5 due to the (basic medium) conditions which favor hydrolysis, whereas with 3-oxopentanedioic acid, a higher efficiency is observed due to the pH decrease from 8 to 5.5.

### 22.4.5.4. *The effect of the NOM concentration*

DBPs are formed through the reaction of a chemical disinfectant (chlorine, chlorine dioxide, chloramines, ozone) with an organic precursor, represented by various NOM components and an inorganic precursor, most often certain halide ions. Any one of these disinfectants can transform (oxidize) a complex NOM molecule into simpler moieties, which are then reactive, with additional chlorine or chloramines acting as a halogen substitution agent.

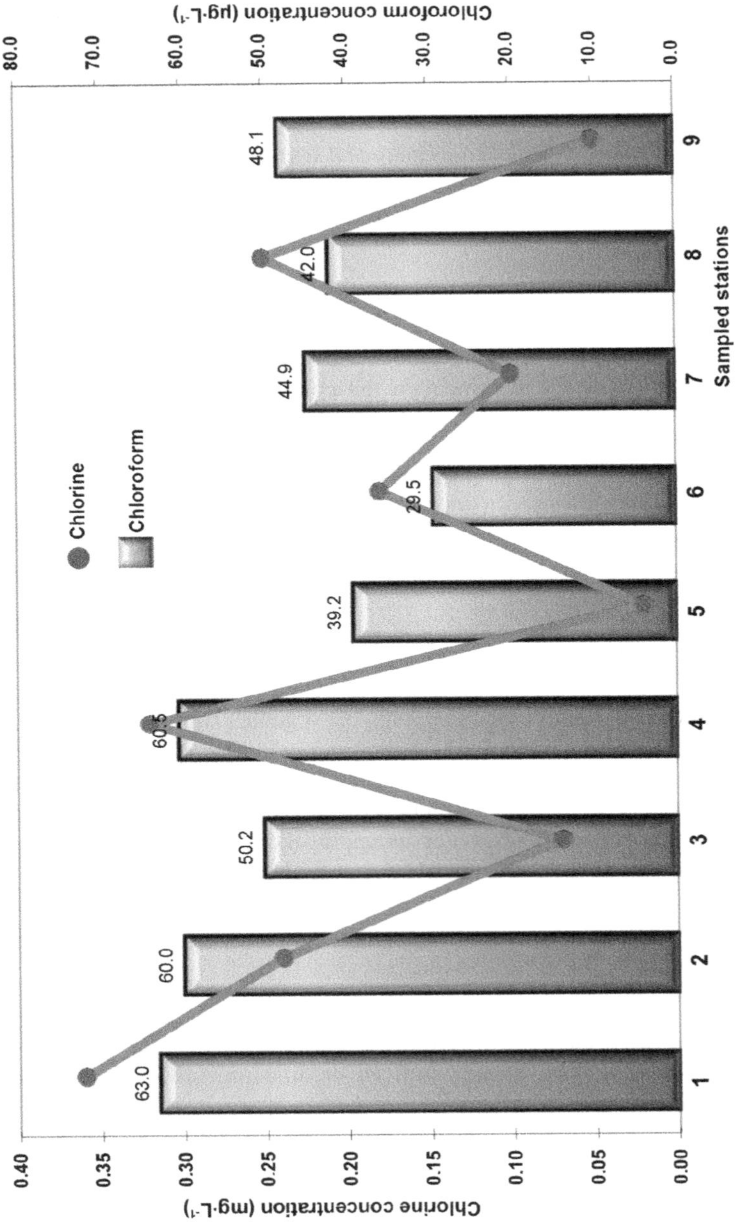


**Figure 22.9.** *Formation of chloroform as a function of chlorine at nine sites. For a color version of this figure, see www.iste.co.uk/gaid/watertreatment5.zip*

Due to its heterogeneous and indefinite nature, the quantity of NOM is measured by surrogate parameters such as total organic carbon (TOC), DOC or UV absorbance at 254 nm.

This last parameter is higher for humic acids due to the higher aromatic content and larger molecular size of this type of compound. Another indicator is specific UV absorbance (SUVA), which results from dividing the UV absorbance by the DOC concentration.

HS have higher SUVAs and CBP formation potentials than the non-humic fraction. Dissolved organic matter (DOM) is composed of a heterogeneous mixture between HS and non-HS. It is mainly HS that act as precursors to DBPs such as THMs and HAAs.

CBP formation is affected by the NOM concentration and characteristics in two ways. First, an increment in the NOM concentration increases the level of CBP precursors, promoting their formation. Then, an increment in the NOM concentration increases water's chlorine demand. A high dosage of chlorine is required to maintain the residual chlorine for network protection. This high dose of chlorine leads to a further increase in the formation of CBP.

Figure 22.9 illustrates the formation of chloroform at some water treatment stations as a function of the chlorine dose. The NOM concentration is variable, set between 1.5 and 3.5 $mg \cdot L^{-1}$.

### 22.4.5.5. *Bromides*

Bromide is an inorganic ion which does not react directly with NOM. However, it can be oxidized by chlorine or ozone into hypobromous acid (HOBr) or hypobromite ($BrO^-$), depending on the pH. By analogy with hypochlorous acid and hypochlorite, hypobromous acid and hypobromite react with NOM, forming brominated DBPs. In water containing bromides, brominated DBPs are formed during chlorination and ozonation. As bromine uses chlorine substitution sites, the formation of chlorine species is reduced:

$$Cl_2 + H_2O \rightarrow HOCl + H^+ + Cl^-$$

$$HOCl \rightarrow OCl^- + H^+$$

$$OCl^- + Br^- \rightarrow OBr^- + Cl^-$$

$$HOCl + Br^- \rightarrow HOBr + Cl^-$$

$$HOCl + HOBr + NOM \rightarrow DBP$$

As bromine (atomic weight 80) is much heavier than chlorine (atomic weight 35.5), the correlated concentration of bromoform is twice that of chloroform. As a result, under given chlorination conditions, an increase in bromides can significantly increase the concentration of THMs. An increase in bromide concentrations also increments the formation of brominated HAAs, while reducing the formation of chlorinated HAAs.

However, HOCl acts as an efficient oxidant, whereas HOBr behaves more efficiently as a halogen substitution agent. As the bromide/TOC ratio increases, the percentage of brominated DBPs rises too. This occurs when the bromide concentration rises or the TOC concentration diminishes.

Thus, chlorination creates hypobromite ($OBr^-$), which in turn are responsible for the formation of brominated derivatives. Figures 22.10 and 22.11 show the potential for HAA formation as a function of the bromide concentration. The formation of HAAs (chlorinated derivatives) is important when the final bromide/chlorine ratio is about 20–30.

The higher the bromide concentration, the lower the production of chloroform in relation to brominated THMs. The chloroform concentration is about 60% of the total THMs for a low bromide concentration and decreases to 20% of the THMs when the bromide concentration is high. Notice that brominated THMs are more harmful than non-brominated THMs.

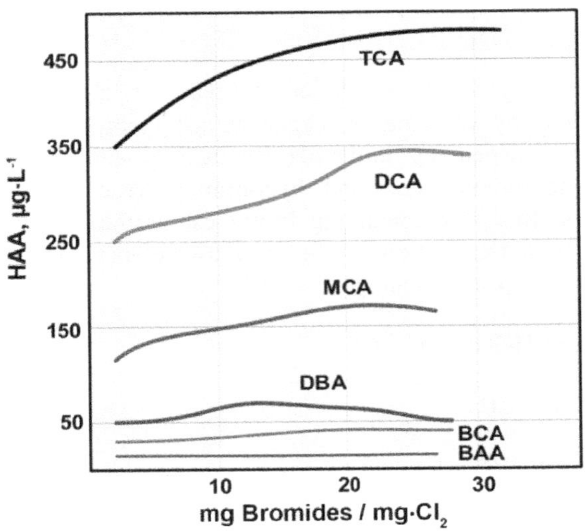


**Figure 22.10.** *Influence of the bromide/chlorine ratio on the formation of HAAs*

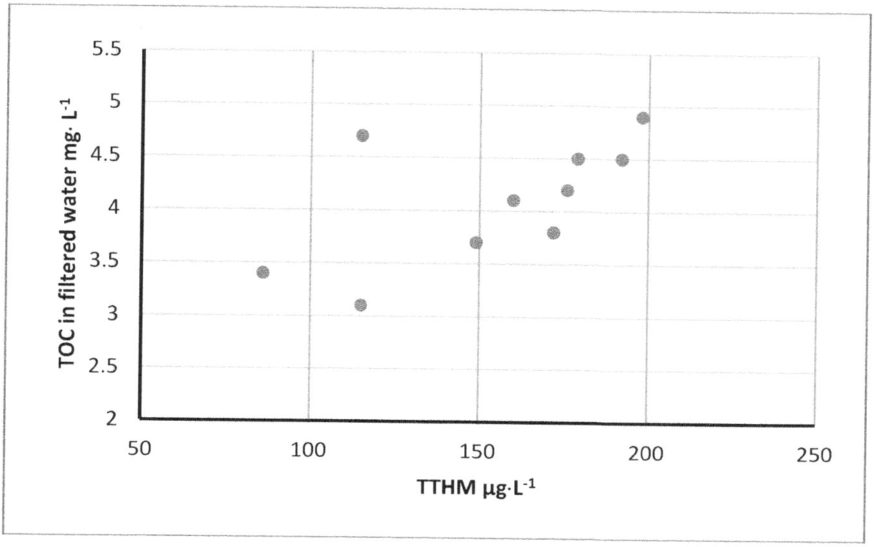


**Figure 22.11.** *TOC versus TTHM (Ouessant), bromides 650 µg·L$^{-1}$*

Figure 22.11 shows the influence of bromides on the potential for THM formation. The operating conditions are the following: T = 20°C, pH = 7.2, chlorine dosage 2 mg·L$^{-1}$, contact time 2 h. Bromides increase DBP formation rates by a factor of 3–10.

### 22.4.5.6. *The effect of the chlorine dose*

Chlorine is one of the two major reagents for the formation of chlorinated by-products. Some CBPs are intermediate products from the chlorination reaction and some are end-products. Intermediate products can be oxidized into end-products, as shown by the following equations:

$$\text{NOM} + \text{HOCl} \rightarrow \text{intermediate products}$$

$$\text{Intermediate products} + \text{HOCl} \rightarrow \text{final products}$$

In general, an increase in the dose of chlorine is responsible for the formation of chlorination end-products in treated water. THMs are typical end-products of chlorination reactions.

Monohalogenated and dihalogenated by-products from chlorination are intermediate products, with the exception of monohalogenated and dihalogenated acetic acids. With a moderate level of residual chlorine, dihalogenated CBPs are formed. A further increase in chlorine dose leads to a rise in trihalo CBPs and a reduction in dihalo CBPs.

An increase in the chlorine dose encourages the formation of THMs, HAAs and many other chlorinated DBPs. However, high doses of chlorine promote the formation of HAAs, rather than THMs. And the formation of trihalogen HAAs is higher than that of di- and monohalogen HAAs.

## 22.5. Regulations

The regulation of THMs and HAAs is based on their use as "surrogates" for the toxicity associated with chlorinated water. Also, it became known that the control of THMs and HAAs also resulted in a reduction in the concentration of many other DBPs as well as unknown chlorination DBPs (based on total organic halogen [TOX] measurements), which may be associated with adverse health effects.

The French government resolution of January 11, 2007 specifies that "the lowest possible value below this value must be targeted without compromising disinfection." Total trihalomethanes refers to the sum of chloroform, bromoform, DBCM and BDCM. The quality limit is set at 150 $\mu g \cdot L^{-1}$.

French regulations require that the consumer's tap – defined as point of compliance or point of use – respects the quality limit of THMs, as is the case for most other regulated chemical pollutants.

In France, the THM monitoring program is characterized by two things:

– a sampling point, which is at the distribution point if the water is treated with chlorine (P2 analyses), and at the consumer's tap, if the free chlorine content measured in the network is greater than 0.5 $mg \cdot L^{-1}$, or at a re-chlorination station, when existent (D2 analyses);

– a sampling frequency, which may vary from once every 10 years (population of less than 50 inhabitants) to once a month (population of more than 300,000 inhabitants), for each of the sampling points.

## 22.6. Predictive models of CBPs

Two types of models are used to predict THM concentrations after chlorination: empirical models and chlorine degradation models (mechanical models). The importance of models is that they can be useful in predicting THM concentration, without resorting to additional tests. Besides, they are sufficiently reliable to assess the effects of operating conditions such as pH change, temperature, contact time and disinfectant doses. The CBP precursors or the CBPs themselves are treated as dependent variables. On the other hand, water quality parameters and process parameters are independent variables. These empirical relationships have practical relevance. Models can be used for controlling the unitary treatment stages, rapidly accessing relevant information. Or they can be used to assess the relative impact of individual parameters on the performances of the treatment station.

Mechanical models are more robust and flexible than empirical models. They provide an overview of the efficiency of modeled unitary stages, acknowledging the characteristics of the reactions between chlorine and NOM. This is something empirical models do not do.

But the implementation of mechanical modeling is considered burdensome due to the complexity of NOM molecules and the reaction pathways between chlorine and NOM. This is even the case when working with computing models, which allow for instantaneous calculations. For operators though, empirical models are easily accessible, enabling them to evaluate THMs very quickly from an Excel sheet.

The predictive models for CBP (THMs and HAAs) concentrations are mostly built using data from laboratory experiments, and sometimes, from the measurements in water distribution networks. The first works have identified the main determinants of THM formation (chlorine dose, pH, TOC, temperature, contact time).

While laboratory tests can offer a good estimate of the general appearance of THMs at the different levels of a treatment chain and at the station's outlet, they are less effective for predicting the evolution of THMs in distribution networks.

There are many formulas for calculating THMs and HAAs at the station's outlet. The following formulas are fairly representative of the results obtained in the field:

– for THMs:

- in the presence of bromides:

$$\text{TTHM } (\mu g \cdot L^{-1}) = 0.15 \ (\text{DOC}^{0.44}) \ (\text{Cl}_2^{\ 0.41}) \ (\text{t}^{0.27}) \ (\text{T}^{1.06}) \ ((\text{pH-2,6})^{0.72}) \ (\text{Br} + 1)^{0.25}$$

- in the absence of bromides:

$$\text{TTHM } (\mu g \cdot L^{-1}) = 0.15\ (DOC^{0.44})\ (Cl_2^{\ 0.41})\ (t^{0.27})\ (T^{1.06})\ ((pH\text{-}2.6)^{0.72})$$

– for HAAs:

- in the presence of bromides:

$$\text{HAA } (\mu g \cdot L^{-1}) = 0.50\ (DOC^{0.935})\ (Cl_2^{\ 0.443})\ (Br^{0.031})\ (T^{0.387})\ (t^{0.178})\ (6.68\ pH^{-0.52})$$

- in the absence of bromides:

$$\text{HAA } (\mu g \cdot L^{-1}) = 0.50\ (DOC^{0.935})\ (Cl_2^{\ 0.443})\ (T^{0.387})\ (t^{0.178})\ (6.68\ pH^{-0.52})$$

with the expression of DOC in $mg \cdot L^{-1}$, chlorine dosage ($Cl_2$) in $mg \cdot L^{-1}$, the contact time (t) in hours, the temperature (T) in °C and bromides (Br) in $\mu g \cdot L^{-1}$.

## 22.7. Removal of THMs and HAAs

Once the treatment chain set in motion, there are a few methods for removing THMs and HAAs.

### 22.7.1. *Aeration*

In the presence of a high concentration of chloroform, tank aeration or stripping removes this compound.

Table 22.6 presents Henry's constants ($atm \cdot m^3 \cdot mol^{-1}$) for different compounds and gases. It shows that THMs like chloroform and bromoform (with high Henry's constants) are well removed by aeration or stripping. However, HAAs are not removed (high Henry's constants).

The aeration or stripping processes are relevant for only a few compounds among the THMs and HAAs. They are chosen depending on water's chemical composition. Air/water ratios between 0.5 and 1 lead to 75–80% chloroform removal efficiencies (Figure 22.12).

| Compound | $K_H$ atm·m$^3$·mol$^{-1}$ |
|---|---|
| **Trihalomethanes (THMs)** | |
| Dichlorobromomethane $CHCl_2Br$ | $1.83 \times 10^{-3}$ |
| Dibromochloromethane $CHBr_2Cl$ | $1.03 \times 10^{-3}$ |
| Chloroform $CHCl_3$ | $3.06 \times 10^{-3}$ |
| Bromoform $CHBr_3$ | $6.29 \times 10^{-4}$ |
| **Haloacetic acids (HAAs)** | |
| Monochloroacetic acid | $9.46 \times 10^{-6}$ |
| Dichloroacetic acid | $8.57 \times 10^{-6}$ |
| Monobromoacetic acid | $6.68 \times 10^{-6}$ |
| Dibromoacetic acid | $4.52 \times 10^{-6}$ |
| Tribromoacetic acid | $3.42 \times 10^{-6}$ |
| **Chlorinated volatile compounds** | |
| Tetrachloroethylene | $1.41 \times 10^{-2}$ |
| Trichloroethylene | $9.89 \times 10^{-3}$ |
| 1,2-Dichloroethane | $1.47 \times 10^{-3}$ |
| Carbon tetrachloride $CCl_4$ | $2.32 \times 10^{-2}$ |
| **Various** | |
| Sulfurous gas $H_2S$ | $9.26 \times 10^{-3}$ |
| Carbon dioxide $CO_2$ | $2.72 \times 10^{-2}$ |
| Ammonia $NH_3$ | $1.6 \times 10^{-5}$ |

**Table 22.6.** *Henry's constants for various compounds and gases at 20°C*

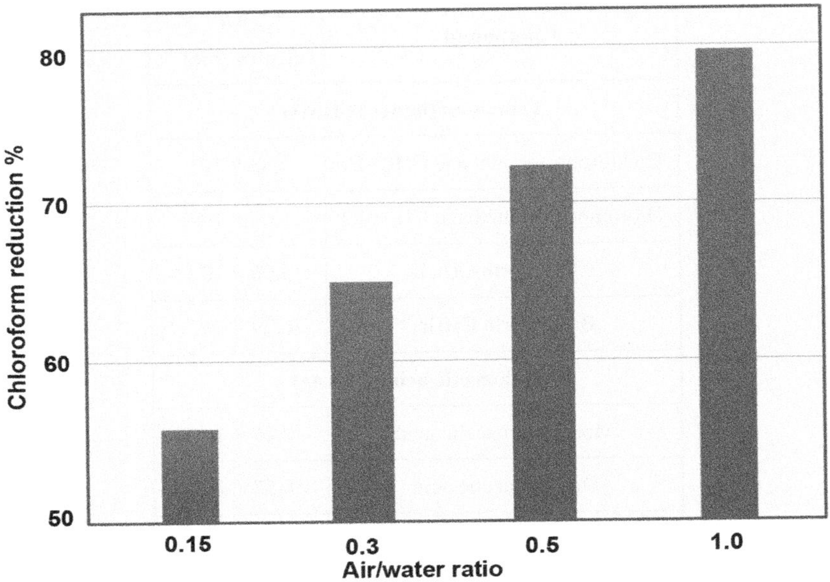


**Figure 22.12.** *Chloroform removal depending on the air/water ratio*

### 22.7.2. *Activated carbon*

While it is possible to implement absorption onto powdered, micrograin or granular activated carbon (GAC), it is powdered activated carbon (PAC) reactors that are the most efficient.

Among the THM compounds, chloroform is the most poorly adsorbed by PAC. The redox reaction between chlorine and the active sites of activated carbon result in a reduction of its adsorption capacity. This means that the addition of PAC into water containing chloroform has little efficiency due to the neutralization of activated carbon sites.

Table 22.7 presents the PAC dosages required to neutralize 100 $\mu g \cdot L^{-1}$ of each THM compound considered separately. The carbon used is a mesoporous activated carbon (Jacobi) with a specific surface of 1000 $m^2 \cdot g^{-1}$.The operating pH is 7.3.

Figure 22.13 illustrates the low adsorbability of chloroform onto activated carbon. It takes 500 $mg \cdot L^{-1}$ of PAC to remove 190 $\mu g \cdot L^{-1}$ of $CHCl_3$, whereas it only takes 42 $mg \cdot L^{-1}$ of PAC to remove the same concentration of $CHBr_3$.

| Compounds | THM concentration ($\mu g \cdot L^{-1}$) | | PAC ($g \cdot m^{-3}$) | Quantities of compounds removed/PAC ($\mu g$ Comp$\cdot mg^{-1}$ PAC) |
|---|---|---|---|---|
| | Inlet | Treated water | | |
| **Chloroform** | 100 | 10 | 300 | 0.30 |
| **Dibromochloromethane** | 100 | 10 | 105 | 0.86 |
| **Bromoform** | 100 | 10 | 20 | 4.5 |
| **Bromodichloromethane** | 100 | 10 | 156 | 0.58 |

**Table 22.7.** *Doses of PAC required for the removal of 90% of each of the THM compounds*

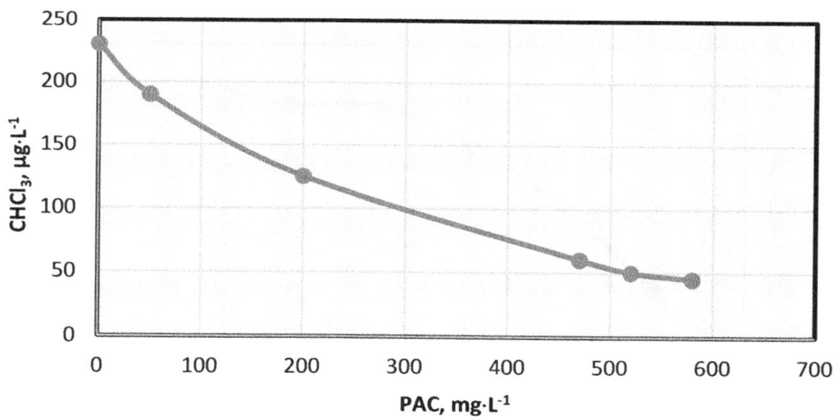


**Figure 22.13.** *Adsorption of chloroform onto powdered activated carbon*

While PAC is undoubtedly efficient, its disadvantage when compared to chloroform is that it requires very high dosages (Table 22.8).

Filtration onto GAC shows rapid kinetics during the first operation periods due to the large mass of available GAC (Figure 22.14). Then, the removed $CHCl_3$/g GAC load (expressed in mg) stabilizes at a 50% reduction, until falling to very low levels, as the activated carbon progressively becomes saturated.

| PAC $mg \cdot L^{-1}$ | THM $\mu g \cdot L^{-1}$ | | | TTHM $\mu g \cdot L^{-1}$ | Removal % |
|---|---|---|---|---|---|
| | $CHCl_3$ | $CHBrCl_2$ | $CHBr_2Cl$ | | |
| 0 | 65 | 9 | 2 | 76 | – |
| 15 | 48 | 7 | 1 | 56 | 27 |
| 65 | 33 | 4 | 1 | 38 | 50 |
| 100 | 30 | 1 | 1 | 32 | 58 |

**Table 22.8.** *Removal of THMs with mesoporous powdered activated carbon*

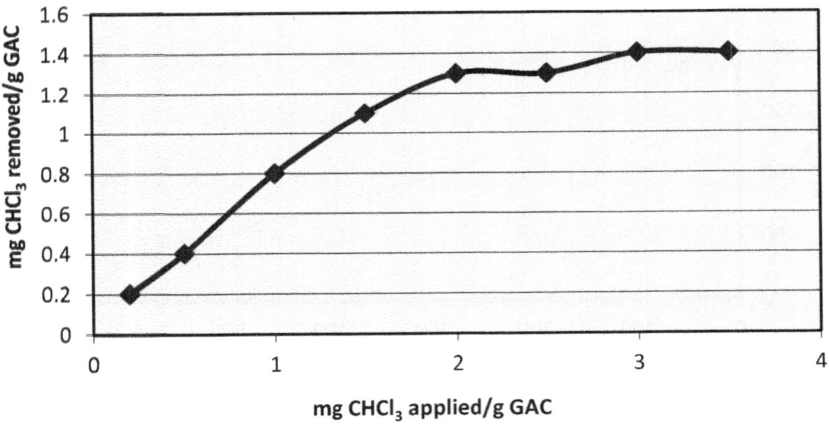


**Figure 22.14.** *$CHCl_3$/g GAC load removed in function of the applied load*

With GAC Picacarb (Jacobi) the k constants ($\mu g \cdot mg^{-1} \cdot L^{1/n} \cdot \mu g^{-1/n}$) and 1/n of the Freundlich equation for THMs are equal to 0.09 and 0.31, respectively.

Table 22.9 presents the Freundlich parameters obtained with mesoporous PAC (Jacobi).

| Compounds | Formulae | Freundlich parameters | | | Final concentrations $\mu g \cdot L^{-1}$ |
|---|---|---|---|---|---|
| | | K $(mg \cdot g^{-1})$ $(L^{1/n} \cdot \mu g^{-1/n})$ | 1/n | Qmax = k $Cout^{1/n}$ $mg \cdot g^{-1}$ PAC | |
| | | *THM* | | | |
| **Chloroform** | $CHCl_3$ | 0.0925 | 0.73 | 2.66 | 60 |
| **Bromoform** | $CHBr_3$ | 0.929 | 0.665 | 19.8 | 100 |
| **Dibromochloromethane** | $CHBr_2Cl$ | 0.585 | 0.636 | 10.9 | 100 |
| **Dichlorobromomethane** | $CHCl_2Br$ | 0.241 | 0.655 | 4.93 | 100 |
| | | **Chlorinated compounds** | | | |
| **Tetrachloroethylene** | $C_2Cl_4$ | 4.05 | 0.46 | 16.06 | 10 |
| **Trichloroethylene** | $C_2HCl_3$ | 2.1 | 0.48 | 8.84 | 10 |
| **Cis 1,2-Dichloroethylene** | $C_2H_2Cl_2$ | 0.202 | 0.59 | 0.79 | 10 |
| **1,1-Dichloroethylene** | $C_2H_2Cl_2$ | 4.9 | 0.54 | 0.40 | 10 |
| **Carbon tetrachloride** | $CCl_4$ | 0.387 | 0.594 | 0.24 | 10 |
| **Chloroethane** | $C_2H_5Cl$ | 0.59 | 0.95 | 5.25 | 10 |
| **Methylene chloride** | $CH_2Cl_2$ | 0.0062 | 1.16 | 0.09 | 10 |
| **1,1-Dichloroethane** | $C_2H_4Cl_2$ | 0.0646 | 0.706 | 0.328 | 10 |
| **1,2-Dichloroethane** | $C_2H_4Cl_2$ | 0.129 | 0.533 | 0.440 | 10 |
| **1,2-Dibromomethane** | $CH_2Br_2$ | 0.888 | 0.471 | 2.62 | 10 |
| **1,2-Dibromoethane** | $C_2H_2Br_2$ | 0.0722 | 0.701 | 0.362 | 10 |
| **Methyl tert-butyl ether (MTBE)** | | 0.218 | 0.479 | 0.656 | 10 |
| | | **Halogenated acetic acids** | | | |
| **Monochloroacetic acid** | $ClCH_2COOH$ | 1.29 | 0.99 | 25.03 | 20 |
| **Dichloroacetic acid** | $Cl_2CHCOOH$ | 1.63 | 0.46 | 6.46 | 20 |
| **Trichloroacetic acid** | $Cl_3CCOOH$ | 11.7 | 0.25 | 24.7 | 20 |
| **Monobromoacetic acid** | $BrCH_2COOH$ | 9.4 | 0.35 | 26.82 | 20 |
| **Dibromoacetic acid** | $Br_2CHCOOH$ | 5.04 | 0.29 | 12.01 | 20 |
| **Tribromoacetic acid** | $Br_3CCOOH$ | – | – | | 20 |
| **Bromochloroacetic acid** | $BrClCHCOOH$ | 3.4 | 0.29 | 8.10 | 20 |

**Table 22.9.** *Freundlich constants for various chlorination by-products*

### 22.7.3. *Biofiltration*

Biofiltration (GAC, Biodagen, sand, etc.) can be an efficient way to remove biodegradable organic matter and water-biodegradable DBPs.

### 22.7.4. *High-pressure membranes*

The rejection of CBPs is influenced by the compounds' physicochemical characteristics (molecular size, solubility, diffusivity, polarity, hydrophobicity and charge), the membrane properties (permeability, pore size, hydrophobicity and charge), the membranes themselves and the operating conditions of the membrane system (flow, transmembrane pressure, etc.).

High pressure membranes (nanofiltration, low-pressure reverse osmosis and reverse osmosis) are efficient in varying degrees and depend on the molecule.

| Pressure (bar) | Pore | Rejection % | |
|---|---|---|---|
| | | DOC | TTHM |
| 5 | NF-270 | 90 | 86 |
| 10 | NF-90 | 90 | 92 |
| 10 | NF-200 | 90 | 85–92 |
| 15 | NF-200 | 90 | 88–96 |

**Table 22.10.** *Performances of nanofiltration in relation to THMs*

Table 22.10 shows the performances of three nanofiltration membranes. Rejection rates are over 85% for THMs. These results are obtained for different applied pressures and for flows ranging from 18 to 20 lmh at 10 bar, and 24 lmh at 15 bar. The temperature is 20°C.

Nanofiltration is effective for the removal of THMs, although with varying rejection rates, depending on the type of compound.

The rejection rate of chloroform is between 80 and 88%, the rejection rate of dichlorobromomethane is between 90 and 94% and the rejection rate of DBCM is between 92 and 96%. The rejections increase with the molecular weight of THMs, especially with bromine atoms which increase the PM of the compound, and therefore the removal efficiency.

## 22.8. The case of nitrosamines and NDMA

Nitrosamines are one of the emerging groups from these DBPs. The nine molecules the most searched for are as follows:

– NDMA, N-nitrosomorpholine (NMOR), N-nitrosomethylethylamine (NMEA);

– N-nitrosodiethylamine (NDEA), N-nitrosodibutylamine (NDBMA), N-nitrosopyrrolidine (NPYR), N-nitrosopiperidine (NPIP), N-nitrosodipropylamine (NDPA) and N-nitrosodiphenylamine (NDPhA).

NDMAs are the most recognizable. These compounds are toxic, even carcinogenic in animals: their carcinogenic potential is by far higher than that of THMs. But nothing has been proven so far as to their effects on human beings.

Until now, nitrosamines have mainly been found in North America (the United States and Canada). The main formation route for NDMAs is due to oxidants, and more particularly, due to monochloramine.

NDMA is the simplest among the dialkylnitrosamines. NDMA ($C_2H_6N_2O$) belongs to a class of chemicals known as N-nitroso compounds (characterized by the presence of the N-nitroso (–N–N=O)) functional group, as well as to the family of amines that additionally have an amine function (–$NR_2$, where R is H or an alkyl group).

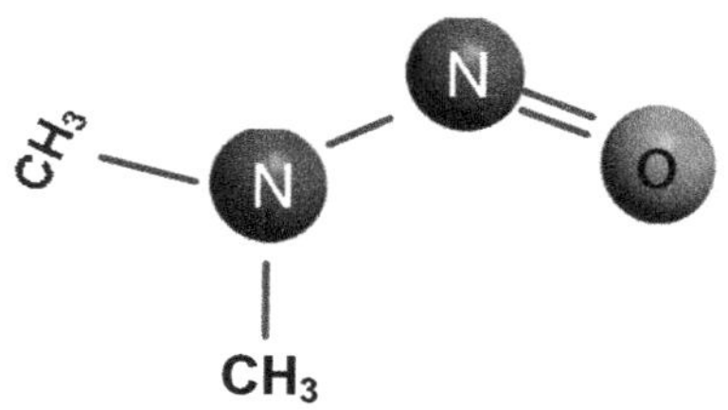


**Figure 22.1 5.** *NDMA chemical structure*

### 22.8.1. *Nitrosation mechanism with HOCl*

One of the possible mechanisms is the reaction between dimethylamine and nitrites with hypochlorous acid (HOCl) as a catalyst:

$$(CH_3)_2\,NH + HOCl \rightarrow (CH_3)_2\,NCl + H_2O \text{ (Dimethylchloramine formation)}$$

$$(CH_3)_2\,NCl + NO_2^- + 2H^+ \rightarrow (CH_3)_2\,NNO + Cl^- + H_2O \text{ (NDMA formation)}$$

As the acidity constants (pKa HOCl = 7.5 and pKa $HNO_2$ = 3.2) show that nitrous acid is a weaker acid than hypochlorous acid, then dimethylchloramine (DMCA) can be oxidized to form NDMA.

A second reaction pathway takes place between HOCl and nitrites as follows:

$$HOCl + NO_2^- \rightarrow NO_2Cl + OH^-$$

$$NO_2Cl + NO_2^- \rightarrow N_2O_4 + Cl^- \text{ (optimized reaction at neutral pH)}$$

$$NO_2Cl + OH^- \rightarrow NO_3^- + H^+ + Cl^-$$

$$DMA + N_2O_4 \rightarrow NDMA$$

In the presence of HOCl and ammonia, NDMA can be formed via monochloramine and dichloramine. In this case, the reaction intermediate is DMCA.

The mechanism is as follows:

$$HOCl + NH_3 \rightarrow NH_2Cl + H_2O \text{ (Monochloramine formation)}$$

$$HOCl + (CH_3)_2NH \rightarrow (CH_3)_2NCl + H_2O \text{ (DMCA formation with HOCl)}$$

$$NH_2Cl + (CH_3)_2NH \rightarrow (CH_3)_2NCl + NH_3 \text{ (DMCA formation with monochloramine)}$$

$$NH_2Cl + (CH_3)_2NH \rightarrow (CH_3)_2NNH_2 + H^+ + Cl^- \text{ (Dimethylhydrazine UDMH formation)}$$

$$(CH_3)_2NNH_2 + 2NH_2Cl + H_2O \rightarrow (CH_3)_2NNO + 2NH_3 + 2H^+ + 2Cl^- \text{ (NDMA formation)}$$

GAC enables a good removal of NMOR but has a short lifespan. Tight nanofiltration membranes (NF-90) grant a removal over 80%. The removal of NDMA is done mainly by UV treatment at doses well above 40 $mJ \cdot cm^{-2}$.

## 22.9. Oxidation by-products related to chlorine dioxide

Although more expensive, chlorine dioxide ($ClO_2$) is often used in water laden with organic matter, because it generates fewer THMs and HAAs compared to chlorine.

$ClO_2$ has a minor role in the substitution reactions with NOM present in the water. With $ClO_2$, up to 30–40% of $ClO_2$ dose applied can be reduced to chlorite ions and 8–10% converted to chlorate ions, whose formation is influenced by the presence of various organic and inorganic constituents in natural water. $ClO_2$ produces fewer organic by-products than chlorine. In addition, the presence of bromides significantly alters the speciation pattern and is also known to increase the DBP concentration. To control the formation of various by-products during the oxidative treatment of source water, it is imperative to know the precursor's characteristics in the source water.

In general, chlorine dioxide is not involved in the direct substitution reaction with the NOM present in water. The possible aqueous chemistry of $ClO_2$ describing the pathways for DBP formation is expressed by the following equations:

$$ClO_2 \rightarrow e^- + ClO_2^- \rightarrow 2e^- + ClO^- \rightarrow 2e^- + Cl^-$$

$$ClO_2 + 2Br^- \rightarrow Br_2 + H_2O \rightarrow HOBr$$

In addition to its high germicidal power and significant persistence, chlorine dioxide also has the advantage of forming few THMs, unlike chlorine and sodium hypochlorite.

While THMs represent the major fraction of by-products from chlorine disinfection, HAAs represent over 80% of the DBP fraction obtained from chlorine dioxide disinfection. This shows that the precursors' reaction mechanism in relation to $Cl_2$ and $ClO_2$ follows different pathways. An increase in temperature and oxidant concentration leads to a decrease in the HAA/THM ratio with chlorine dioxide, whereas it is less affected with chlorine.

Disinfection with chlorine dioxide produces little TOX, even fewer THMs and no chlorophenols: the potential risks related to organo-chlorinated compounds (except HAAs), as well as tastes and odors, are thus avoided. Chlorination produces more THMs than HAAs, whereas chlorine dioxide produces more HAAs.

The only by-products formed following a treatment with chlorine dioxide are chlorites and, to a lesser extent, chlorates and chlorides. While the number of these DBPs may be small, their formation is almost systematic and not negligible: over 20% of the quantity of chlorine dioxide introduced is transformed into chlorites.

In comparison with chlorine, the benefits of $ClO_2$ are as follows: a higher oxidizing power, a longer persistence in the distribution network practically independent of pH, and the non-proliferation of haloforms or chlorophenols. However, the settings and efficiencies of chlorine dioxide generators should be closely monitored: in devices operating with a strong excess of chlorine, the formation risks of organo-chlorinated compounds can be significant.

The main undesirable by-products resulting from the use of chlorine dioxide are chlorites ($ClO_2^-$) and chlorates ($ClO_3^-$). There are two main origins for chlorates:

– The first is within the generators. The residual hypochlorous acid due to excess chlorine applied in the settings can react with chlorine dioxide, forming chlorates (which is an additional reason to limit excess chlorine).

– The second is the reaction of ozone when injected upstream. The action of ozone on chlorine dioxide and chlorites quickly triggers the formation of chlorates.

Neutral and alkaline conditions favor the formation of acid radicals, and under high pH conditions $ClO_2$ tends to react by exchanging an electron to produce a chlorite ion:

$$2ClO_2 + 2OH^- \leftrightarrow ClO_2^- + ClO_3^- + H_2O \text{ (acidic pH)}$$

$$ClO_2 + e^- \leftrightarrow ClO_2^- \text{ (basic pH)}$$

When injecting chlorine dioxide at the treatment station outlet before introduction into a distribution network, the evolution of residual dioxide and its oxidation by-products can be schematized as shown in Table 22.11.

| | |
|---|---|
| **Chlorine dioxide ($ClO_2$)** | Very rapid decrease in the residual after injection (traces after only a few hours) |
| **Chlorites $ClO_2^-$** | Very rapid increase related to the destruction of chlorine dioxide to reach over 20% of the injected dioxide content<br>Fairly slow decrease associated with the reactivity of chlorites, which are then transformed into chlorides (after a few days) |
| **Chlorates $ClO_3^-$** | Constant values between 10 and 15% of the injected chlorine dioxide value |

**Table 22.11.** *The evolution of residual dioxide and its oxidation by-products*

Due to the rapid transformation of chlorine dioxide into chlorites, the application of this value limits the use of chlorine dioxide at the post-final disinfection stage, and this should be done in low doses (0.3–0.5 mg·L$^{-1}$ maximum).

There are currently no regulatory constraints for chlorates. Their formation should be limited because they also produce undesirable health effects.

The following recommendations may be considered in order to reduce chlorite and chlorate levels:

– regularly check the settings and production efficiency of generators;

– in particular, check the pressure of chlorinated water, as well as the flow rate of the chlorite and gaseous chlorine solution. Analyze the generated dioxide solution: concentration of dioxides, chlorites and chlorates;

– do not apply too high chlorine overdoses: a generator maintained in good operating condition results in efficiencies over 90% with an excess of chlorine of 15%;

– do not store chlorine dioxide solutions for too long (in generators equipped with a storage tank before injection). It is advisable not to exceed a half-day. The generator should be started up at least twice a day. It is possible to adjust the concentration of the stored solution by increasing the flow rate of dilution water (which should not decrease below 2 $g \cdot L^{-1}$ in $ClO_2$);

– protect generators from heat and sunlight;

– do not produce and store chlorine dioxide solutions that are too concentrated. The higher the concentration, the greater the degradation reactions by dismutation or due to the reaction with hypochlorous acid.

## 22.10. Ozonation by-products

Oxidation is defined as the transfer of one or more electrons from an electron donor (reductant) to an electron acceptor (oxidant) having a higher affinity for electrons. The electron transfer results in the chemical transformation of both the oxidant and the reductant and, in some cases, producing chemical species with an odd number of valence electrons. These species, known as radicals, tend to be highly unstable and, therefore, highly reactive because one of their electrons is unpaired. Oxidation reactions that produce radicals tend to be followed by additional oxidation reactions between the radical oxidants and other reactants (both organic and inorganic), until thermodynamically stable oxidation products are formed.

Ozone is the most powerful oxidant available to treat water. However, its stability in water is low, which is why ozone is not suitable for maintaining its disinfecting conditions during distribution. It is all the more effective when pH and temperature are low. The incidence of pH is much lower than for chlorine, however. On the other hand, ozone does not produce undesirable chlorine compounds.

Although ozone forms only a few halogenated DBPs (bromoforms and brominated AAs) when these participate in oxidation–reduction reactions with NOM, it also leads to a variety of organic and inorganic by-products.

Acetaldehyde formed during pre-ozonation was found to react during post-chlorination to form chloroacetaldehyde, which in the presence of a sufficient amount of free chlorine rapidly reacted to form chloral hydrate. In addition, pre-ozonation was found to increase chloropicrin (trichloronitromethane) formation post chlorination.

Table 22.12 shows the main known by-products associated with ozonation.

| Organic by-products of ozonation | |
|---|---|
| **Aldehydes**<br>Formaldehydes<br>Acetaldehyde<br>Glyoxal<br>Methyl glyoxal | **Aldo and ketoacids**<br>Pyruvic acid |
| **Acids**<br>Oxalic acid<br>Succinic acid<br>Formic acid<br>Acetic acid | **Brominated by-products**<br>Bromates<br>Bromoforms<br>Brominated acetic acids<br>Bromopicrins<br>Brominated acetonitriles |

**Table 22.12.** *Main organic by-products formed during ozonation*

Ozone does not form chlorites. On the other hand, it can generate chlorates (which are undesirable for health) due to the rapid reaction with chlorites and chlorine dioxide.

In virtue of its very high oxidizing power, ozone is likely to modify the structure of a part of organic matter. Ozone oxidizes TOC and transforms it into biodegradable dissolved organic carbon (BDOC). As BDOC (including aldehydes, ketones, organic acids, alcohols) serves as a substrate for heterotrophic bacteria, it can cause problems associated with bacterial revival in the network. This is the reason why GAC filters are increasingly used downstream of ozonation towers: the bacterial biomass attached to them uses the BDOC substrate resulting from the ozone treatment. This practice corresponds to the use of ozone in interozonation.

Ozone treatment can generate bromate ions ($BrO_3^-$) that are formed by the oxidation of bromides ($Br^-$). The oxidation reaction of the bromide ion ($Br^-$) into the hypobromite ion ($BrO^-$) and bromite ion ($BrO_2^-$), and then into the bromate ion ($BrO_3^-$), takes place due to a direct reaction with ozone.

Brominated by-products include the bromate ion, bromoform, brominated acetic acids and acetonitriles, bromopicrin, and cyanogen bromide (if ammonia is present).

The bromate formation mechanism involves both the radical pathway and the direct oxidation pathway (Figure 22.16).

Under these conditions, the production of water containing less than 10 $\mu g \cdot L^{-1}$ of bromates is not easy to guarantee all year round.

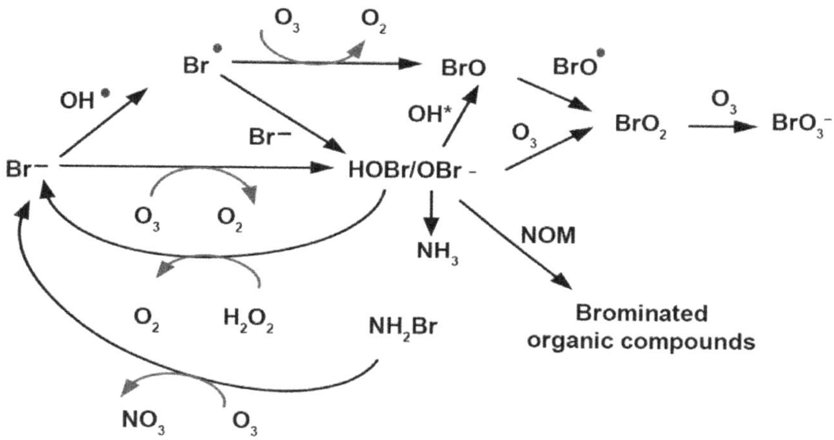


**Figure 22.16.** *Bromate ion (molecular and radical) formation mechanisms*

Veolia's experience has shown that the most favorable conditions for the production of bromates during the ozonation stages are as follows:

– the presence of $Br^-$ even at very low levels can lead to the formation of bromates;

– an increase in pH leads to an increment in the formation of bromates;

– an increase in temperature leads to an increment in the formation of bromates; when temperature rises from 7 to 22°C, the bromate concentration doubles. This risk starts at temperatures 10–12°C;

– the presence of ammonia seems to reduce the formation of bromates thanks to the bromamine formation parallel reaction, but this solution is far from being satisfactory;

– the formation of bromates is correlated with the Ct (the product of the dissolved ozone residual × contact time);

– the risk of bromate formation remains low if the ozone content/TOC is lower than 0.4 $g \cdot g^{-1}$;

– the treatment rate: an ozone overdose (maintaining too high a residual after disinfection, the ozone demand having already been satisfied) can lead to a significant increase in the production of bromates;

– contact time: too long a contact time has similar effects to an ozone overdose (attention should be paid to the stations' settings that are scheduled for long contact times of more than 8 min required for pesticide degradation; a trade-off should be found to ensure their removal without generating too many bromates);

– radical "traps": carbonate ions, for example, promote the formation of bromates.

The main advantage of ozone for controlling the formation of DBPs is that it enables the application of free chlorine at lower doses for network protection, something which reduces the formation of DBPs. Nevertheless, the formation of by-products during ozonation reduces the interest of this advantage.

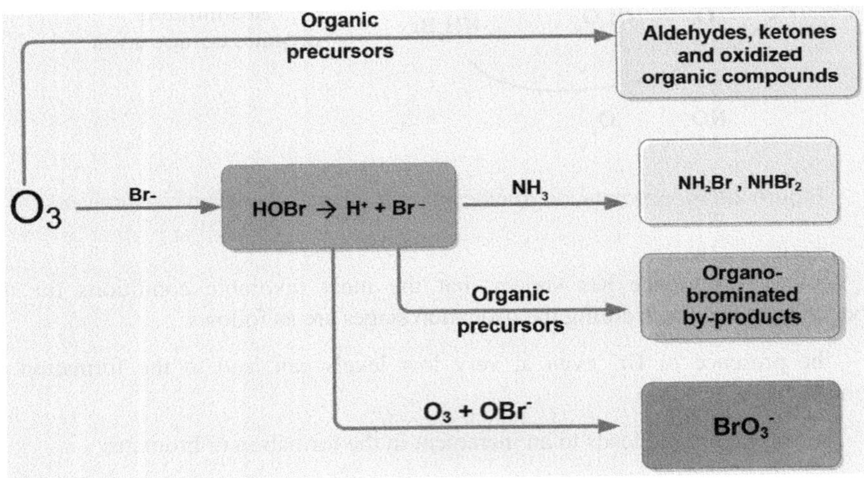


**Figure 22.17.** *Formation of brominated by-products*

## 22.11. *Recommendations*

Disinfection/oxidation and treatment practices may be optimized to minimize, within limits, the formation of both the regulated and emerging DBPs. Pre-oxidation with chlorine, chlorine dioxide or ozone may destroy or transform the precursors for NDMA. However, pre-ozonation can increase the potential formation for trihaloacetaldehydes and halonitromethanes (HNMs).

Emerging DBPs of potential health concern are being identified in disinfected drinking water. Because all chemical disinfection processes produce DBPs, it is important to include a wide range of potential DBPs in treatment studies to determine how best to minimize the formation of as many DBPs as possible, recognizing that not all will be minimized (and some may be increased).

Exposure to DBPs is highly prevalent because of the use of some disinfectant or oxidant at drinking water treatment plants in most parts of the world. The challenge for scientists, engineers, utility staff, regulators, toxicologists and epidemiologists is to be aware of both emerging DBPs and emerging sources of DBP precursors and how their formation and control may be similar to and/or different from that of currently regulated DBPs.

## 22.12. References

Amy, G. and Sohn, J. (1999). Scientific basis for modeling trihalomethanes (THMs) in the Paris-area distribution systems operated by Compagnie Générale des Eaux. Report, Boulder.

Amy, G.L., Chadik, P.A., Chowdhury, Z.K. (1987). Developing models for predicting trihalome-thane formation potential and kinetics. *J. Am. Water Works Assoc.*, 89–96.

Amy, G.L., Siddiqui, M., Ozekin, K., Zhu, H.W., Wang, C. (1998). Empirically based models for predicting chlorination and ozonation by-products: Trihalomethanes, haloacetics acids, chloral hydrate, and bromate. Report EPA-815 R-98 005, University of Colorado, Boulder.

Ates, N., Yilmaz, L., Kitis, M., Yetis, U. (2009). Removal of disinfection by-product precursors by UF and NF membranes in low-SUVA waters. *J. Membr. Sci.*, 328, 104–112.

Barrett, S.E., Krasner, S.W., Amy, G.L. (2000). Natural organic matter and désinfection by products: Characterization and control in drinking water – An overview. *ACS Symp. Ser.*, 72, 485–495.

Bellar, T.A., Lichtenberg, J.J., Kroner, R.C. (1974). The occurence of organohalides in chlorinated drinking water. *J. Am. Water Works Assoc.*, 66, 703–710.

Berrahal, Y. (2019). Evaluation de la matière organique dans les eaux de surface des barrages de l'ouest d'Algérie et évolution des trihalométhanes et le plomb dans le réseau d'eau potable. PhD Thesis, Université de Djillali Liabes, Sidi Bel Abbés.

Bond, T., Henriet, O., Goslan, E.H., Parsons, S.A., Jefferson, B. (2009). Disinfection by-product formation and fractionation behavior of natural organic matter surrogates. *Environ. Sci. Technol.*, 43, 5982–5989.

Bond, T., Goslan, E.H., Parsons, S.A., Jefferson, B. (2011). Treatment of disinfection by-product precursors. *Environ. Technol.*, 32, 1–25.

Boyce, S.D. and Hornig, J.F. (1983). Reaction pathways of trihalomethane formation from the halogenation of dihydroxyaromatic model compounds for humic acid. *Environ. Sci. Technol.*, 17, 202–211.

Chang, E.E., Chiang, P.C., Chao, S.H., Lin, Y.L. (2006). Relationship between chlorine consumption and chlorination by-products formation for model compounds. *Chemosphere*, 64, 1196–1203.

Chang, H., Chen, C., Wang, G. (2011). Identification of potential nitrogenous organic precursors for C-, N-DBPs and characterization of their DBPs formation. *Water Res.*, 45, 3753–3764.

Chen, W.J. and Weisel, C.P. (1998). Halogenated DBP concentrations in a distribution system. *J. Am. Water Works Assoc.*, 90(4), 151–163.

Chen, B. and Westerhoff, P. (2010). Predicting disinfection by product formation potential in water. *Water Res.*, 44, 3755–3762.

De la Rubia, A., Rodriguez, M., Leon, V.M., Prats, D. (2008). Removal of natural organic matter and THM formation potential by ultra- and nanofiltration of surface water. *Water Res.*, 42, 714–722.

Deborde, M. and von Gunten, U. (2008). Reactions of chlorine with inorganic and organic compounds during water treatment – Kinetics and mechanisms: A critical review. *Water Res.*, 42, 13–51.

Deborde, M., Rabouan, S., Gallard, H., Legube, B. (2004). Aqueous chlorination kinetics of some endocrine disruptors. *Environ. Sci. Technol.*, 38, 5577–5583.

Dickenson, E.R.V., Summers, R.S., Croué, J.-P., Gallard, H. (2008). Haloacetic acid and trihalo-methane formation from the chlorination and bromination of aliphatic β-dicarbonyl acid model compounds. *Environ. Sci. Technol.*, 42, 3226–3233.

Dubowski, Y., Greenberg-Eitan, R., Rebhun, M. (2018). Removal of trihalomethane precursors by nanofiltration in low-SUVA drinking water. *Water*, 10, 1370–1380.

El-Shahat, M.F., Abdel-Halim, S.H., Hassan, G.A. (2001). Factors influencing the formation of trihalomethanes in drinking water treatment plants. *Bull. Environ. Contam. Toxicol.*, 67, 549–553.

Fayad, N.M. (1993). Seasonal variations of THMs in Saudi Arabian drinking water. *J. Am. Water Works Assoc.*, 85, 46–50.

Gallard, H. (2002). Chlorination of natural organic matter: Kinetics of chlorination and of THM formation. *Water Res.*, 36(1), 65–74.

Gallard, H. and von Gunten, U. (2002). Chlorination of natural organic matter: Kinetics of chlorination and of THM formation. *Water Res.*, 36, 65–74.

Geriesh, M.H., Blake, K.D., El-Rayes, A.E (2008). Problems of drinking water treatment along Ismailia Canal Province. *Egypt. J. Zhejiang Univ. Sci.*, B9(3), 232–242.

Golfinopoulos, S.K. and Arhonditsis, G.B. (2002). Multiple regression models: A methodology for evaluating trihalomethane concentrations in drinking water from raw water characteristics. *Chemosphere*, 47(9), 1007–1018.

Hoyer, O. (1998). Measures to reduce or avoid THM formation during water treatment and distribution. *Water Supply*, 16(3–4), 143–149.

Hua, G. and Reckhow, D.A. (2007). Characterization of disinfection by-product precursors based on hydrophobicity and molecular size. *Environ. Sci. Technol.*, 41(9), 3309–3315.

Kiso, Y., Sugiura, Y., Kitao, T., Nishimura, K. (2001). Effect of hydrophobicity and molecular size on rejection of aromatic pesticides with nanofiltration membranes. *J. Membr. Sci.*, 192, 1–10.

Koch, B., Krasner, S.W., Sclimenti, M.J., Schimpff, W.K. (1991). Predicting the formation of DBPs by simulated distribution system. *J. Am. Water Works Assoc.*, 83(10), 62–70.

Korshin, G.V., Wu, W.W., Benjamin, M.M., Hemingway, O. (2002). Correlations between differential absorbance and the formation of individual DBPs. *Water Res.*, 36, 3273–3282.

Koudjonou, B., Lebel, G.L., Dabeka, L. (2008). Formation of halogenated acetaldehydes, and occurence in Canadian drinking water. *Chemosphere*, 72, 875–881.

Koyuncu, I. and Yazgan, M. (2001). Application of nanofiltration and reverse osmosis membranes to the salty and polluted surface water. *J. Environ. Sci. Health A*, 36, 1320–1333.

Krasner, S.W., McGuire, W.H., Jacangelo, J.G., Patania, N.L., Reagan, K.M., Aieta, M.E. (1989). The occurence of disinfection by-products in US drinking water. *J. Am. Water Works Assoc.*, 81(8), 41–53.

Krasner, S.W., Weinberg, H.S., Richardson, S.D., Pastor, S.J., Chinn, R., Sclimenti, M.J. (2006). Occurrence of a new generation of disinfection by-products. *Environ. Sci. Technol.*, 40(23), 7175–7185.

Le Roux J. (2011). Mécanisme de formation des nitrosamines et sous-produits halogénés lors de la chloramination de contaminants organiques azotés émergents. PhD Thesis, Université de Poitiers, Poitiers.

Lee, C., Schmidt, C., Yoon, J., von Gunten, U. (2007). Oxidation of N-nitrosodimethylamine (NDMA) precursors with ozone and chlorine dioxide: Kinetics and effect on NDMA formation potential. *Environ. Sci. Technol.*, 41, 2056–2063.

Li, X. and Zhao, H.B. (2006). Development of a model for predicting trihalomethanes propagation in water distribution systems. *Chemosphere*, 62(6), 1028–1032.

Liang, L. and Singer, P.C. (2003). Factors influencing the formation and relative distribution of haloacetic acids and trihalomethanes in drinking water. *Envir. Sci. Technol.*, 37(13), 2920–2928.

Lin, Y., Chiang, P., Chang, E.E. (2006). Reduction of disinfection by-products precursors by nano-filtration process. *J. Hazard. Mater.*, B137, 324–331.

Malliarou, E., Collins, C., Graham, N., Nieuwenhuijsen, M.J. (2005). Haloacetic acids in drinking water in the United Kingdom. *Water Res.*, 39(12), 2722–2730.

Mitch, W.A. and Sedlak, D.L. (2002). Formation of N-nitrosodimethylamine (NDMA) from dimethylamine during chlorination. *Environ. Sci. Technol.*, 36, 588–595.

Montiel, A.J. (1983). *Les halométhanes dans l'eau – Formation et élimination*. CIFEC, Paris.

Mouly, D., Joulin, E., Rosin, C., Beaudeau, P., Zeghnoun, A., Olszewski-Ortar, A., Munoz, J-F. (2008). Les sous-produits de chloration dans l'eau destinée a la consommation humaine en France – Campagnes d'analyses dans quatre systémes de distribution d'eau et modélisation de l'évolution des trihalométhanes. Report, Institut de veille sanitaire, Saint-Maurice.

Muellner, M.G., Wagner, E.D., McCalla, K., Richardson, S.D., Woo, Y.T., Plewa, M.J. (2007). Haloacetonitrils versus acides haloacetiques. *Environ. Sci. Technol.*, 41, 645–651.

Navalon, S., Alvaro, M., Garcia, H. (2008). Carbohydrates as trihalomethanes precursors. Influence of pH and the presence of Cl- and Br- on trihalomethane formation potential. *Water Res.*, 42, 3990–4000.

Neveu, A. (2014). Retour d'expériences sur le risque THM de l'eau de ferel. Campagnes de mesures 2013–2014. In *5^e^ rencontres de la Vilaine*. Safège.

Nieminski, E.C., Chaudhuri, S., Lamoreaux, T. (1993). The occurrence of DBPs in Utah drinking waters. *J. Am. Water Works Assoc.*, 85(9), 98–105.

Nieuwenhuijsen, M.J., Toledano, M.B., Eaton, N.E., Fawell, J., Elliott, P. (2000). Chlorination disinfection by-products in water and their association with adverse reproductive outcomes: A review. *Occup. Environ. Med.*, 57(2), 73–85.

Rapinat, M. and Welte, B. (2001). Dossier : les sous-produits de la desinfection. *Technique Sci. Methods*, 6, 36–79.

Reckhow, D.A., Singer, P., Malcolm, R.L. (1990). Chlorination of humic material: By-product formation and chemical interpretations. *Environ. Sci. Technol.*, 24(11), 1655–1664.

Rodriguez, M.J. and Serodes, J.B. (2001). Spatial and temporal evolution of trihalomethanes in three water distribution systems. *Water Res.*, 35(6), 1572–1586.

Rodriguez, M.J., Serodes, J.B., Levallois, P. (2004). Behavior of trihalomethanes and haloacetic acids in a drinking water distribution system. *Water Res.*, 38(20), 4367–4382.

Sadiq, R. and Rodriguez, M.J. (2004). Disinfection by-products (DBPs) in drinking water and predictive models for their occurrence: A review. *Sci. Total Environ.*, 321(1–3), 21–46.

Singer, P. C. (1994). Control of disinfection by-products in drinking water. *J. Environ. Eng.*, 120, 727–774.

Summers, R.S., Knappe, D.R.U., Snoeyink, V.L. (2011). Adsorption of organic compounds by activated carbon. In *Water Quality & Treatment*, 6th edition, Edzwald, J.K. (ed.). American Water Works Association, Denver.

Tung, H.-H., Unz, R.F., Xie, Y.F. (2006). HAA removal by GAC adsorption. *J. Am. Water Works Assoc.*, 98(6), 107–112.

Urano, K., Wada, H., Takemasa, T. (1983). Empirical rate equation for thialomethane formation with chlorination of humic substances in water. *Water Res.*, 17(12), 1797–1802.

Uyak, V., Koyuncu, I., Cakmakci, M., Toroz, I. (2008). Removal of trihalomethanes from drinking water by nanofiltration membranes. *J. Hazard. Mater.*, 152, 789–794.

Visvanathan, C., Marsono, B.D., Basu, B. (2003). Removal of THMP by nanofiltration: Effects of interference parameters. *J. Membr. Sci.*, 225, 91–103.

Weinberg, H.S., Glaze, W.H., Krasner, S.W., Sclimenti, M.J. (1993) Formation and removal of aldehydes in plants that use ozonation. *J. Am. Water Works Assoc.*, 85(5), 72–85.

Weisel, C.P., Kim, H., Haltmeier, P., Klotz, J.B. (1999). Exposure estimates to disinfection by-products of chlorinated drinking water. *Environ. Health Perspect.*, 107(2), 103–110.

Westerhoff, P., Chao, P., Mash, H. (2004). Reactivity of natural organic matter with aqueous chlorine and bromine. *Water Res.*, 38, 1502–1513.

Williams, D.T., LeBel, G.L., Benoit, F.M. (1995). Étude nationale sur les sous-produits de désinfection chlores dans l'eau potable au Canada. Report, Government of Canada.

# 23

# Sludge Treatment

The sludge extracted from drinking water stations contains little biodegradable matter. A direct consequence of this is that no stabilization or hygienization treatment is required. On the other hand, this low organicity endows it with a poor rheological quality in terms of texture: a solid, stackable state is reached only at the cost of a significant addition of lime.

Furthermore, the volumes generated by the drinking water stations (DWS) are low and are often discharged directly into the natural environment. However, design is subject to uncertainty due to variations in the quality of raw water (flood phenomena), particularly at the level of the buffering structures.

This is done on the assumption that the (self-cleaning) network and the wastewater station (load and instantaneous flow) can accept and treat the additional sludge flow, the dirty water from filters and low-pressure membranes, the eluates from resin denitration, organic sludge or the concentrates from membrane treatments and that these are all discharged into the sewage.

In principle, pasty or solid sludge can be discharged without any particular constraints on agricultural land plots (known under the more complimentary term, "settling soil"). This condition is only obtained by means of a lime treatment, granting soil amelioration and a better rheological quality.

The sludge from DWS does not constitute hazardous waste and can be received in existing landfills or new storage facilities for household waste, provided that it respects a minimum of 30% dryness. Agricultural devolution implies de facto liming, except in the case of softening sludge, which is already rich in $CaCO_3$.

## 23.1. Choosing a treatment chain

There is no treatment in the event of direct discharge. On the other hand, agricultural devolution, landfilling or conditioning as an industrial additive require a volume reduction treatment and possibly lime conditioning in order to give sludge a "soil amelioration" status and thus facilitate its handling and storage, making it shovelable and even stackable.

There are some cases in which the thickening of softening sludge (i.e. lime sludge from water softening) reaches dryness levels of 300 $g \cdot L^{-1}$, enabling the direct discharge to landfill, with no subsequent dewatering. However, piling may be required (even at 300 $g \cdot L^{-1}$, the sludge may not be stackable).

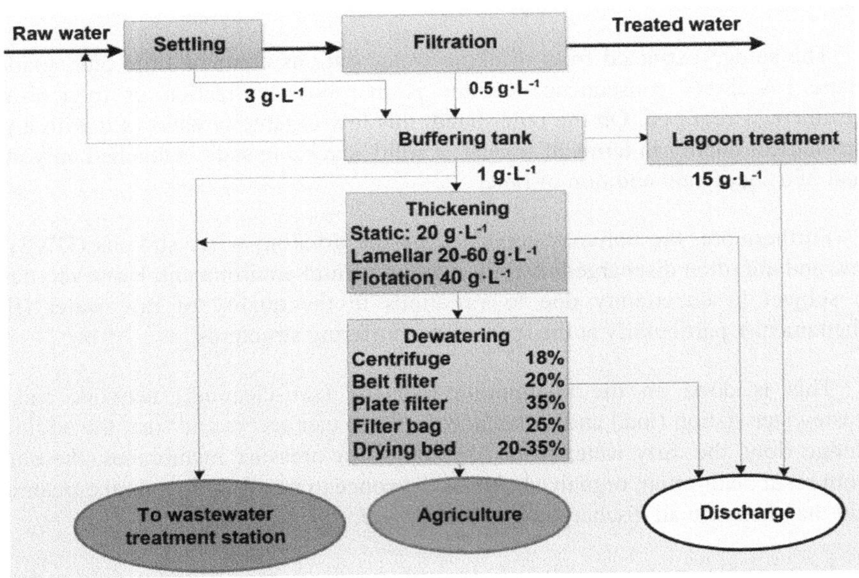


**Figure 23.1.** *Main drinking water sludge treatment processes (hydroxide sludge from surface water). For a color version of this figure, see www.iste.co.uk/gaid/watertreatment5.zip*

Several technologies are used for separating the solid and liquid phases of the sludge extracted at drinking water stations. The majority of such technologies are either based on differences in density, or on the physical retention of particles on a filter. Each technology is adapted to treat a specific sludge range, depending on its texture and concentration. Table 23.1 shows an overview of common devices and solid fractions indicating inlet and outlet values.

## 23.2. Characteristics of drinking water sludge

The characterization of DWS sludge is estimated from laboratory tests measuring specific parameters such as dryness limit, specific resistance and so on, and making it possible to predict the performances and consumables at each stage. In most of the cases studied, as sludge is not readily available, estimates are made based on the existing literature or on data obtained at similar stations.

### 23.2.1. *The quantity of sludge produced*

Water treatment sludge typically comprises the following:

– naturally-occurring colloidal and other particulate matter (e.g. algae, Fe and Mn particles);

– dissolved natural organic matter (NOM) (e.g. humic acids, fulvic acids);

– precipitated water treatment chemicals;

– oxide precipitates of inorganic species dissolved in the raw water (e.g. Fe, Mn);

– addition of powdered activated carbon or clay, sand or other solid mineral substances.

It is deduced from the following formula:

$$P/V = \{[SS] + 0.07H + k\cdot D + X\} + \{1.58\cdot[Mn] + a\cdot[Mn] + 1.91\cdot[Fe] + [\text{Ca hardness Cin} - \text{Ca hardness Cout}]\}$$

where:

– a: coefficient related with the type of chemical oxydant used;

– P: mass of recovered sludge in $g\cdot j^{-1}$;

– V: volume of treated raw water in $m^3\cdot j^{-1}$;

– SS: SS concentration in raw water in $g\cdot m^{-3}$;

– H: apparent color of raw water in Hazen units (HU);

– D: coagulant dose in $g\cdot m^{-3}$;

– k: coagulant fraction found in the sludge;

– X: other added products (losses in microsand, insolubles, PAC, etc.) in $g\cdot m^{-3}$.

And, in the case of iron removal, manganese removal, softening treatments:

– Fe: iron concentration in raw water in $g \cdot m^{-3}$;

– Mn: manganese concentration in raw water in $g \cdot m^{-3}$;

– oxidant dosage: oxidant dosage rate applied (g oxidant per g Mn) in $g \cdot g^{-1}$;

– Ca hardness Cin: raw water calcium hardness in $mg \cdot L^{-1}$ $CaCO_3$;

– Ca hardness Cout: treated water calcium hardness in $mg \cdot L^{-1}$ $CaCO_3$.

As the SS parameter is not always known, the following approximation can be used:

$$\text{SS concentration } [mg \cdot L^{-1} \text{ or } g^{-3}] = 1.5 \times \text{NTU}$$

The k coefficient depends on the type of reagent used (Table 23.1).

| **Reagents** | **k values** |
|---|---|
| Crystallized hydrated ferric chloride 60% $FeCl_3$ | 0.40 |
| Crystallized ferric chloride 99% $FeCl_3$ | 0.65 |
| Ferric chloride solution 39-41% (596 $g \cdot L^{-1}$) | 0.27 |
| WAC 10% $Al_2O_3$ | 0.15 |
| Aluminum sulfate “lumps” 17.2% (645 $g \cdot L^{-1}$) $Al_2O_3$ | 0.27 |
| Aluminum sulfate solution 8.25% $Al_2O_3$ | 0.17 |
| Clairtan 41–44% $FeClSO_4$ | 0.24 |
| Ferric sulfate 43% $Fe_2(SO_4)_3$ | 0.23 |
| Ferrous sulfate 54.7% $FeSO_4$ | 0.38 |

**Table 23.1.** *Value of k depending on the type of reagent used*

Table 23.2 provides a quick estimate of the amount of sludge produced. Sludge can be extracted from different places along the water treatment chain.

In general, the main extraction levels are:

– settler or flotation unit;

– powdered activated carbon reactors;

– filters (dirty water);

– low-pressure membranes (dirty water);

– saturators;

– thickeners.

It is advisable to optionally enrich this sludge with the brines made from the eluates resulting from the regeneration of the ion exchange resins and the concentrates from the high-pressure membrane treatments. These brines represent a negligible flow of dry matter, but can be quite voluminous.

| Water origin | Resources | Sludge production | Types of sludge |
|---|---|---|---|
| Surface waters | Medium quality river | 25–45 g $DM{\cdot}m^{-3}$ raw water | Hydroxides |
| | River during winter flood; borehole in a karst environment under storm events | 65–150 g $DM{\cdot}m^{-3}$ raw water | Hydroxides |
| | Dead river surfaces, abundant plankton | 80–95 g $DM{\cdot}m^{-3}$ raw water | Hydroxides |
| | Dam water, clean lake, river close to the source | 2–15 g $DM{\cdot}m^{-3}$ raw water | Hydroxides |
| | River at the mouth | 70–140 g $DM{\cdot}m^{-3}$ raw water | Hydroxides |
| Borehole waters | Ferruginous borehole waters, physico-chemical treatment | 1.9 g $DM{\cdot}g^{-1}$ Fe removed | Hydroxides |
| | Ferruginous well water, biological treatment | 2.2 g $DM{\cdot}g^{-1}$ Fe removed | Organic |
| | Nitrated borehole waters, biological treatment | 1.6 $DM{\cdot}g^{-1}$ $NaNO_3$ removed | Organic |
| | Hard borehole waters | 2 $mg{\cdot}mg^{-1}$ $CaCO_3$ removed | Carbonates |
| | Borehole in a karst environment during a violent storm event (a few hours) | 200–800 g $DM{\cdot}m^{-3}$ raw water | Hydroxides |

**Table 23.2.** *Quantities of sludge produced (DM: dry matter)*

### 23.2.2. *Sludge concentration estimate at different stages of the chain*

During a conventional settling-filtration chain, 90–95% of suspended solids are recovered at the settling stage, and 5–10% at the filtration stage. Depending on purge dilution, the ratio between the settled and filtered sludge volumes can vary from 1 to 5. There is no difference in quality between settling and filtration sludge. Only the concentrations are different.

On a lime saturator, the recovered SS represents about 15% of the $Ca(OH)_2$ weight used.

The volume of extracted raw sludge represents 2–5% of the volume of water produced. On average, 20–100 g of sludge per $m^3$ of treated water are recovered for surface waters (excluding flood periods).The volume flows are calculated depending on the mass balance at each stage. The orders of magnitude regarding the concentrations are shown in Table 23.3.

| Sludge origin | SS concentration $g \cdot L^{-1}$ |
|---|---|
| Corridor settler, with no scraper | 0.5–5 |
| Scraped corridor settler | 2–5 |
| Sewage drain corridor settler | 10–60 |
| Cyclofloc | 1–10 |
| Multiflo | 2–30 |
| Actiflo | 0.5–20 |
| Spidflow® | 25–40 |
| Lamella water softener | 50–300 |
| Catalytic water softener | 200–500 |
| Powdered activated carbon reactor | 1–20 |
| Filter washwaters | 0.1–0.6 |
| Water backwashing for MF/UF membranes | 0.2–0.8 |
| Saturator purges | 15–30 |

**Table 23.3.** *Extracted sludge concentration per processing stage*

### 23.2.3. *Sludge quality: physical and chemical properties*

The knowledge of sludge quality is essential to estimate not only thickening and dewatering performances but also its ability to be transported and stored.

The knowledge of the physical and chemical properties of sludge requires a number of analyses relating to:

– dry matter (DM) and volatile matter (VM) (≈ OM, organic matter) or their complement MM (mineral matter) in order to predict the physical properties;

– the hydroxides or calcium salt content (which can also be calculated by the mass balance);

– the Ca/Mg ratio, notably in the case of softening sludge;

– the silica content, especially in the case of centrifugation.

In broad terms, a distinction can be made between:

– hydrophobic-type sludge (rich in carbonates, silts, oxides, silica) → easy to treat;

– hydrophilic-type sludge (rich in hydroxides, organic matter [OM]) → difficult to treat. Aluminum hydroxides are less favorable than iron hydroxides.

By analogy with soils, the $CaCO_3$ content is useful for classifying sludge as specified in Table 23.4.

| $CaCO_3$ levels | 0–10% | 10–30% | 30–70% | 70–90% | 90–100% |
|---|---|---|---|---|---|
| **Soil/sludge types** | Clay | Marl clay | Marl | Marl limestone | Limestone |

**Table 23.4.** *Sludge classification depending on their $CaCO_3$ content*

The dryness limit and the specific resistance provide reliable information as to the performances to be expected in terms of dewatering. In order to avoid problems related to transport and storage, it is necessary to consider:

– *dry matter (DM)*: over 150 $g \cdot L^{-1}$ approximately, sludge is no longer transportable with the usual technologies (eccentric rotor pumps). Particles settle easily and sludge is only pumped using standard air-lift technology, membrane piston pump or hidrostal-type centrifugal screw pump, or equivalent;

– *organic matter (OM)*: drinking water sludge is often poor in organic compounds (<20% of DM), which explains its liquid behavior, even at very high concentrations. The knowledge of OM, associated with that of hydroxides, helps to predict sludge behavior. If the sludge contains both hydroxides and OM, the sludge is hydrophilic and its treatability, is less easy;

– *bulk density (ρ)*: for a liquid or pasty sludge of dryness S (expressed as a fraction), the value of ρ is obtained by the following formula:

$$\rho = 1000\,(1 - S) + \rho MM\,(S \times MF) + 850\,(S \times VF) \text{ in } (kg \cdot m^{-3})$$

where:

– MF and VF: mineral and volatile (organic) fractions of the dry matter, respectively;

– ρMM: mineral matter density, that is: ρMM = 2,300 $kg \cdot m^{-3}$ for hydroxides and ρMM = 2,700 $kg \cdot m^{-3}$ for carbonates. A completely dry hydroxide sludge has a bulk density of 1200–1500 $kg \cdot m^{-3}$ (taking into account the vacuum degree);

– the *dynamic viscosity (μ)*: a sludge having VF = 26% and a dryness S greater than 2% (20 $g \cdot L^{-1}$), transported in a pipe of diameter D (in m) at speed u (in m/s), follows a Bingham behavior ("toothpaste" type):

$$\mu = \mu o + D/(6 \times u) \times ro \text{ in } (Pa \cdot s)$$

μo and ro are also linked to dryness S by the following relations:

$$\mu o = 2.5 \times 10^{-3} \exp [18.7 (S - 0.02)] \text{ in (Pa·s)}$$

$$ro = 2.31 \exp [34.4 (S - 0.02)] \text{ in (Pa)}$$

– and for $S \in [0; 0.02]$ or (0 to 20 $g \cdot L^{-1}$), we have: $\mu = 0.001 + 2$ S in (Pa·s). A lower volatile matter content results in a lower viscosity. A temperature higher than room temperature also results in a lower viscosity;

– the *grain size*: large particles can cause transportation (deposits) or abrasion (silica) problems. The silica level may be requested in the case of centrifugation: a maximum value of 5% $SiO_2$/DM may be required.

## 23.3. Handling and storage: shovelable and stackable nature

The graph in Figure 23.2 provides an estimation of the behavior of sludge in accordance with the organicity criterion: at 30% dryness, hydroxide sludge is pasty, but softening sludge is always liquid. Most sludge is not shovelable below 20% dryness.

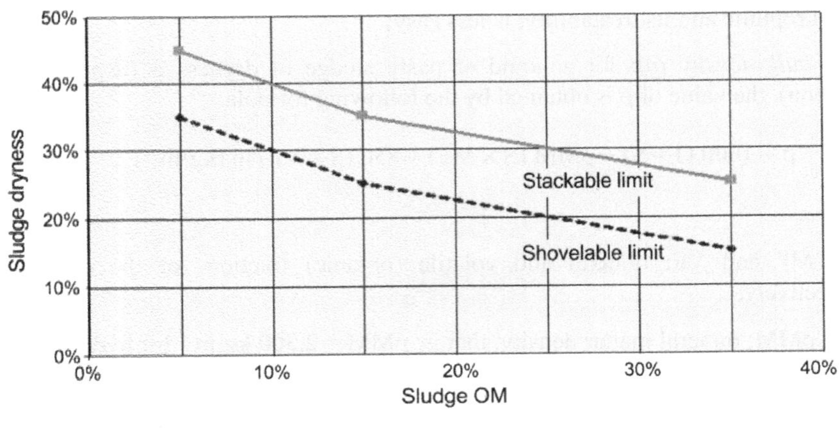


**Figure 23.2.** *Drinking water sludge handling limits*

Using the soil mechanics terminology, these are known as Atterberg limits, that is to say:

– the liquid limit wL separating the liquid state from the plastic state: above such limit sludge becomes shovelable;

– the plasticity limit wP separating the plastic state from the solid state: above such limit, sludge becomes stackable;

– the shrinkage limit wR, separating the solid state without shrinkage from the solid state having shrunk;

– the plasticity index IP, defined as the difference between the liquid limit and the plasticity limit. It is experimentally associated in virtue of the Casagrande relation:

$$IP = (wL - wP) = 0.73 \times (wL - 20)$$

## 23.4. Different classes of sludge

This quality depends on the origin of treated water and the treatment applied. The different classes of sludge which can be identified are as follows.

### 23.4.1. *Hydroxide sludge*

This mainly comes from the treatment of surface water. Also knows as clarification sludge, it is obtained by settling or the retention of suspended solids and colloids precipitated with iron or aluminum salts. Depending on the quality of the resource (suspended solids in raw water), it may contain more or less easily decantable matter such as marl, silt, alluvium or clay.

Hydroxide sludge can be sub-divided into four subgroups:

– highly hydrophilic sludge from resources poor in alluvium (dam, river close to the source);

– slightly hydrophilic sludge, from resources moderately rich in alluvium ("medium" river);

– moderately hydrophilic sludge, from resources rich in alluvium (flooding river, rivers at the mouth, etc.);

– very hydrophilic sludge, from resources rich in OM.

### 23.4.2. *Softening sludge*

Hydrophobic, lime softening sludge generally results from the precipitation of carbonates from hard borehole waters, and sometimes from surface waters.

### 23.4.3. *Metal species sludge treatment*

This refers to the physicochemical manganese and iron removal sludge obtained by the oxidation of dissolved ions and the formation of the corresponding hydroxides. These are very hydrophilic.

### 23.4.4. *Biological sludge*

Resulting from denitrification or biological iron removal, this sludge has similar features to both biological sludge from wastewater treatment stations (denitrification sludge) and sludge from the physicochemical iron removal group. They are very hydrophilic. This type of sludge is still anecdotal.

### 23.4.5. *The case of mixed sludge*

Hard surface waters may be subject to softening, but this scenario is quite rare. Conversely, borehole sludge may contain hydroxide sludge as a result of reagent adhesion. Therefore, the station generates a mixed sludge by mixing. Being aware of the respective sludge proportions makes it possible to know the characteristics of the mixture and to estimate the expected performances.

| Types of sludge | Origin | Composition | Nature |
|---|---|---|---|
| **Clarification sludge (hydroxide sludge)** | Surface waters (+ adhesion of borehole→ water and mixed sludge) | Hydroxides (Fe + Al) > 20%<br>OM > 20% | Mildly to strongly hydrophilic |
| **Softening sludge** | Hard borehole waters (+ hard surface waters → mixed sludge) | $CaCO_3$> 55%<br>Hydroxides (Fe + Al) < 5%<br>OM < 10% | Not hydrophilic (hydrophobic) |
| **(Physicochemical) iron removal sludge** | Borehole waters | $CaCO_3$ = 20–40%<br>Fe hydroxides > 40% | Mixed hydrophilic |
| **Physicochemical manganese removal sludge** | Borehole waters | Al hydroxides > 5–10% | Hydrophilic or hydrophobic |
| **Biological sludge** | Denitrification or iron removal on biofilters | OM > 70% | Highly hydrophilic |

**Table 23.5.** *Simplified composition of sludge types*

The composition of the main types of sludge can be summarized as shown in Table 23.5.

In terms of treatability, the following classification can be established:

> Softening sludge > clarification sludge > iron removal sludge > biological sludge

## 23.5. Sludge composition depending on the characteristics of raw water

### 23.5.1. *Surface water sludge*

This sludge is of the hydrophilic type. Depending on the origin of raw water, the hydroxide and OM contents may vary considerably. The treatability is all the stronger as the OM and hydroxide contents are low and have a high $CaCO_3$ content.

| Resources | Units | Algal barrier water | Dam water (clear water) | Normal river | Flooding river |
|---|---|---|---|---|---|
| **Stagnant character** | | ++++ | +++ | ++ | + |
| **Raw water SS** | $mg \cdot L^{-1}$ | 5 to 70 | 1 to 10 | 30 to 70 | >250 |
| **Raw water turbidity** | NTU | 10 to 30 | < 5 | 20 to 40 | >100 |
| **Raw water TOC** | $mg \cdot L^{-1}$ | 3 to 20 | 1 to 3 | 3 to 6 | 5–10 |
| **OM** | DM % | 30 to 50 | 20 to 30 | 20 to 30 | 10–15 |
| **Hydroxides** | DM % | > 65 | 30 to 80 | 20 to 25 | 10–30 |
| **CaO** | DM % | | 5 to 10 | | 15–35 |
| **Clays (<2 μm)** | DM % | – | – | – | 25–40 |
| **Silts (2–20 μm)** | DM % | – | – | – | <60 |
| **Total $SiO_2$** | %zMM | – | 20–40 | – | – |

**Table 23.6.** *Example of surface water sludge composition*

1) 2,000 algal units/L correspond to 1 $mg \cdot L^{-1}$ SS.

2) the term normal river covers a broad diversity of water, including river water during algal blooms. Because of their more or less abundant alluvial support, these are easier to treat than sludge from dam waters, even clean waters.

Raw water characteristics depend on its origin:

– organic water: contains algae, plankton and/or humic substances;

– dam water: clear water laden with algae at some moments of the year:

  - hydrophilic sludge (strong hydroxide content);

– flooding river: silty, clayey, rich in alluvium, mouth:

  - hydrophobic sludge (weak hydroxide content).

### 23.5.2. *Treatment sludge with coagulants (Fe or Al)*

Since particle size is smaller, treatability is reduced. Note that this sludge, as settling sludge, contains iron hydroxides. But, unlike the latter, which are amorphous and laden with suspended solids, the hydroxides formed from iron precipitation in underground aquifers are composed of relatively pure crystals (entirely inorganic, and therefore, more difficult to treat).

### 23.5.3. *Borehole sludge*

Borehole water sludge comes from the physicochemical removal of iron, manganese and other heavy metals (arsenic, selenium, antimony, nickel, etc.), as well as from the biological removal of iron, manganese, nitrates, ammonia and organic carbon.

| **Treatment** | **Unit** | **Softening** | **Physicochemical removal of iron, manganese and other heavy metals** | **Denitrification, iron removal, biological manganese removal** |
|---|---|---|---|---|
| **Dosage applied (raw water)** | – | – | 2050 $mg \cdot L^{-1}$ Fe | – |
| **OM** | DM % | <5 | – | >70 |
| **Hydroxides** | DM % | <5 | $Fe(OH)_3 > 40$ | <5 |
| **$CaCO_3$** | DM % | 50–90 | 20–40 | – |

**Table 23.7.** *Composition of borehole water sludge*

## 23.6. Thickening of drinking water sludge

### 23.6.1. *Function and criteria for choosing a thickener*

Concentration, clarification and storage are the three functions that thickening should fulfill.

#### 23.6.1.1. *Concentration*

Excluding the flooding periods and specific softening sludge, drinking water sludge has a fairly mediocre capacity to thicken and to dehydrate in terms of concentration (about 20 $g \cdot L^{-1}$ underflow).

#### 23.6.1.2. *Clarification*

A second specificity is related to drinking water processes: the thickener must have a very good separation capacity in order to enable direct discharge to the supernatant medium (35 $mg \cdot L^{-1}$ SS or 20 NFU being the upper limit). As well, an anionic polymer-type flocculant is often required to obtain these performances. Another consequence of the addition of polymers is a considerable reduction in the thickening time, to such extent that the concentration limit is approached after a few tens of minutes and depends little on the type of thickener, unless it is a lamellar (Actydin) with a very high load (flood, etc.), for which the sludge retention time (SRT) can become limiting.

#### 23.6.1.3. *Storage function*

On surface waters, the need to absorb peaks due to floods requires a high sludge buffering capacity (several days or several times the average load). The very low fermentability observed on drinking water sludge (except highly organic algal-type sludge) favors such storage either under specific storage capacities, or directly in the thickener.

SRT influences the storage capacity and is calculated by considering that the average concentration in the sludge blanket is approximately two-thirds of the concentration withdrawn. Sludge loads can vary greatly in the treatment of drinking water sludge, and SRT may span a few tens of minutes (example of a lamella treating a flood episode) to several months.

#### 23.6.1.4. *Buffering storage*

Most drinking water sludge is made up of filter washwater, which should first be collected and homogenized before treatment. The buffering tank can be of the *storer-homogenizer* type, for example, the water collected is homogenized using an impeller. All of the inlet flow goes back to the thickeners.

The design depends on the filtered water volumes and washing frequencies, on the one hand, and on the dewatering operation, on the other hand. As the number of filters increases with the station's capacity, the flow to be treated tends to be continuous, and the relative storage capacities, lower. The graph in Figure 23.3 gives a first idea of the volumes needed, which should be checked in all cases.

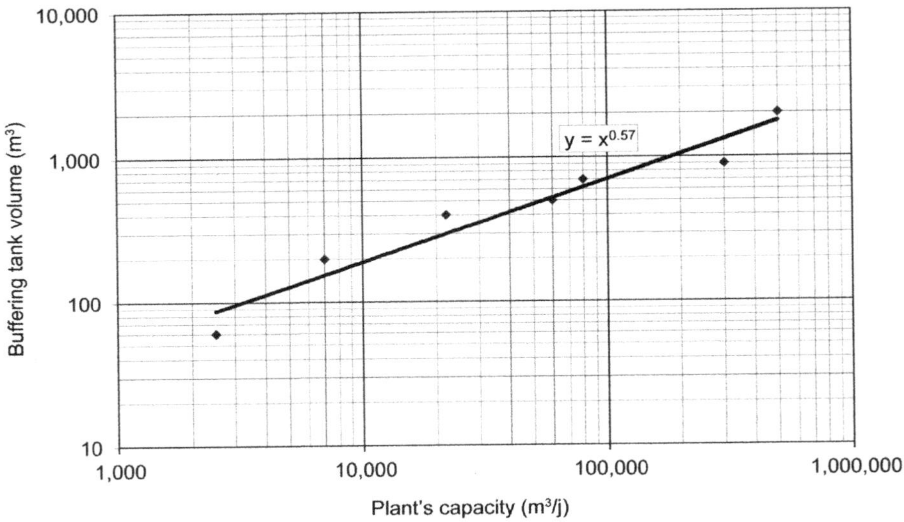


**Figure 23.3.** *Pre-designing a buffering capacity*

#### 23.6.1.5. *Storer-settler*

The water admitted into the storer is separated from sludge by settling, overflows return upstream of the treatment chain or are discharged into sewage. Concentrated underflows are sent to the thickening area.

The storer-settler is commonly referred to as wash water recovery tank (WWRT). It is divided into three identical tanks. Each tank operates under the following cycle:

– reception of the washwater corresponding to 2–3 washing cycles;

– particle settling for 2 h (up to 3 h, depending on sludge decantability);

– discharge of overflows and of sludge having concentrated for 20–40 min;

– hydraulic loading rate: 0.4–0.5 $m \cdot h^{-1}$ without polymer or 0.8 $m \cdot h^{-1}$ with polymer.

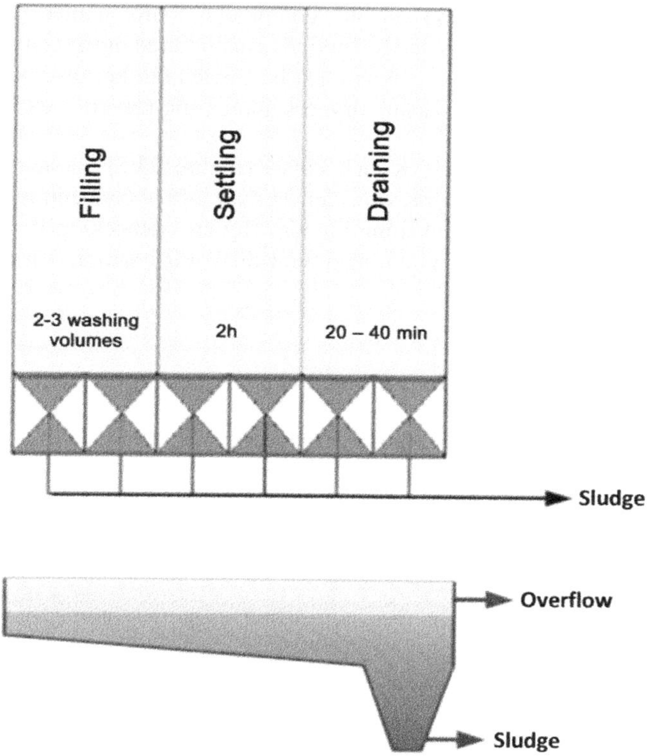


**Figure 23.4.** *Block diagram of a WWRT (Veolia reference)*

### 23.6.2. *Thickener design*

Unlike a wastewater station, the thickener design takes into account the most penalizing case found over the reference year and divides it into shorter periods in order to judge the buffering capacity and determine the operating balance sheet. For example, for river water:

– calm period: moderately organic sludge;

– flood period: strong increase in the mass to be treated, but sludge generally presents good treatability (abundance of alluvium);

– algal bloom: organic sludge, quite difficult to treat.

In general, the design period for the equipment is the winter flood period because, in some cases, the "bad sludge" periods encountered during an algal bloom can be limiting. In any case, simulations must be carried out because performances decrease during these times (decreasing dryness, increasing reagents, etc.).

If sludge is present, a simple Kynch curve (Figure 23.5) for each reference period is of great help because it provides an estimate of the concentration obtained from overflows and from the thickening kinetics (by cross-checking it with data on sludge types). Most sludge attains its maximum thickening within 3 h of SRT and then continues very slowly.

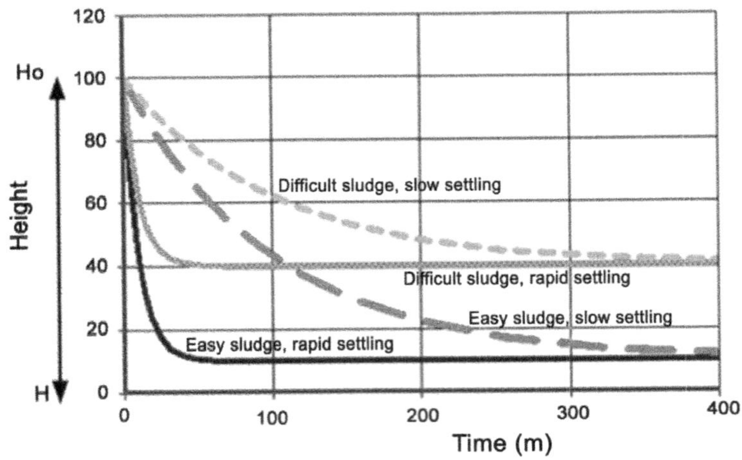


**Figure 23.5.** *Kynch curves for different types of sludge. For a color version of this figure, see www.iste.co.uk/gaid/watertreatment5.zip*

The *hydraulic loading rate* $C_h$ ($m \cdot h^{-1}$) is given by:

$$Ch = \frac{Q}{S}$$

where:

– Q: inlet sludge flow ($m^3 \cdot h^{-1}$);

– S: thickener surface area ($m^2$).

The *floor load* $C_r$ (kg DM·$m^{-2}$·$j^{-1}$) is:

$$Cr = \frac{QCs}{S}$$

where:

– Cs: sludge concentration (g·$L^{-1}$);

– Q: flow expressed in $m^3$·$j^{-1}$.

The *hydraulic retention time* (h) is:

$$ts = \frac{V}{Q} = \frac{SH}{Q}$$

where:

– V: thickener volume ($m^3$);

– H: thickener water level (m);

– Q: flow expressed in $m^3$·$h^{-1}$.

### 23.6.2.1. *Conventional and lamellar thickening*

These structures are mainly intended for sludge from surface waters, but in some cases they may be applied to certain well waters. Equipped with a harrow over at least two-thirds of the sludge bed height, they consume little energy and ensure problem-free sludge storage. They require a regular feed, which justifies the presence of a buffering tank upstream.

The presence of a lamellar clarification compartment above the sludge bed offers excellent separation efficiency, and therefore, a higher load on the floor.

#### 23.6.2.1.1. Sludge from surface water

Table 23.8 presents the design values to be taken into account for physicochemical sludge drawn from an Actiflo, Multiflo, a corridor settler, Cyclofloc, eventually mixed with filter washwaters (the first waters, exclusively), representing less than 30% by volume.

The prerequisites are an optimal physicochemical treatment (coagulant and flocculant) and a sludge layer at least 3 m high.

| Types of raw water | Organic waters | | Dam waters | | Normal rivers | | Flooding rivers | |
|---|---|---|---|---|---|---|---|---|
| Thickener types | C | L | C | L | C | L | C | L |
| Thickener inlet concentrations, $g \cdot L^{-1}$ | 0.5–1.0 | | 1–2 | | 2–10 | | 10–50 | |
| Coagulants: $Al_2SO_4$ or $FeCl_3$, $mg \cdot L^{-1}$ | 150 | | 100 | | 100 | | 50 | |
| Flocculants: polymers, $mg \cdot L^{-1}$ | 0.5–1.0 | | 0.5–1.0 | | 0.5–1.0 | | 0.5–1.0 | |
| Hazen velocity, $m \cdot h^{-1}$ | 1.0 | 1.0 | 1.0 | 1.0 | 1.0 | 1.0 | 1.0 | 1.5 |
| Floor loads, $kg \cdot m^{-2} \cdot j^{-1}$ | – | 250 | – | 250 | – | 600 | – | 1,500 |
| Underflow sludge concentrations $g \cdot L^{-1}$ | 10 | | 15 | | 30 | | 60-100 | |
| Overflow SS concentrations, $mg \cdot L^{-1}$ | 30 | | 30 | | 30 | | 30 | |

**Table 23.8.** *Performance of conventional (C) and lamellar (L) thickeners*

**Figure 23.6.** *Sludge from a flooded river in Asia (Veolia site). For a color version of this figure, see www.iste.co.uk/gaid/watertreatment5.zip*

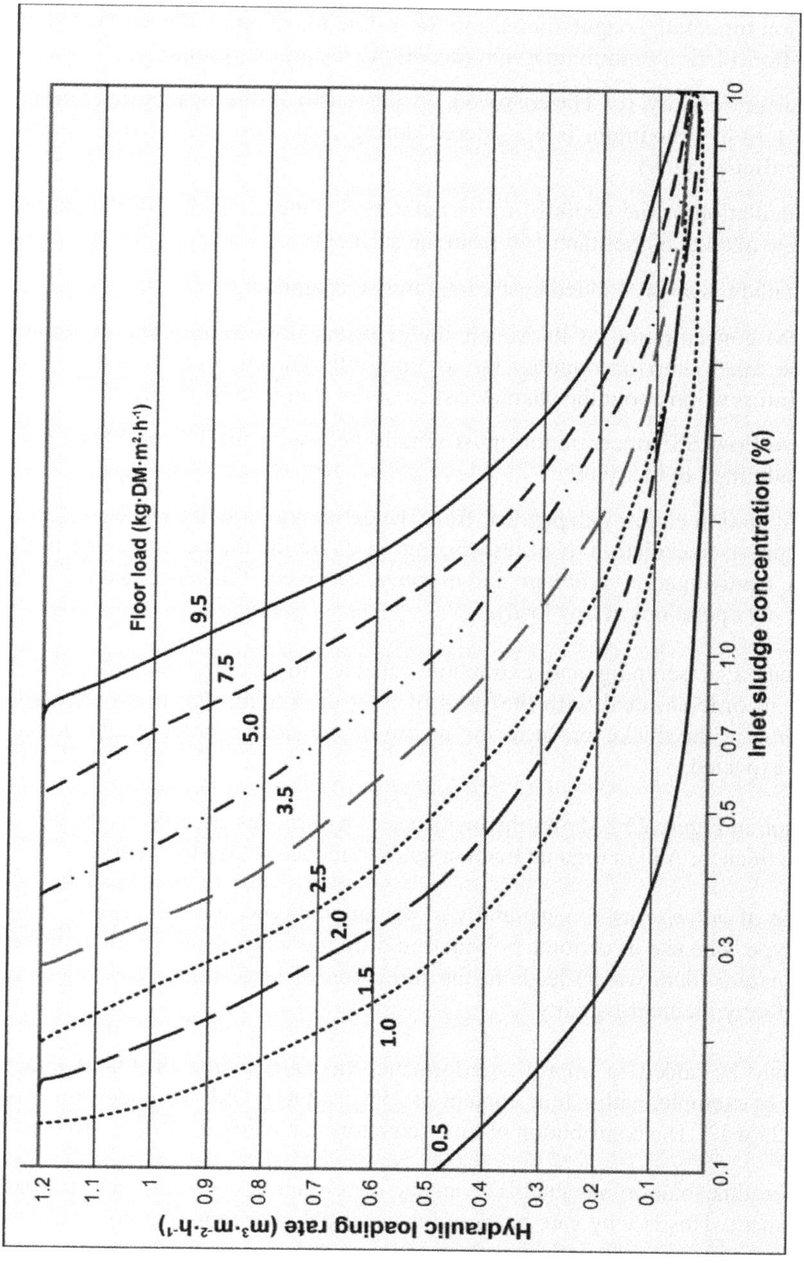


**Figure 23.7.** *Floor loads ($kg{\cdot}m^{-2}{\cdot}h^{-1}$) as a function of the hydraulic loading rate ($m{\cdot}h^{-1}$) and inlet sludge concentration (%), applicable to static thickeners for hydroxide sludge*

Comments on the main parameters:

– retention time with coagulation: 2 min or in-line injection (followed by a static mixer) and flocculation: minimum 6 min (according to Jar-test results);

– the mirror velocity (or Hazen speed, in this case) is the design-determining parameter: 1 $m \cdot h^{-1}$ maximum is considered during peak flow (hence the possible need for a buffering tank);

– the calculation of the depth (4 m by default) is done according to the storage needs (sludge blanket higher than 1 m from the surface);

– a flocculant is always added upstream, rarely a coagulant;

– the DM concentration of thickened sludge (underflow) represents the default value to be taken into account as far as currently known, due to the lack of additional information about the characteristics of the sludge to be thickened;

– the overflow SS concentration must remain below 35 $mg \cdot L^{-1}$ on 95% of the samples if admitted to the river;

– a recirculation circuit independent from the extraction circuit must be set up in order to improve flocculation, if necessary, and to dilute the thickened sludge in the event of a centrifugation problem and/or on excessively thickened sludge (e.g. during start-up operations at low load).

In all cases, a scraping and extraction capacity functioning at 150 $g \cdot L^{-1}$ is guaranteed in order to deal with the risks of over-thickening due to an excessive retention time in the thickener, or to the nature of the sludge which might thicken better than expected.

The graph in Figure 23.8 shows the incidence of hydroxides and OM on the ability of sludge to thicken. A high organic fraction tends to reduce sludge thickening.

The type of *polymer* used is generally of the anionic type, but can also be of the non-ionic type. The use of cationic polymer on drinking water sludge with a marked hydroxide nature (dam water) leads to the formation of large flocs, resulting in an overflow of very mediocre quality.

Lime may be added to improve performance (a contact time of 15–30 min is required). For example, with a lime content of 25% $Ca(OH)_2$/DM, the underflows rise from 10 to 25 $g \cdot L^{-1}$.The contribution of lime saturators can partly fulfill this mission.

However, this solution has the disadvantage of resolubilizing aluminum upstream of the treatment. This is why care is taken for water pH after remixing not to exceed 7.4 in the case of a treatment with aluminum salts, and 8.5 if iron salts are used.

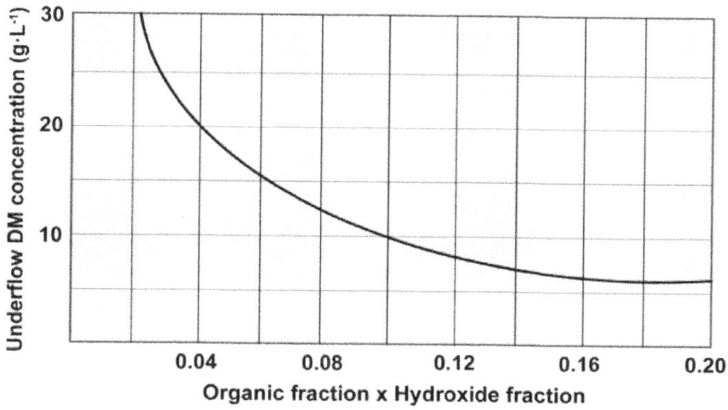


**Figure 23.8.** *Influence of the product (OF × HF) on the thickened sludge concentration*

#### 23.6.2.1.2. Softening sludge

Sludge rich in $CaCO_3$ thickens very well. The overflows obtained can vary from 60 (partial softening) to 300 $g \cdot L^{-1}$, depending on the treatment.

The floor load admitted on a conventional thickener can exceed 200 kg $DM \cdot m^{-2} \cdot j^{-1}$, but thickening is not essential given the inlet concentration.

#### 23.6.2.1.3. Denitrification sludge or biological iron removal

The thickening capacity of sludge is less satisfying. Rarely does it exceed 25 $g \cdot L^{-1}$.

#### 23.6.2.1.4. Physicochemical iron removal sludge

Performances are much lower for filter washwater sludge (absence of SS in raw water): the underflows will concentrate between 5 and 10 $g \cdot L^{-1}$.

### 23.6.3. *Implementation of thickeners*

#### 23.6.3.1. *Conventional thickener*

The main design characteristics of the conventional OTV thickener:

– the diameter range of thickeners extends from 4 to 30 m;

– velocity in the feed line: 0.5 $m \cdot s^{-1}$;

– minimum height of the clarification zone: 1 m.

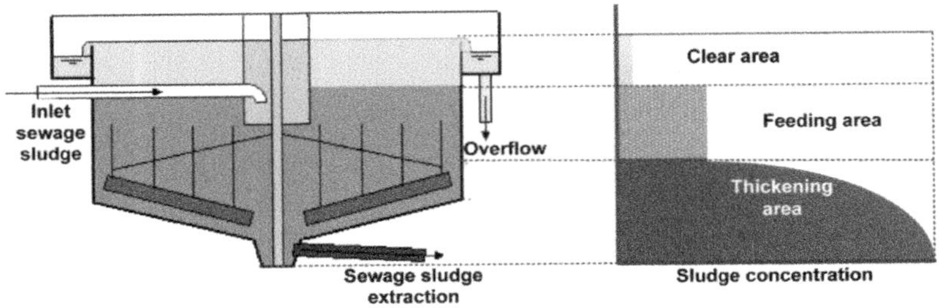


**Figure 23.9.** *OTV-type conventional thickener. For a color version of this figure, see www.iste.co.uk/gaid/watertreatment5.zip*

a)

b)

**Figure 23.10.** *(A) Settler-thickener for softening sludge. (B) Settler-thickener for softening sludge. For a color version of this figure, see www.iste.co.uk/gaid/watertreatment5.zip*

The structures can be concrete and steel for small diameters.

#### 23.6.3.2. *Lamellar thickener*

In order to prevent the lamellae from being blocked by thickened sludge, the chosen lamella plates should be spaced wide enough (80–100 mm) to avoid clogging with an alveolar PVC or stainless steel material. The lamellae length is 1–1.5 m.

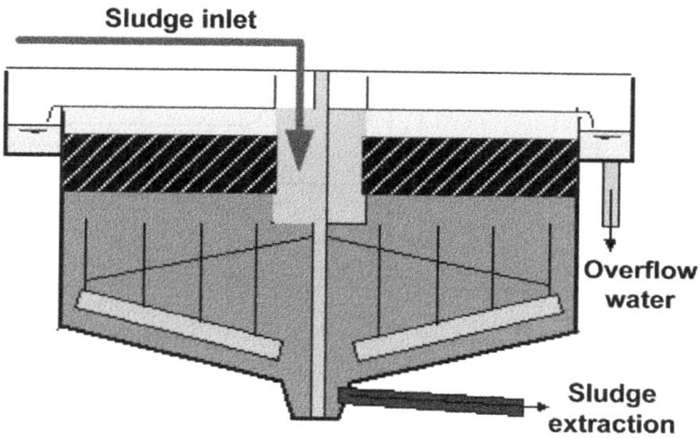


**Figure 23.11.** *Diagram of a lamellar thickener. For a color version of this figure, see www.iste.co.uk/gaid/watertreatment5.zip*

#### 23.6.3.3. *Actidyn lamellar thickener (Veolia)*

The Actidyn lamellar thickener is composed of a flocculation tank which, if necessary, can be stirred with a polymer injection. The retention time in the flocculation tanks is less than 15 min because too long a retention time in the maturation tank is detrimental to the proper formation of flocs, which tend to be torn apart.

The flocculated water is then introduced into the settling tank under the plates forming the lamellar modules and circulates bottom-down (Figure 23.12). The clarified water is collected in the upper part of the tank by recovery collectors, which discharge the water into the waste. These collectors are equipped with notched spillways granting an appropriate flow distribution within the structure.

Sludge is deposited on the lamellae, then slides toward the bottom of the structure and is collected by bottom scraping that transports it into a central sludge well. Afterward, the sludge is withdrawn by extraction pumps and sent to sludge treatment.

The structure's depth and the SRT in the thickener make it possible to attain concentrations of the order of 25–40 $g \cdot L^{-1}$. Actidyn works with mirror velocities of

8–15 $m \cdot h^{-1}$ and floor loads greater than 100 kg $DM \cdot m^{-2} \cdot j^{-1}$.The sludge thickening rate is a function of the thickening zone height under the lamellae and under the settling zone.

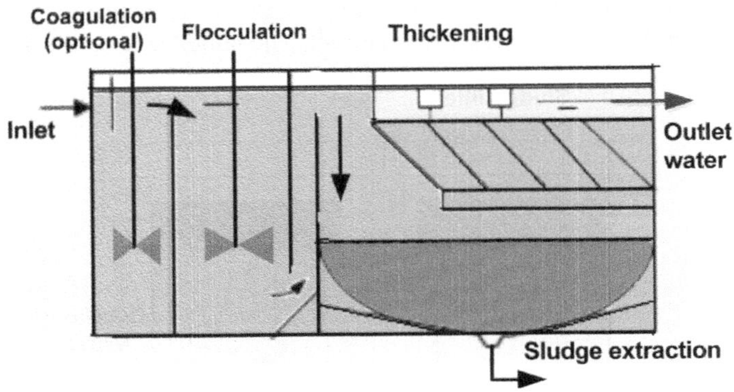


**Figure 23.12.** *Diagram of the Actidyn process. For a color version of this figure, see www.iste.co.uk/gaid/watertreatment5.zip*

To facilitate maintenance operations, Actidyn has lamellae cleaning equipment including a blower which sends air into a ramified network fixed under the structure's joists. The cleaning of the lamellae is done by bubbling. Sliding along the lamellae, the air bubbles loosen the sludge which then falls to the bottom of the structure. Lamellae cleaning is necessary to avoid the accumulation of sludge on the lamella plates.

**Figure 23.13.** *Actidyn settler-thickener. For a color version of this figure, see www.iste.co.uk/gaid/watertreatment5.zip*

The mirror velocity is the parameter to be retained. As for the sludge blanket maximum height, and therefore, the total height of the structure, it is related to the minimum retention time, thus obtaining high concentrations in the underflow.

A coagulant is added upstream of flocculation when one wishes to obtain overflow water with low turbidity (<10 NTU).

| Parameters | Characteristics |
|---|---|
| Contact time in the maturation tank | 10–15 min |
| Mirror settling velocity | 8–15 $m \cdot h^{-1}$ |
| Hazen velocity | 0.53–1.0 $m \cdot h^{-1}$ |
| Lamellae length | 1–1.15 m |
| Total height of the settling/thickening zone | 5–7 m |
| Floor load | Up to 1,500 $kg \cdot m^{-2} \cdot j^{-1}$ |

**Table 23.9.** *Technical characteristics of the Actidyn process*

#### 23.6.3.4. *Lagoon thickening*

The sludge lagoon treatment is generally chosen as a dewatering procedure. Nevertheless, it can be considered as a thickener if the feed is not regular, for example, in the event of a flood followed by a long period of rest. Polymers are not necessary.

The implementation conditions are the following:

– superficial velocity: <0.2 $m \cdot h^{-1}$;

– water depth: about 2.5 m;

– underflow concentration: about 1.5 times the concentration at the thickeners' outlet, depending on the settling time;

– overflow concentration: <35 $mg \cdot L^{-1}$.

The underflows are recovered by a hydrostal-type submersible pump guided from the surface by a maneuverable float.

The recovery of clear water is done by a perforated collector. The sludge blanket must be at a distance greater than 30 cm below the pipeline's lower generatrix (V = 0.8 $m \cdot s^{-1}$ in the pipeline, 1.2 $m \cdot s^{-1}$ at the openings).

### 23.6.4. *Flotation*

The flotation phenomenon that occurs when sludge is introduced into a graduated cylinder is illustrated in Figure 23.14.

Four areas can be identified:

– a zone of clear water in the lower area of the cylinder which can be likened to the same zone that exists in a flotation unit;

– an intermediate zone in which the residual particles are free or flocculated. These particles float while maintaining the same relative position in relation to one another;

– a heterogeneous flotation zone where the particles become agglomerated and are pushed upwards by the ascending flow;

– a sludge compression zone which is located in the upper part of the flotation unit where the particles form a solid structure with strong physical interactions.

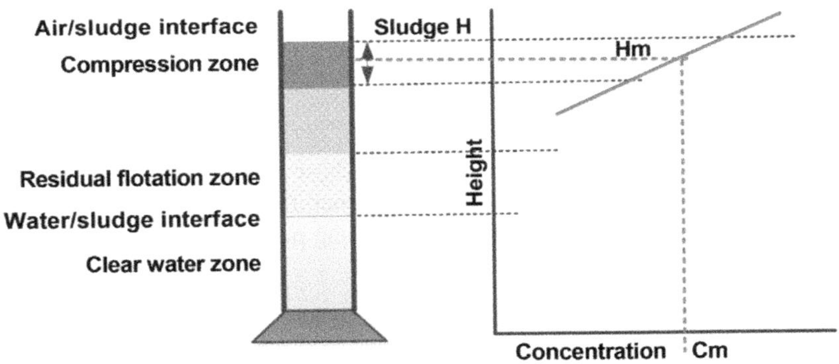


**Figure 23.14.** *Classification of zones during sludge flotation*

The surface of the flotation unit is given by:

$$S = \frac{Q}{v} = \frac{Qt}{H}$$

where:

– S: flotation unit surface area ($m^2$);

– Q: flow rate through ($m^3 \cdot h^{-1}$);

– V: flow velocity ($m \cdot h^{-1}$);

– t: retention time in the thickening zone (h);

– H: total height (m).

The sludge average concentration is obtained by plotting the tangent to the curve's inflection point corresponding to a height value in the flotation unit's sludge zone.

Figure 23.15 summarizes the points considered to determine the sludge average concentration, which is then deduced from:

$$\frac{Ho}{Hb} = \frac{Cb}{Cin}$$

where:

– Ho and Hb: total height of the flotation unit and compression zone, respectively;

– Cin and Cb: inlet sludge concentration and in the compression zone.

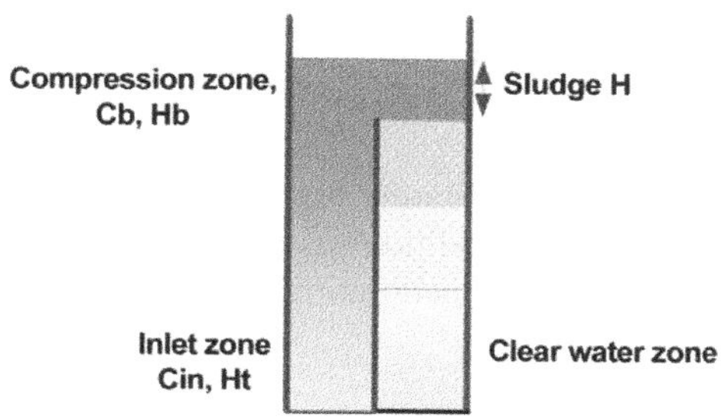


**Figure 23.15.** *Sludge concentration estimate in the compression zone*

For example, for a float 4 m high and a sludge inlet concentration of 5 $g \cdot L^{-1}$, the sludge average concentration equals:

$$Cb = \frac{HoCin}{Hb} = \frac{4 \times 5}{0.5} = 40 \text{ g} \cdot \text{L}^{-1}$$

Consider that the sludge height Hb is 0.5 m. This procedure is chosen when sludge settles with difficulty (it is hydrophilic), in other words, sludge rich in OM, in particular plankton or hydroxide-rich micro-algae (over 40%). Contrary to this, when hydroxides are lower than 20%, the static thickener is more interesting.

Sludge flotation preferably applies to filter washwater sludge (highly organic water), as well as iron removal sludge. This technique is not used for softening sludge or for any other hydrophobic sludge.

The process is generally of the indirect type: the quality of the underflows is good and the energy cost moderate, insofar as recirculation remains low. Pressurization is related to a low flow rate and is energetically less costly than working on a direct flotation unit, which requires pressurizing the entire flow to be treated. The flotation unit can operate discontinuously for a few hours a day. The flow must be as regular as possible, in terms of load and volume.

| | | |
|---|---|---|
| **Hydraulic loading rate** | $m^3 \cdot m^{-2} \cdot h^{-1}$ | <5 |
| **Mass loading rate** | kg DM$\cdot m^{-2} \cdot h^{-1}$ | 5–13 |
| **Inlet SS** | $mg \cdot L^{-1}$ | 300–3,000 (<8,000) |
| **Polymer** | $mg \cdot L^{-1}$ | <2 |

**Table 23.10.** *Some operating characteristics*

## 23.7. Drinking water sludge dewatering

Sludge dehydrated with polymer rarely reach the "stackable sludge" status, even at high dryness (>30%). This condition is only obtained by means of a lime treatment, which translates as soil amelioration, making the land apt for agricultural purposes.

The conditioning of thickened sludge with lime attains high dryness levels during mechanical dewatering. The addition of a ferric chloride-type coagulant is unnecessary for this type of sludge.

### 23.7.1. *Plate filter*

#### 23.7.1.1. *Description*

The process involves pressing sludge laden with solid matter with a high-pressure injection (from 5 to 15 bar) between two plates. These plates are dressed with a fixed filter cloth and suspended vertically between compression plates or compacting membranes.

When the filter is closed, the plates touch one another and form chambers intended to receive limed sludge. These are injected at high-pressure through the plates' central orifice, filling the chambers. The liquid part – the filtrate – crosses the filter cloths and is discharged by a drain. Then, the rubber compacting membranes are inflated with water from a hydraulic power station and compact the sludge in the chamber, pursuing dewatering until the formation of a final compacted cake.

**Figure 23.16.** *Plate filter (small drinking water station). For a color version of this figure, see www.iste.co.uk/gaid/watertreatment5.zip*

When the filter press opens plate after plate, the cakes fall successively and are directed toward storage silos. Compared to other sludge dewatering techniques, this one requires more manpower, in particular for monitoring the removal of filters (the other sequences of the "pressing" cycle can be automated).

This constraint can be much less severe with the choice of fully automatic filter presses, which have very satisfactory performances.

Lime conditioning is commonly used (20–35% of suspended solids), reaching dryness levels of between 30 and 40% for surface water sludge. For highly mineral sludge, dryness can reach 40%.

#### 23.7.1.2. *Plate filter performances on hydroxide sludge*

Most of the time, this technology is used with lime conditioning, due to the above-mentioned reasons. The addition of sawdust is also being considered.

On hydroxide sludge (surface waters: rivers or dams), the dryness obtained is generally between 30 and 40%. Conditioning with $Ca(OH)_2/SS$ is around 20–30%.

Characteristics are as follows:

– cycle duration (pressing + filter removal): 2 h 30 min;

– cake density obtained: 1.15–1.20 kg·dm$^{-3}$;

– plate spacing: 35 mm.

Operating conditions are as follows:

– productivity: 2–5 kg DM·m$^{-2}$·h$^{-1}$;

– working pressure: 10–25 bar.

When recycling is chosen, the filtrates and washwater can be sent back upstream of the thickener, or to the dirty water homogenization tank. Sometimes, to limit the alkalinization of the filtrates discharged into the receiving environment (in the case of a small river with a low flow), the liming rate is restricted to less than 15%. This translates into a deterioration in cake dryness and a substantial extension of the cycle duration. Switching to polymer is then recommended.

For a pressing time of around 4 h, the data are shown in Tables 23.11 and 23.12.

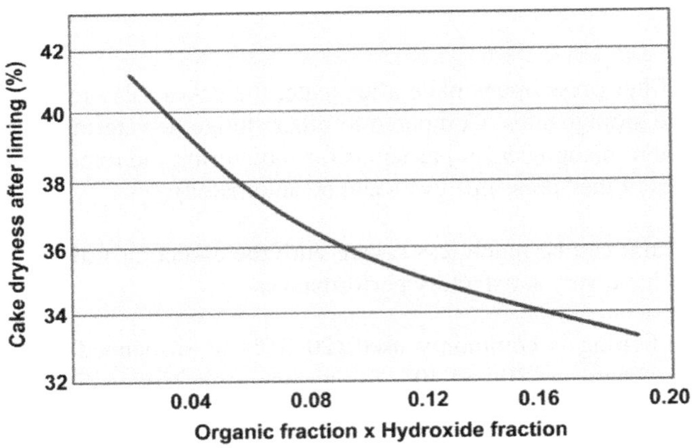


**Figure 23.17.** *Influence of sludge quality on liming efficiency*

| | Range |
|---|---|
| **Polymer rate (kg/tDM)** | 3–8 |
| **Dryness achieved (%)** | 15–22 |

**Table 23.11.** *Polymer conditioning*

| | Polymers alone | Polymers + lime 5–10% | Lime + $FeCl_3$ |
|---|---|---|---|
| **Cake thicknesses** | 25 mm | 25–30 mm | 30–40 mm |
| **Powder polymers** | 3–8 kg/tDM | 2–3 kg/tDM | 0 |
| **Lime** | 0 | 50–100 kg $Ca(OH)_2$/tDM | 150–400 kg $Ca(OH)_2$/tDM |
| **Ferric chloride** | 0 | 0 | 0–70 kg $FeCl_3$/tDM |
| **Final dryness** | 20–28% | 30–35% | 35–40% |
| **Final pH** | 7 | 8–9 | >11 |
| **Cycle duration** | 4–6 h | 4–5 h | 3.5–4.5 hours |
| **Membrane cycle duration** | 3 h | 3 h | 2–3 h |

**Table 23.12.** *Performances observed depending on the reagents added*

#### 23.7.1.3. *Performances on softening sludge*

Very high dryness values, up to 70%, or even higher, can be observed for this type of sludge, unless they contain too much magnesium. For this type of sludge, there is no conditioning, neither mineral nor organic. Cake thickness can reach 50 mm for a cycle of the same order, that is, 2 h to 2 h 30 min.

### 23.7.2. *Centrifugation*

Under the effect of the centrifugal force, suspended solids experience accelerated settling and are compacted on the wall of the centrifuge tank. A centrifuge is made of a horizontal cylindro-conical bowl rotating at high speed, and a helical extraction screw in its interior, rotating at a different speed. The extraction screw discharges the settled sludge toward the conical end, while the liquid phase is evacuated toward the other end.

The design of the centrifuge will be adapted to the sludge feed rate, the suspended solids concentration and the final conditioning (higher or lower dryness).

Centrifugation performances significantly improve with the use of a thickener, which should be defined in accordance with the settling test.

The choice of thickening by flotation is only possible provided that the filter washwater has not been treated. Otherwise, static thickening should be chosen. It is necessary to regulate the flow rates and concentrations for the centrifuge's smooth functioning. In this view, it is advisable to provide buffering tanks for the settler purges and filter washwaters. Poly-electrolyte, or lime and poly-electrolyte conditioning should be systematically provided.

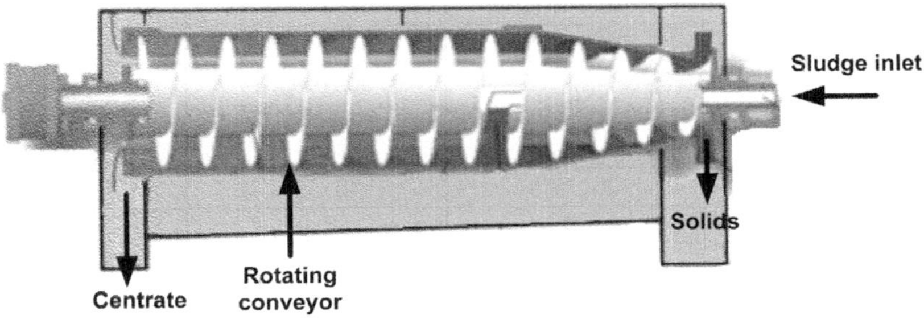


**Figure 23.18.** *Operation diagram of a centrifuge. For a color version of this figure, see www.iste.co.uk/gaid/watertreatment5.zip*

Applying the principle according to which "a sludge that thickens well, dewaters well", the correlation displayed on the graph of Figure 23.19 may be useful as a first approximation.

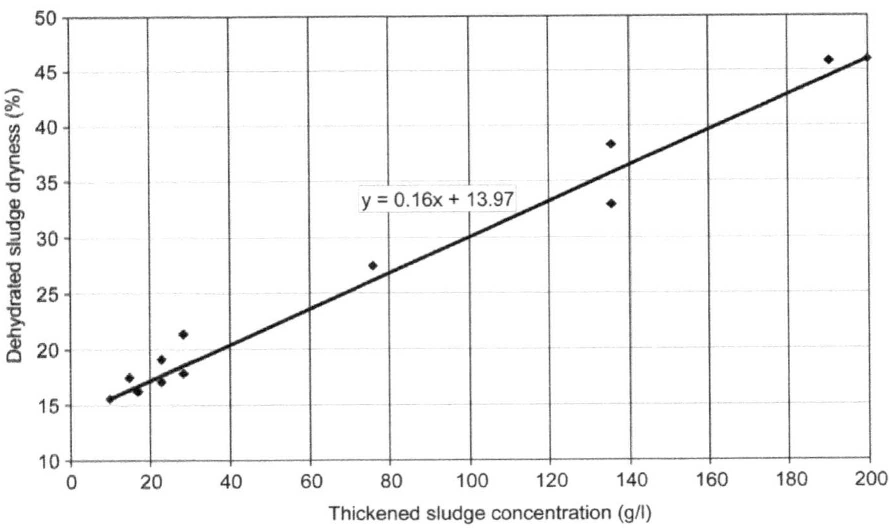


**Figure 23.19.** *Relationship between thickened sludge concentration and dryness*

There are different types of centrifuges: conventional and high-performance ones which result in higher dryness, but only in the presence of polymers.

Performances are related to the increase in bowl speed, retention time in the bowl, or the technology of the screw.

The main causes of malfunction are found in the conveyor screw, whose state of wear or state of fouling must be regularly checked, together with the polymer injection equipment. For proper operation, it is essential to clean the centrifuge after each shutdown, and to check all the equipment. Needless to say, equipment maintenance requires qualified operators.

#### 23.7.2.1. *Surface water sludge centrifugation*

Sludge dryness is all the greater when ferric chloride is used rather than aluminum sulfate for the treatment, when the turbidity of raw water is high and the hydroxide content is low (Table 23.13).

| Types of water | Dam | Normal river | Alluvium-rich river |
|---|---|---|---|
| **Inlet sludge concentration ($g \cdot L^{-1}$)** | 16 (10–25) | 19 (10–30) | 150 (100–200) |
| **Polymers (kg/TDM)** | 7 (5–9) | 6 (3–8) | 2 (1–3) |
| **DM cake (%) with ferric chloride** | 18 (15–20) | 20 (15–25) | 30 (25–35) |
| **DM cake (%) with aluminum sulfate** | 15 (12–18) | 17 (15–20) | 28 (20–30) |
| **Centrate SS ($mg \cdot L^{-1}$)** | <1,000 | <500 | <500 |

**Table 23.13.** *Centrifugation performances depending on sludge origin*

#### 23.7.2.2. *Softening sludge centrifugation*

The high $CaCO_3$ content ensures very good performances, as shown in Tables 23.14 and 23.15.

| | Range |
|---|---|
| **Dryness achieved (%)** | 30–55 |
| **Polymers (kg/tDM)** | 0–3 |

**Table 23.14.** *Centrifugation performances on borehole water sludge*

#### 23.7.2.3. *Iron removal sludge centrifugation*

| | Range |
|---|---|
| **Dryness achieved (%)** | 15–20 |
| **Polymers (kg/tDM)** | 2–5 |

**Table 23.15.** *Centrifugation performances on iron removal sludge*

### 23.7.3. *Belt filters*

Sludge dewatering is obtained by applying progressive pressure to a continuously moving sludge bed under a pressure system. Low- and medium-pressure belt filters (4 and 5 bar) are the most commonly used.

Medium-pressure dewatering is carried out in four stages:

– sludge flocculation with a mixture of polymers in a tank serving both as a flocculator and a buffering tank;

– gravity draining on a mesh fabric (Ø 600 μm);

– pre-pressing the sandwiched sludge, which has been pre-pressed around a roller;

– high-intensity pressing for progressive sludge dewatering.

The pre-pressing stage is non-existent in the low-pressure belt filter. Belt filter dewatering is used in medium production units and for sludge recovery in agriculture. Dewatering by high-pressure belt filters makes it possible to obtain two dryness points compared to the medium-pressure filter, a gain which does not justify an additional investment cost.

Sludge must contain a maximum of 40% hydroxides and its structure should not be too colloidal. For clarification of sludge concentrations below 15 $g \cdot L^{-1}$, operations become too difficult. It is necessary to carry out a preliminary sludge thickening and choose this option in accordance with the settling test. The choice of thickening by flotation is only possible provided that the wash water is not treated.

To ensure feed regularity, buffering tanks on the settler purges and on the filter washwaters should be considered.

To obtain good performance, the sludge entering the system must have a low compressibility. To this end, conditioning with a polyelectrolyte is done during the system's first phase.

This technique is not recommended for biological iron and manganese removal because sludge adhesion problems are not uncommon. Belt filters are well suited to treat softening sludge. For hydroxide sludge, the melting risk is greater and the facility requires closer monitoring.

The dryness obtained varies between 15 and 50% and the filtrate concentration is lower than 30 mg·L$^{-1}$ of SS. Dewatered sludge can then be limed to reach the 30% dryness required for admission into landfill, or for being recycled in agriculture.

The performances to be expected are shown in Table 23.16.

| | Dams | Normal rivers | Alluvium-rich rivers | Softening |
|---|---|---|---|---|
| **Inlet sludge concentration (g·L$^{-1}$)** | 16 (10–25) | 19 (10–30) | 150 (100–200) | 200 (100–300) |
| **Lime/DM (%)** | 30 | 30 | – | 0 |
| **DM cake (%)** | 15–22 | 20–25 | 25–40 | 25–50 |
| **Filtrates (mg·L$^{-1}$)** | <30 | <30 | <30 | <30 |

**Table 23.16.** *Performance of belt filters on sludge of different origins*

## 23.7.4. *Filter bags*

This system is halfway between thickening and dewatering. It is responsible for sludge conditioning. It is presented as a compact device. Previously flocculated sludge is introduced into porous bags, where it experiences gravity drainage.

This technique is interesting for low levels of sludge production. The parameter governing the design is the average dryness obtained after draining.

**Figure 23.20.** *Example of filter bags. For a color version of this figure, see www.iste.co.uk/gaid/watertreatment5.zip*

It is necessary to carry out a preliminary sludge thickening, and to choose this option in accordance with the settling test. To ensure feed regularity, buffering tanks on the settler purges and on the filter washwaters should be considered. It is also advisable to systematically provide for conditioning using a poly-electrolyte.

This technique applies to sludge having been conditioned at the thickener. It is very economical in terms of investment but requires regular handling with a hoist, which limits its use at DWS, producing less than 100 kg $DM{\cdot}d^{-1}$.

Bags with a volume of 60–800 L (large-type bags) are filled in first. The bags are drained and then, after about a week has elapsed, they are placed on a drying area. The removal of dry bags is generally done once or twice a year. The preferred destination is the landfill.

Several daily fillings are possible: an 800 L bag of sludge at 15 $g{\cdot}L^{-1}$ represents a load of 12 kg DM/bag. Handling can also be done by hand truck on 60 L bags and by forklift on 800 L bags.

Dry contents of 60% for 60 L bags and 50% for 800 L bags can be reached for softening sludge. Conversely, some settling sludge barely reaches 8–10% at the end of the dewatering phase, in which case dryness can be improved during the storage phase.

Table 23.17 provides some indications on the dryness and the treatment capacity of a bag.

| | **Clean rivers** | **Average quality rivers** | **Flooding rivers** | **(Physicochemical) Fe sludge** | **Biological Fe sludge** | **Denitrification sludge** | **Softening sludge** |
|---|---|---|---|---|---|---|---|
| **Dryness after draining (%)** | 12 | 13 | 15 | 10 | 18 | 15 | 15–25 |
| **Processing capacities for one bag (kg DM)** | 48 | 52 | 60 | 40 | 72 | 72 | 240 |

**Table 23.17.** *Dryness and treatment capacities per filter bag for different types of sludge*

### 23.7.5. *Drying beds*

After a thickening stage or passing through a buffering tank, sludge is sent to a filtering body where the water percolates. The supernatant is discharged to the surface.

Drying is done by evaporation. This method requires the operation of at least two drying beds in parallel, one performing the drying while the other one is filled.

#### 23.7.5.1. *Description*

The treatment chain includes:

– a buffering basin collecting washwater or liquid sludge and ensuring it mixes up;

– sludge conditioning with poly-electrolytes: they reduce the clarification time on the drying bed (4 h) and restore a good quality supernatant;

– a thickening where conditioned sludge settles; the supernatant water is discharged on the surface toward the natural environment;

– a filtration and drying phase whereby sludge is sent to draining beds through which the water percolates.

Drying is obtained by water evaporation. Sludge reveals characteristic fissures enabling the rapid discharge of rainwater, which limits the risk of sludge rehydration. The end product is chunky and solid.

##### 23.7.5.1.1. Drying Bed Features

From top to bottom, the draining bed contains:

– 10 cm thick sand, with a grain size between 0.3 and 1 mm;

– gravel 20 cm thick, with a grain size between 5 and 15 mm;

– gravel 10 cm thick, with a grain size between 15 and 25 mm with plastic or brick drains spaced every 2.5 m;

– compacted soil on a slight slope of 1.5%;

– a geotextile protecting the drainage layer.

The design should consider the surface area to be used during the period of maximum sludge production, which also corresponds to the period for which the most sludge needs to be stored.

The minimum bed area is 75 $m^2$ (15 m × 5 m) and the liquid height on the draining bed should measure 1 m. A loading mass flow of the order of 50 kg $DM{\cdot}m^{-2}$ is recommended in order to obtain a shovelable sludge with a 20–25% dryness within 10 weeks.

**Figure 23.21.** *Thickening and drying (Veolia site). For a color version of this figure, see www.iste.co.uk/gaid/watertreatment5.zip*

**Figure 23.22.** *Thickening and drying in covered sites (Veolia site). For a color version of this figure, see www.iste.co.uk/gaid/watertreatment5.zip*

### 23.7.5.2. *Thickened sludge drying beds*

Preferably, the drying beds are preceded by a thickening stage in order to obtain short-duration filling-drying cycles. When the feed is made from the buffering tank, thickening takes place directly in the beds. Beds require manpower for the removal of dried sludge, but this maneuver is periodic, generally every 30–60 days. Conditioning is important: drying is adequate for sludge treated with $FeCl_3$, whereas aluminum salts are much less favorable.

The sludge layer to be applied is in the order of a few tens of centimeters. At least two structures must be provided in order to operate in two filling-drying phases. Conditioning with poly-electrolytes makes it possible to restore a good quality supernatant within 4 h.

The operation is divided into several phases:

– thickened sludge transfer toward the bed;

– sludge settling for 4 h;

– supernatant extraction via a shovel valve toward a recovery sump; the supernatant, still very charged, must be reintroduced into the thickener;

– opening of the drain valve;

– drainage and evaporation for a week before a new filling.

After a number of fillings, depending on the acceptable height of sludge, the bed is left until it completely dries up, while the second bed is filled.

### 23.7.5.3. *Non-thickened sludge drying beds*

The sludge layer to be applied can reach 1.5–2 m. At least, two structures must be provided in order to operate alternately in two filling-drying phases, with a drying period of 70 days for each phase:

– Settling and thickening phase: sludgy water is admitted into the structure, where it settles naturally. The supernatant is permanently discharged by pouring through the wall valves (it is possible, for example, to lower the valves by 5 cm every 30 min).

– The sludgy water feed stops when the filling time contemplated by the design has been reached. The DN 200 drain is insulated and filled with water during this phase in order to avoid any filtration or clogging phenomena.

– Drainage and drying phase: the body of water is lowered very slowly ($\approx$10 $cm \cdot h^{-1}$) by the drainage system to avoid sludge re-suspension. Emptying comes to a stop when the body of water reaches about 10 cm above the gravel. From this height, the concentration is sufficient for adequate drying and recovery by a backhoe loader. Once the drying cycle is complete, sludge is scraped off. The drains are cleaned and filled with raw water.

The structure's floor is covered by a bed of gravel about 10 cm thick, onto which the sludge is deposited. The DN 200 drain is placed onto the bed's material along the structure's longitudinal axis and coated with sand or gravel. It is necessary to provide valves at the high level to act as a spillway, as well as valves at the low level for emptying the basin. The longitudinal dikes have concrete slopes. The transverse dikes have an access ramp for backhoe loaders. The floor slope in the direction of the longitudinal axis is 1%.

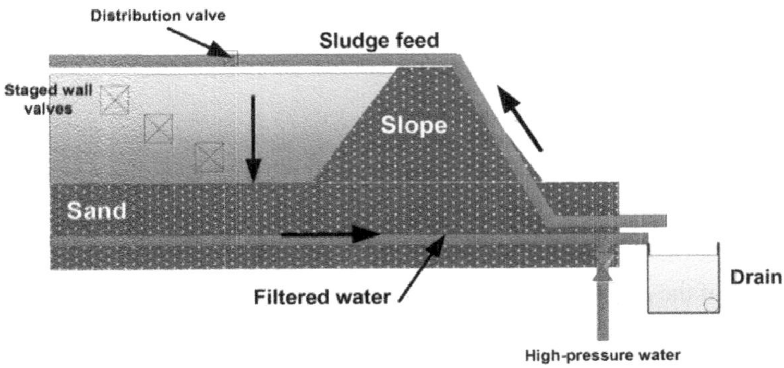


**Figure 23.23.** *Cross-section of a drying bed. For a color version of this figure, see www.iste.co.uk/gaid/watertreatment5.zip*

"Wall" solutions only apply to already thickened sludge. If the sludge layer is more voluminous (unthickened sludge), the dike solution is mandatory.

It must take into account the climate, the desired dryness and the load applied. At the beginning of the drying phase, the height of the thickened sludge spread over the drain is around a few tens of centimeters.

**Figure 23.24.** *Filling a drying bed. For a color version of this figure, see www.iste.co.uk/gaid/watertreatment5.zip*

The data expected for 2–3 month cycles (60–70 days) are shown in Table 23.18.

| Climates | Dryness to be obtained % | Loads applied kg DM·m$^{-2}$ per cycle |
|---|---|---|
| **Oceanic** | 15–20 | 30 |
| **Continental** | 20–25 | 50 |
| **Mediterranean** | 40 | 70 |
| **Arid** | 60 | 160 |

**Table 23.18.** *Loads applied in function of the geographical areas*

If the climate is very humid, a cover should be provided. Furthermore, frost is very favorable to dewatering because it enables the release of bound water. For proper management, the operation is weekly, which means that the volume of thickened sludge to be spread corresponds to the production of 1 week. The sun and the wind increase dewatering, as well as frost does. During the growth of ice crystals, solids become compressed and depleted in water.

While the bed cover guarantees independence from climatic conditions, it reduces direct evaporation due to solar radiation, and increases air currents.

Sludge is recycled for agricultural spreading or sent to landfill when the dryness reaches 30%. This condition limits the application of such a method in oceanic or continental areas.

### 23.7.6. *Sludge lagoon treatment*

The lagoon treatment is considered as a rectangular settler. The lagoon sludge dewatering technique ensures sludge thickening, with discharge of the supernatant into the natural environment, storage of the settled materials and finally their drying.

The design depends on the mode of operation. It must be perfectly defined for a good exploitation. The key design parameter is the average sludge concentration in the thickening zone after storage and before drying.

#### 23.7.6.1. *Main characteristics*

The main characteristics are as follows:

– the lagoon is fed with non-thickened sludge;

– the lagoon fulfills two functions: thickening during filling and dewatering during drying;

– the implementation of a lagoon requires the presence of impermeable ground;

– the cleaning of the lagoon at the end of the drying cycle will preferably be done at the end of the summer.

The cake obtained tends to be very heterogeneous. The sludge can be very dry on the surface (about 65%) and still be in a semi-liquid state 20 or 30 cm deep.

The design principle of a lagoon is similar to that of a drying bed: the difference only lies in the cycle duration, which is 1 year instead of 70 days and the non-discharge and non-percolation of overflows: all the water withdrawn is evaporated.

| | Oceanic climate | Continental climate | Mediterranean climate |
|---|---|---|---|
| **Sludge concentrations at the end of the thickening phase** | 1.5 times the quantity obtained on a static thickener after 1 month thickening | | |
| **Dryness obtained after 1 year** | 15% | 25% | 40% |

**Table 23.19.** *Dryness obtained in function of geographical area*

Data:

– initial liquid depth:1.5–2.5 m which corresponds to a good compromise between groundwork ease and embankment;

– sludge layer height: 0.7–1.4 m.

Sludge is usually destined to the landfill. A sludge with an initial dryness of 4% is treated over an average cycle of 1 year. To obtain a dryness of between 20 and 30%, a dewatering period of at least 3–12 months (depending on climatic conditions) is required, without the addition of new sludge.

The loading mass flow is between 35 and 40 kg $DM \cdot m^{-3} \cdot year^{-1}$, whereas the thickened sludge concentration is between 30 and 50 $g \cdot L^{-1}$.

It is necessary to monitor soil quality. It must be impermeable so as not to contaminate the groundwater. Favorable climatic conditions and large available surfaces are required. The process operates directly on settler purges. On the other hand, a buffering tank must be provided for the filter washwaters.

The simple lagoon treatment, without bottom drainage, involves 1-year cycles: filling plus thickening for 1 year, then drying for 1 year.

The lagoon must accept a capacity equal to two volumes of successive filter washes. It is important to respect a water layer height of approximately 30 cm in order to obtain a proper settling of the supernatant. Insofar as this condition is respected, the operation of this process is attractive due to its simplicity, its flexibility, its reduced monitoring during operation, its adaptation to all types of sludge, as well as its low power consumption.

If the lagoon is fed by sequential purges, it is important to stop the supernatant discharge during filling in order to avoid the appearance of turbulence, which degrades the quality of the supernatant. The removal of water is mainly done by withdrawal, and not by evaporation, unlike the drying bed. The main operational task is the annual cleaning of sludg in the lagoon. The amount of sludge extracted depends on effluent concentration.

This process is mainly used on simple treatment low-flow stations which do not include a coagulation stage or biological treatment stations.

Clarified overflow water can be discharged into the natural environment, provided that regulatory constraints have been followed.

## 23.8. Advantages and drawbacks of the different sludge dewatering treatments

| | **Advantages** | **Disadvantages** | **Applications** |
|---|---|---|---|
| **Lagoon treatment** | Thickening + storage + dewatering<br>Not labor intensive | Surface area, slowness<br>Climatic hazards<br>Sludge removal | All types of sludge<br>Climate with low humidity<br>Two lagoons in parallel on clay soil |
| **Drying beds** | Same as lagoon + solid sludge | Climatic hazards<br>Surface area | Climate with low humidity |
| **Filter bags** | Very archaic and simple | Handling<br>Very small capacities | Small size DWS (a few hundred $m^3$/day) |
| **Plate filters** | Very good dryness | Cost, labor intensive<br>Reagents | Land not available |
| **Belt filters** | Fairly good dryness | Not applicable to organic or sticky sludge | Too humid climate<br>Dryness requirement |
| **Centrifugation** | Good dryness | Cost, energy, polymers | – |

**Table 23.20.** *Advantages and drawbacks of the different sludge dewatering treatments*

## 23.9. References

Bustos, M.C., Concha, F., Bürger, R., Tory, E.M. (1999). Sedimentation and thickening: Phenomenological foundation and mathematical theory. In *Mathematical Modeling: Theory and Applications*, Lowen, R. (ed.). Kluwer Academic, Dordrecht.

Fitch, B. (1993). Thickening theories – An analysis. *AIChE J.*, 39(1), 27–36.

Guibelin, E. (2001). Note process : les boues d'eau potable. Report OTV.

Harbour, P.J., Anderson, N.J., Aziz, A.A., Dixon, D.D., Hillis, P., Scales, P.J., Stickland, A.D., Tillotson, M. (2004). Fundamental dewatering characteristics of potable water treatment sludges. *J. Water Suppl.*, 53(1), 29–36.

de Krestser, R.G., Usher, S.P., Scales, P.J., Boger, D.V., Landman, K.A. (2001). Rapid filtration measurement of dewatering design and optimization parameters. *AIChE J.*, 47(8), 1758–1769.

Landman, K.A., White, L.R., Buscall, R. (1988). The continuous-flow gravity thickener: Steady state behavior. *AIChE J.*, 34(2), 239–252.

Landman, K.A., White, L.R., Eberl, M. (1995). Pressure filtration of flocculated suspensions. *AIChE J.*, 41(7), 1687–1700.

Nicholson, J.E., Joseph, C., Goldbach, J.C. (1991). Belt press dewatering of alum sludge proves effective. *Water/Engineering & Management – Scranton Gillette*, 138(12), 28–31.

Nielsen, H.L., Carns, K.E., Deboice, J.N. (1973). Alum sludge thickening and disposal. *J. Am. Water Works Assoc.*, 65(6), 385–394.

Pan, J.R., Huang, C., Chuang, Y.-C., Wu, C.-C. (1999). Colloids and surfaces – Physicochemical and engineering aspects. *Dewatering Characteristics of Algæ-Containing Alum Sludge*, 150(1–3), 185–190.

Pan, J.R., Huang, C., Cherng, M., Li, K-C., Lin, C.-F. (2003). Correlation between dewatering index and dewatering performance of three mechanical dewatering devices. *Adv. Environ. Res.*, 7(3), 599–602.

Qian, F., Scales, P.J., Dixon, D.R., Newcombe, G., Ho, L. (2011). Dewatering of cyanobacteriarich sludge. *OzWater'11*, 1–9.

Schwoyer, W.L. and Luttinger, L.B. (1973). Dewatering of water-plant sludges. *J. Am. Water Works Assoc.*, 65(6), 399–403.

Stickland, A.D. (2005). Solid–liquid separation in the water and wastewater industries. PhD Thesis, University of Melbourne, Melbourne.

Talmage, W.P. and Fitch, E.B. (1955). Determining thickener unit areas. *Indust. Eng. Chem./Am. Chem. Soc.*, 47(1), 38–41.

Tiller, F.M. and Chen, W. (1988). Limiting operating conditions for continuous thickeners. *Chem. Eng. Sci.*, 43(7), 1695–1704.

Tiller, F.M. and Hsyung, N.B. (1993). Unifying the theory of thickening, filtration, and centrifugation. *Water Sci. Technol.*, 28(1), 1–9.

Thompson, P.L. and Paulson, W.L. (1998). Dewaterability of alum and ferric coagulation sludges. *J. Am. Water Works Assoc.*, 90(4), 164–170.

Verrelli, D.I. (2008). Drinking water treatment sludge production and dewaterability. PhD Thesis, University of Melbourne, Melbourne.

Zhao, Y.Q. (2004). Settling behaviour of polymer flocculated water-treatment sludge II: Effects of floc structure and floc packing. *Sep. Purif. Technol.*, 35(3), 175–183.

# 24

# The Treatment Chain: Conception and Design

The goal of any drinking water treatment is to remove all the organic and mineral compounds that may harm the consumer's health, restrict their comfort of use, or even degrade the production and distribution equipment, effects which indirectly affect the consumer.

In particular, energy savings should be evaluated, as well as the chemicals used and the sludge treatment.

The engineering of a drinking water treatment station represents an important choice in terms of the number and content of the treatment stages. This conceptualization must ensure the achievement of drinking water quality goals. Proper conception must meet a number of technical considerations, associated with economic investment costs.

Often, the simplest solution comes down to using the average of the data concerning water qualities. Surface waters, as well as groundwater, show significant variations throughout the year and the design based on an average parameter cannot work when these parameters reach their maximum value.

The conception of a treatment chain generally follows a succession of various stages, from the definition of the general and specific goals of water quality to the requirements imposed by the implementation of the most appropriate solution. This must be a compromise between the various parameters: quality goals, project specifications, customer requirements and economic costs.

Virtually all water sources require treatment before distribution to ensure that they do not pose a health risk to the consumer. Water's immediate health-related risks are primarily due to microbiological contamination. The absence of microbiological contamination is a key factor for human health. Health risks due to chemical contamination (with the exception of a few substances, such as cyanide or high-dose arsenic) show a longer-term trend. Substances affecting the appearance, smell or taste of water can make it disagreeable, leading to its rejection for esthetic reasons. Microorganisms associated with particles and water turbidity can also pose a health risk because turbidity makes disinfection more difficult.

Treatment processes are based on the removal of all particulate and dissolved contaminants that may harm the consumer's health.

The treatment chain comprises a certain number of stages arranged in a series, thus creating multiple barriers. This multi-barrier approach is an important notion because it provides the basis for effective water treatment, treating it to a suitable quality level for each of the processes scheduled downstream:

– The first stage is to identify the most relevant raw water parameters which need to be treated. Their removal constitutes the treatment's backbone.

– The second stage is to plan the treatment chain respecting the specifications, while integrating new (technical and economic) processes which will make all the difference compared to the competition.

– The third stage draws up a preliminary economic assessment of this configuration. Then, a potential alternative – the most reasonable one according to the new processes available on the market – is examined. The costs and scope of these potential solutions are estimated according to an initial design, which contemplates structure size, energy required, consumption of chemical reagents as well as the question of liquid and solid discharges.

The potential advantages and drawbacks of a particular solution are then listed with a view to making the decision to evaluate them in further detail or not.

A particular alternative solution is abandoned when its potential advantages do not suffice to make a clear difference against the basic solution.

Throughout all these stages, the qualitative approach combining expert systems with heuristic rules and the numerical approach integrating simulations and optimization methods are used. These two approaches are complementary and lead to a highly competitive solution.

Drinking water stations are intended to be operated for decades. For this reason, it is essential for the proposed treatments not to induce exorbitant operating costs, which have an impact on the cost per cubic meter of water produced.

Depending on the evolution of standards, a drinking water station must be upgraded and improvements must be made to the existing processes. Otherwise, additional treatment stages can be proposed (addition of new equipment, decommissioning of old equipment, readjustment of hydraulic profiles, etc.).

Innovative processes can be an important choice factor for the customer insofar as they can anticipate better, more reliable and constant water qualities. The example of Actiflo has demonstrated this on many occasions.

The expert system intended to support engineers in the conception of water treatment stations, and to help them improve their creativity and efficiency, is recognized as useful and often decisive. It is based on strong technical knowledge, on databases drawn from numerous operating stations and on the R&D results obtained by research centers (Veolia Environment Research and Innovation). These data have made it possible to develop modeling software adapted to various applications, serving as invaluable tools for decision support.

Database access by engineers allows them to:

– quickly assess the compliance of their alternative proposals with the basic solutions;

– provide strong arguments for the most optimized solution;

– access specific documents and articles.

Because of this, technical alternatives can be quickly generated, respecting the evaluation criteria for each case and the appropriate parameters, finally selecting the best alternative that meets technical and economic performances.

When conceptualizing and designing a drinking water plant, the design engineer interacts with engineers from other disciplines (electrical, hydraulic, civil engineering, etc.) in order to summon their opinion on his proposals. Often, the winning solution is the most economical one, while meeting customer specifications in a safe, reliable and robust manner.

## 24.1. The treatment chain

The study of a treatment chain involves defining the sequence of the various treatment stages best suited to the physicochemical and microbiological quality of raw water, and acknowledging its variations.

For this, manufacturers use design and performance prediction software. This software is often developed from laboratory tests, industrial pilot results or obtained by accumulation over many years of study on various sites, as well as from operating feedback from drinking water plants.

When possible, the calibration of IT tools can be based on laboratory tests intended to qualify and quantify the most appropriate reagents for a given site.

Finally, for complex cases and in order to minimize uncertainties, pilot tests can be implemented. Thus, the type of treatment chain chosen results from the combination of technical (best adapted technology), economic (operating investment costs) and environmental considerations.

The conception of a treatment chain necessarily requires the prior availability of a set of analytical data regarding the resource's physicochemical and microbiological quality. A large number of contract documents overlook this aspect, which obliges the design engineer to make assumptions which may prove to be risky.

For example, the specifications recommend compliance with drinking water standards, without indicating all the data relating to the quality of raw water. In addition, the information provided often concerns spot samples that do not reflect the variations in the quality observed throughout the year, or even during a single season.

## 24.2. The definition of a treatment chain

Potable water comes from different sources as follows:

– groundwater from deep aquifers;

– surface water (streams, ponds, artificial reservoirs, springs, canals).

Whether of surface or underground origin, water has variable physical, chemical and microbiological characteristics, which require the choice of treatment chain to be always examined on a case-by-case basis. For example, in the west and center of France, reservoir waters (which are often not very mineralized) contain high concentrations of organic matter, iron, manganese and pesticides. These waters also present variable concentrations of ammonium ions, depending on the season. This example shows that, for this type of water, a special study must be carried out before defining the most optimized treatment chain.

A treatment chain implements specific processes for the removal of particulate pollution (mineral and organic matter in suspension, in particular micro-organisms) and for the removal of dissolved substances by means of physical, chemical, physicochemical or biological mechanisms.

Deep water from protected aquifers generally has a physicochemical quality often requiring simple treatments. Sometimes a simple disinfection treatment is necessary. The engineer recommends setting up a disinfection unit in all cases as a simple precautionary measure, even if the raw water is drinkable. It can be done with chlorine or chlorine dioxide. Disinfection by ultraviolet radiation is possible, but should be followed by the addition of a residual oxidant to maintain water quality in the distribution network.

In the presence of elements such as iron, manganese, arsenic, a specific treatment is implemented, involving either physicochemical processes (coagulation–flocculation–filtration, filtration on various adsorbent materials or inducing catalytic reactions, membranes), or biological processes (iron, manganese removal).

In the case of the presence of nitrates and/or pesticides, the removal of nitrates can be done by the biological pathway (biological filter) or by the physicochemical pathway (ion exchange). The removal of pesticides is obtained by adsorption onto activated carbon. Other treatments are proposed such as the implementation of membrane techniques (nano-filtration, reverse osmosis) which, depending on the type of membrane, enable the simultaneous removal of nitrates and pesticides.

In the case of karstic waters, which are characterized by strong variations in turbidity, processes including microfiltration or ultrafiltration membranes are the most effective.

Conventional coagulation–flocculation–settling–filtration treatments can always be applied, but respecting very strict operating conditions, due to the sudden variations in turbidity.

Surface water is characterized both by variations in the quality of raw water and by variations in the water flow to be treated. It usually represents very large volumes to be treated.

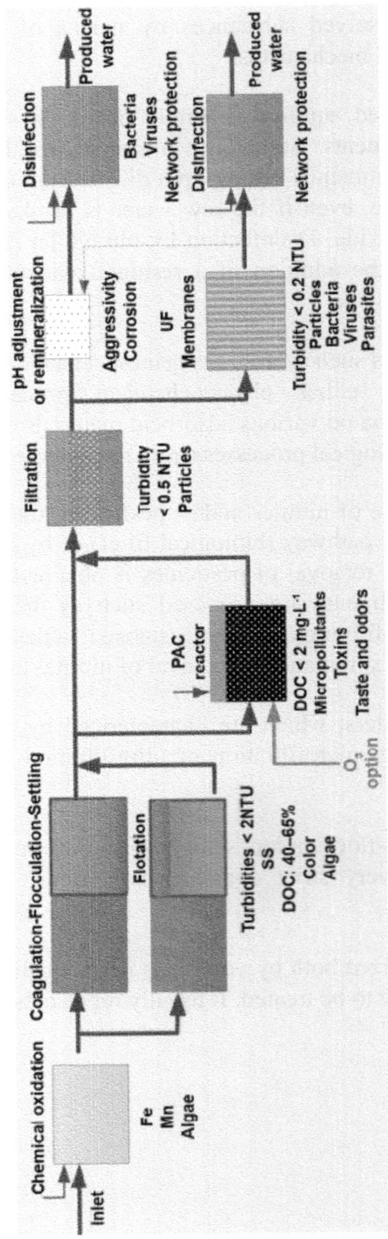


**Figure 24.1.** *Example of a surface water treatment chain. For a color version of this figure, see www.iste.co.uk/gaid/watertreatment5.zip*

Surface water is composed of a variety of families of organic and inorganic substances, among which the most salient are:

– chemical substances whose presence in drinking water is critical to health;

– chemicals which do not affect health at the levels normally found in drinking water;

– substances and parameters in drinking water which may give rise to consumer complaints.

Microbiological parameters are also relevant.

The conception of a treatment chain must not only take into account all these parameters to obtain treated water complying with the standards in force, but also acknowledge variations in the quality of raw water.

## 24.3. The stages of a treatment chain

Table 24.1 summarizes the different possible treatments for each water contaminant. Depending on the case, contaminants can be removed through the physical, chemical, physicochemical or biological pathway, or using low-pressure (LP or UF) or high-pressure (NF, LPRO or RO) membranes.

| Parameters to be reduced or removed | Technologies |
|---|---|
| Turbidity | Direct filtration (fast or TGV filter): monolayer or dual media coagulation-filtration<br>Clarification: coagulation-flocculation-settling (corridor, Cyclofloc, Actiflo, Multiflo, Clariflocculator) or flotation (Spidflow®) filtration<br>Low-pressure membranes: microfiltration or ultrafiltration, with and without upstream coagulation |
| Bacteria<br>Parasites<br>Viruses | Disinfection: chlorination (gaseous chlorine and Na and Ca hypochlorite), chlorine dioxide, ozone, low and medium-pressure ultraviolet rays, chloramines (depending on country authorization) |
| Algae | Chemical pathway: coagulation–flocculation--flotation (Spidflow®), oxidation–coagulation–flocculation–settling (Actiflo, Multiflo), ozonation–flotation (Ozoflot, Flottazone)<br>Low-pressure membrane pathway (if limited concentration): microfiltration (possible coagulation upstream), ultrafiltration (possible coagulation upstream) |

| Parameters to be reduced or removed | Technologies |
|---|---|
| Natural organic matter (humic substances)<br>Color | The chemical pathway: coagulation–flocculation–dual media filtration (if low inlet concentrations), coagulation-advanced flocculation–settling or flotation–filtration<br>The physicochemical pathway: powdered activated carbon reactors (Actiflo Carb, Multiflo Carb, Opacarb MF), micrograin activated carbon reactors (Filtraflo® Carb, Opacarb FL, Oparcarb MG), granular activated carbon filtration with one or two filters in series<br>Biological pathway: ozonation and GAC filtration<br>The high-pressure membrane pathway: "open" and "tight" nanofiltration, low-pressure reverse osmosis, reverse osmosis |
| Micropollutants: pesticides, pharmaceutical residues, endocrine-disrupting compounds, industrial products | The physicochemical pathway: powdered activated carbon reactors (Actiflo Carb, Multiflo Carb, Opacarb MF), micrograin activated carbon reactors (Filtraflo® Carb, Opacarb FL, Oparcarb MG), granular activated carbon filtration with one or two filters in series<br>The high-pressure membrane pathway: "open" and "tight" nanofiltration, low-pressure reverse osmosis, reverse osmosis |
| Perfluorinated substances<br>Perchlorates | The physicochemical pathway: powdered activated carbon reactors (Actiflo Carb, Multiflo Carb, Opacarb MF), micrograin-activated carbon reactors (Filtraflo® Carb, Opacarb FL, Oparcarb MG), granular-activated carbon filtration with one or two filters in series, ion exchange resins<br>The high-pressure membrane pathway: "open" and "tight" nanofiltration, low-pressure reverse osmosis, reverse osmosis |
| Softening | Chemical pathway: softening with lime and soda: (Actiflo Soft, Multiflo Soft)<br>Catalytic softening with lime or soda : (Actina, Saphira), ion exchange resins |
| Nitrates | The biological pathway: denitrification: Biodenit<br>The physico-chemical pathway: denitrification on ion exchange resins: Ecodenit<br>The high-pressure membrane pathway: "tight" nanofiltration, low-pressure reverse osmosis, reverse osmosis |
| Taste and odors<br>Algal toxins | The physico-chemical pathway: powdered activated carbon reactors (Actiflo Carb, Multiflo Carb, Opacarb MF), micrograin activated carbon reactors (Filtraflo® Carb, Opacarb FL, Oparcarb MG), granular activated carbon filtration with one or two filters in series, ozonation<br>The high-pressure membrane pathway: "open" and "tight" nanofiltration, low-pressure reverse osmosis, reverse osmosis |
| Iron<br>Manganese | The biological pathway: biological iron removal, biological manganese removal<br>The physical pathway: Fe: aeration (cascade-air bubble induced oxidation in a tank-Aerator) filtration<br>The physicochemical pathway: Fe: chemical oxidation ($O_3$, chlorine, potassium permanganate, chlorine dioxide), Mn: chemical oxidation ($O_3$, potassium permanganate, chlorine dioxide)<br>The catalytic pathway: catalytic iron removal on $MnO_2$, catalytic manganese removal on $MnO_2$ |

| Parameters to be reduced or removed | Technologies |
|---|---|
| Carbon dioxide<br>Methane<br>Hydrogen sulfide | The physical pathway: aeration (cascade, air bubble induced oxidation in a tank), stripping |
| Heavy metals: arsenic, antimony, selenium, nickel, chromium, zinc | Oxidation–coagulation–settling or direct filtration<br>Ferric oxyhydroxide filtration<br>Manganese dioxide filtration<br>Chemical precipitation at high pH values<br>Activated alumina<br>Oxidation–coagulation–microfiltration or ultrafiltration membranes<br>Nanofiltration<br>Low-pressure reverse osmosis<br>Reverse osmosis |
| Control of disinfection by-products: (THMs, halogenated acetic acids, etc.) | The physicochemical pathway: coagulation-advanced flocculation, powdered activated carbon reactors (Actiflo Carb, Multiflo Carb, Opacarb MF), micrograin-activated carbon reactors (Filtraflo® Carb, Opacarb FL, Oparcarb MG), granular-activated carbon filtration with one or two filters in series<br>Biological pathway: ozonation and GAC filtration<br>The high-pressure membrane pathway: "open" and "tight" nanofiltration, low-pressure reverse osmosis, reverse osmosis |
| Aggressivity and corrosion control | pH control and adjustment (lime, lime water, soda)<br>Remineralization ($CO_2$-lime or limestone filters)<br>Chloride and sulfate control |

**Table 24.1.** *Methods for reducing or removing various contaminants present in water*

## 24.4. The renovation of water treatment plants

For all local authorities, water-related issues are decisive because it is necessary to simultaneously respond not only to a deterioration in the quality of raw water, but also to consumption increase, changes in the regulations (national or European), as well as the concern to distribute better quality water.

To this end, it is crucial to redevelop, modify or supplement the existing stations, in view of adapting them to the new goals in mind. Very often, the age of the treatment plant does not allow for a flow increase, or for the compliance with new standards, due to different requirements when it was built back in the day. In this sense, it is difficult for an operator to guarantee a water production compliant with the new flow requirements, or to ensure a better quality of treated water.

It is the responsibility of the authorities to reconsider the case of such station in order to back up the modifications suggested. The manufacturer puts in practice the actions that will help meet the new requirements, following a threefold approach: the diagnosis of the existing station, the proposal of modifications to be implemented and the guarantee of the quality of treated water.

In France, and around the world, OTV-Veolia has repeatedly contributed to the modernization of existing processes by proposing more comprehensive treatments, without stopping water production during the process.

### 24.4.1. *Adaptation of new goals*

The newly assigned goals may include:

– a flow increase without modifying the initial quality of treated water;

– an overall improvement in the quality of treated water;

– a specific improvement of certain already existing parameters;

– better performances in relation to a deterioration in the quality of raw water;

– adaptation to new standards.

The first stage of an intervention includes a complete diagnosis of the existing facility. This diagnosis provides a general assessment of:

– the station;

– the electrical and mechanical equipment (pumps, valves, etc.);

– the structures (concrete, steel, etc.);

– automation and instrumentation;

– the reagent and cleaning stations (filter washing);

– the power stations available.

Table 24.2 summarizes some causes of structural malfunction.

| Process | Malfunction causes |
|---|---|
| Chemical reagents | Choice of reagent<br>Choice of dose and pH<br>Power line maintenance<br>Dosing pumps-diverse equipment<br>Reagent dilutions |
| Coagulation | Types of agitation<br>Number of mixers<br>Stirring speeds<br>Reaction time<br>Coagulant dosages |
| Flocculation | Reaction time<br>Stirring speeds<br>Number of reaction tanks<br>Flocculant dosages |
| Settling | Flow velocities<br>Sludge accumulation quantities and velocities<br>Sludge extraction<br>Loads applied |
| Filtration | Filtration velocities<br>Cycle durations<br>Hydraulic parameters<br>Washing conditions<br>Washing equipment |

**Table 24.2.** *Causes of structural malfunctions*

### *24.4.2. The choice of treatment technologies*

Once the station's diagnosis has been established and the treatment targets assigned, it is possible to choose the most suitable technologies to achieve such goals. It is obvious that the same technology can result in different removal performances, depending on its position in the chain, on whether it is used in pre-treatment or post-treatment. While this performance may be optimal in relation to a substance, it may nonetheless induce other undesired phenomena. For example, chlorine disinfection is effective at low pH values, although this same pH can lead to corrosion phenomena.

In the case of renovation, it is imperative to consider all the implications and consequences of the introduction of a new reagent or a new structure.

Among the most common treatment stages improving water quality are:

– *Chemical pre-oxidation*, using an oxidant such as chlorine, chlorine dioxide or ozone. This stage is necessary for the removal of algae, tastes and odors, iron and manganese oxide. Depending on the organic matter content present in water, dioxide or ozone may be preferred over chlorine in order to prevent the formation of THMs. Special attention must be paid to the formation of bromates.

– One of the decisive operating parameters during the *coagulation–flocculation* stage is the judicious choice of reagents. Depending on the local context and reagent availability (ferric chloride, aluminum sulfate, polymer) and the quality of the raw water, the use of reagents is adapted to obtain optimal performances. For example, with aluminum sulfate, acidification is often necessary in order to coagulate within the optimum pH zone. This is done by adding acid or by overdosing aluminum sulfate when acid is not available. This always implies that at the end of the treatment, water must return to its calco-carbonic equilibrium. The reagent injection order and their location is also a determining point for reaching satisfying performances at this stage.

Apart from the coagulation–flocculation chemistry, it is important to properly monitor the mixing and homogenization conditions in the reaction tanks. Mechanical agitation is the ideal method, but baffle mixing, if these already exist, can be a good palliative.

Regarding *settling*, renovation can also be decided on the basis of the existing facility. Various options are available:

– a corridor (horizontal) settler if the velocity suffices, only modifying the sludge extractor, for example;

– modify a corridor (horizontal) settler by creating a lamellar zone in order to increase the flow velocity;

– build an additional structure if no solution is possible due to an expected excessive flow.

The improvement of *sand filtration* also depends on the existing structure. Regarding the material, there are many possibilities such as increasing sand height whenever possible, modifying the sand's grain size or setting up a dual media filter in order to increase cycle durations.

One may optionally inject a polymer upstream to increase flocculation within the filters.

But it is also important to intervene on filter washing conditions, hydraulic distribution, type of filtering floor, etc. Hydraulic conditions are an essential operating parameter for sand filters and the change from one material to another does not suffice to increase this filter's performances. This stage will play an essential role in guaranteeing an efficient treatment, since it is responsible for retaining the residual suspended matter (which affects turbidity), as well as a good part of the microbiological contaminants.

Currently, there are particle counters on filtered water which serve as an alarm in relation to turbidity and microbiological contaminants. Particles hide bacteria which can be improperly removed, disinfected or inactivated during ozonation or chlorination at the final disinfection stage.

Moreover, the existing relation between the number of particles (>5 μm) and turbidity at the filtration outlet makes it possible to intervene on the washing frequencies and to ensure permanent quality water.

The presence of micropollutants such as pesticides or a high concentration of organic matter in raw water requires an additional treatment including *zonation* and *activated carbon*.

The purpose of ozonation is to remove residual tastes and odors to trigger pre-disinfection, to oxidize pesticides and organic matter by making it more biologically assimilable by bacteria which thrive on activated carbon.

Intermediate ozonation is significant, and a residual of 0.4 $g^{-3}$ must be maintained for 4 min to kill all viruses.

*Activated carbon* acts as an adsorbent of pesticides and other by-products or can constitute a biological support for the bacteria in charge of removing a good part of the total organic carbon.

Often, it is the combined action of these two functions that is sought. Depending on the availability of one of these two stages, the treatment chain is adapted accordingly. Otherwise, a new stage is incorporated in the frame of the treatment renovation.

Finally, *disinfection* by chlorine, chlorine dioxide or ozone is carried out taking into account the existing situation, the new pre-oxidation stage that may need to be implemented, the network's condition, the type of microbial germs to be removed, etc.

In general, efforts are made to maintain the disinfection pH at around 7.5, and always below 8. The regularity of this parameter is essential to prevent any disinfection defects, or complaints from consumers, who may perceive certain regulatory adjustments.

These first intervention elements on the treatment chain make it possible to apprehend the complexity of treatment processes. Permanent interactions exist and should seriously be taken into account in order to avoid reaching the opposite of what was initially intended.

| Parameters | Processes | | | | | | | | | | | | | | | |
|---|---|---|---|---|---|---|---|---|---|---|---|---|---|---|---|---|
| | aeration | Coag-floc Settling or flotation | Filtration Direct | Filtration $MnO_2$ | precipitation | Chemical oxidation | Activated carbon | Activated alumina | Resins | Biological treatment | Ultrafiltration | Microfiltration | Nanofiltration | LP and HP Reverse Osmosis | Desalination | Disinfection |
| Microbiology | | | | | | | | | | | | | | | | |
| Coliforms | | | | | | | | | | | XXX | XXX | | | | XXX |
| Viruses | | | | | | | | | | | XXX | | | | | XXX |
| Parasites | | | | | | | | | | | XXX | XXX | | | | XXX |
| Particulate pollution | | | | | | | | | | | | | | | | |
| Turbidity | | XX | X | | | | | | | | XXX | XXX | | | | |
| Suspended solids | | XX | X | | | | | | | | XXX | XXX | | | | |
| Algae | | XXX | X | | | | | | | | XX | XX | | | | |

| Parameters | Processes | | | | | | | | | | | | | | | |
|---|---|---|---|---|---|---|---|---|---|---|---|---|---|---|---|---|
| | aeration | Coag-floc Settling or flotation | Filtration Direct | Filtration $MnO_2$ | precipitation | Chemical oxidation | Activated carbon | Activated alumina | Resins | Biological treatment | Ultrafiltration | Microfiltration | Nanofiltration | LP and HP Reverse Osmosis | Desalination | Disinfection |
| **Organic substances** | | | | | | | | | | | | | | | | |
| Humic substances | | xx | | | | | xx | | xx | x | | | xxx | xxx | xxx | |
| Color | | xxx | | | | x | xx | | | | | | xxx | xxx | xxx | |
| Dissolved organic carbon (DOC) | | xx | | | | | xx | | xx | x | | | xxx | xxx | xxx | |
| MIB and geosmin | | | | | | | xx | | | | | | xxx | xxx | xxx | |
| Toxins | | | | | | x | xx | | | | | | xxx | xxx | xxx | |
| Tastes and odors | | | | | | xx | xx | | | | | | xxx | xxx | xxx | |
| Primary chlorination by-products | x | | | | | | xx | | | | | | xxx | xxx | xxx | |

| Parameters | Processes | | | | | | | | | | | | | | | |
|---|---|---|---|---|---|---|---|---|---|---|---|---|---|---|---|---|
| | aeration | Coag-floc Settling or flotation | Filtration Direct | Filtration $MnO_2$ | precipitation | Chemical oxidation | Activated carbon | Activated alumina | Resins | Biological treatment | Ultrafiltration | Microfiltration | Nanofiltration | LP and HP Reverse Osmosis | Desalination | Disinfection |
| **Metals** | | | | | | | | | | | | | | | | |
| Iron (2+) | xx | O | O | xxx | xxx | xxx | | xx | | | O | O | | | | |
| Manganese (2+) | | | | xxx | xxx | xxx | | xx | | | O | O | | | | |
| Arsenic (3+) | | O | O | xx | o | xxx | | x | x | | | | | xxx | | |
| Arsenic (5+) | | xxx | xxx | xx | xx | | | xxx | | | | | | xxx | xxx | |
| Selenium (4+) | | xx | | | | | | xx | xx | | | | | xxx | | |
| Selenium (6+) | | xxx | xxx | | | | | xx | xx | | | | | xxx | | |
| Nickel (2+) | | | | x | xx | | | | x | | | | xxx | xxx | xxx | |
| Fluorides | | | | | xxx | | | xx | xx | | | | xxx | xxx | xxx | |
| Chrome (3+) | | xx | | | xx | | | | xx | | | | xxx | xxx | | |
| Chrome (6+) | | | | | | | | | xx | | | | xxx | xxx | | |

| Parameters | Processes | | | | | | | | | | | | | | | |
|---|---|---|---|---|---|---|---|---|---|---|---|---|---|---|---|---|
| | aeration | Coag-floc Settling or flotation | Filtration Direct | Filtration $MnO_2$ | precipitation | Chemical oxidation | Activated carbon | Activated alumina | Resins | Biological treatment | Ultrafiltration | Microfiltration | Nanofiltration | LP and HP Reverse Osmosis | Desalination | Disinfection |
| **Miscellaneous compounds** | | | | | | | | | | | | | | | | |
| Ammonia | | | | | | XX | | | | XXX | | | | XX | XX | |
| Nitrites | | | | | | XX | | | | XXX | | | XXX | XXX | XXX | |
| Nitrates | | | | | | | | | | XXX | | | XXX | XXX | XXX | |
| Boron | | | | | | | | | XX | | | | X | XXX | XXX | |
| Silica | | | | | XXX | | | | | | | | | | | |
| **Micropollutants** | | | | | | | | | | | | | | | | |
| Pesticides | | | | | | X | XXX | | | | | | XXX | XXX | XXX | |
| Drug residues | | | | | | X | XXX | | | | | | XXX | XXX | XXX | |
| Endocrine-disrupting compounds | | | | | | X | XXX | | | | | | XXX | XXX | XXX | |
| Industrial waste | | | | | | | XXX | | | | | | XXX | XXX | XXX | |

**Table 24.3.** *Adequate processes for the removal of various compounds*

COMMENTS ON TABLE 24.3.–

– O: associated with chemical oxidation;

– aeration: Aeralator, Degaz;

– coag-floc-settling: Actiflo, Multiflo, Cyclofoc, Corridor settler, Clariflocculator;

– coag-floc-flotation: Spidflow®, Spidflow® Filter, Ozoflo, Flottazone, conventional flotation;

– filtration: Filtraflo® family (monolayer, dual media, three-layer);

– $MnO_2$ filtration: Mangaflo, Mangaflo+;

– biological treatment: Biodenit, Selenium+, Filtraflo®;

– precipitation (lime, soda): Actina, Actiflo Soft, Multiflo Soft, Saphira;

– chemical oxidation: chlorine, hypochlorite, chlorine dioxide, potassium permanganate, ozone, Metclean (for metals);

– activated carbon: Actiflo Carb, Multiflo Carb, Filtraflo® Carb, Opacarb, Opacarb FL, Opacarb MG, Opacarb MF, Filtraflo® GAC (mono, dual media and three-layer), Opaline C, Opaline S, Opaline B, occasional addition of PAC in Actiflo, Multiflo and Cyclofoc;

– resins: Ecodénit, Opalix, Opadem;

– ultrafiltration: Opaline C, approved market membranes;

– microfiltration: approved market membranes;

– nanofiltration: approved market membranes;

– low-pressure and high-pressure reverse osmosis: approved market membranes;

– desalination: approved market membranes, Barrel, Hyprode;

– disinfection: chlorine, hypochlorite, chlorine dioxide, ozone, chloramines, ultraviolet rays.

## 24.5. References

Crittenden, J.C., Trussell, R.R., Hand, D.W., Howe, D.W., Tchobanoglous, G.G. (2005). *Water Treatment Design: Principles and Design*, 2nd edition. John Wiley and Sons, New York.

Dinka, M.O. (2018). Safe drinking water: Concepts, benefits, principles and standards, water challenges of an urbanizing world. *IntechOpen*. doi:10.5772/intechopen.71352.

Domenico, P.A. and Schwartz, F.W. (1990). *Physical and Chemical Hydrology*. John Wiley and Sons, New York.

Edzwald, J.K. (2011). *Water Quality & Treatment, American Water Works Association*, 6th edition. McGraw Hill, New York.

Gerba, C.P. (2009). Drinking water treatment, In *Environmental Microbiology*, Maier, R.M., Pepper, I.L., Gerba, C.P. (eds). Elsevier, Amsterdam.

Hannoun, I.A., Boulos, P.E., List, E.J. (1998). Using hydrauling modeling to optimize contact time. *J. Am. Water Works Assoc.*, 90, 77–87.

Legube, B. (2015). *Production d'eau potable*. Dunod, Paris.

Lisle, J. (2000). Drinking water treatment and regulations in the United States. *Lab. Med.*, 31(9), 492–496.

Masschelein, W.J. (1992). *Processus unitaires du traitement de l'eau potable*. CEBEDOC, Liège.

Mechouk, C. and Pichard, T. (eds) (2021). Réglementation et traitement des eaux destinées à la consommation humaine. Commission eau potable de l'ASTEE.

Novotny, V. (2003). *Water Quality: Diffuse Pollution and Water Shed Management*, 2nd edition. John Wiley and Sons, New York.

Okun, D.A. (1977). Best available source. *J. Am. Water Works Assoc.*, 83(3), 30–32.

Pyke, G.W., Becker, W.C., Head, R., O'Melia, C.R. (2003). *Impacts of Major Point and Non-Point Sources on Raw Water Treatability*. Water Research Foundation, Denver.

Schmoll, O., Howard, G., Chilton, J., Chorus, I. (2006). *Protecting Groundwater for Health: Managing the Quality of Drinking Water Sources*. World Health Organization/IWA Publishing, London.

Sharma, S. and Bhattacharya, A. (2017). Drinking water contamination and treatment techniques. *Appl. Water Sci.*, 7, 1043–1067.

Todd, D.K. and Mays, L.W. (2005). *Groundwater Hydrology*, 3rd edition. John Wiley and Sons, New York.

Wu, J., Cao, M., Tong, D. (2021). A critical review of point-of-use drinking water treatment in the United States. *Clean Water*, 4(40), 1–25.

# 25

# The Future of Water

## 25.1. The major elements of the future of water

Alternative resources such as treated wastewater are a good example of how it is possible to preserve and recycle the planet's limited resources. The water, energy, nutrients and metals that are present in wastewater are all increasingly viable parts of a cyclical economy. The technology for converting wastewater into drinking water is rapidly evolving and in the right direction. The safety and quality of this water can be assured. All that is needed is to change the people's mentalities to guarantee its acceptance because products must be judged on their quality not on their origin. The future will be inventive and add value by transforming recovered resources into valued commodities for consumers.

The concept of "One Water" is gaining increasing importance around the world. This trend assesses water as a whole, and not as individual compartments separating drinking water, wastewater or gray water. The relevance of water for health adds a new dimension to this concept, which could be expressed as "One Water, One Heath".

This includes the detection and risk management of known and emerging contaminants in water resources. Future technical and scientific developments will focus on having more frequent information updates on contaminants and pathogens, and on developing specific protocols particularly adapted to new techniques, facilitating their automation.

A multi-barrier approach could be systematically implemented by region, department and country. The purpose of this approach is to reduce the risk of drinking water contamination, while increasing the practicality and efficiency of corrective controls and preventive measures.

This approach focuses on three major elements:

– the enhanced protection of water resources;

– the treatment of drinking water;

– the drinking water distribution network.

The approach is presented as an integrated system involving various procedures, processes and tools which—when put together—prevent or reduce drinking water contamination from source to tap to reduce risks to public health. Brought together, they provide the assurance that drinking water will be safe to drink. The approach also ensures the long-term viability of distribution networks.

Such approach will be viable with the participation of all stakeholders (public institutions and private sector, associations). By means of associations, the public will not only become sensitized to issues concerning drinking water quality – gaining understanding on the quality of resources and the available measures to mitigate pollution, health risks, the results of monitoring and control activities – but will also engage in public consultations involving decision-making processes.

Figure 25.1 shows the evolution of water treatment since the 19th century. At the beginning of the 20th century (1900s), only turbidity and color reduction, followed by disinfection, were implemented in drinking water stations. Throughout this century, the development of analytical tools and the risks averted in relation to many organic substances led the authorities to restrict standards, and designers to modify treatment chains. Many processes were developed to reduce the turbidity of the water produced, remove organic matter and pesticides almost entirely and finally proceed to a strengthening of disinfection by increased performances against bacteria, viruses and parasites.

The end of the 20th century and the beginning of the 21st century saw the emergence of differentiating technologies with the use of low-pressure membranes (microfiltration and ultra-filtration) and (high-pressure (nanofiltration)) membranes applied to conventional surface water. This was revolutionary in itself, notwithstanding the fact that membranes like reverse osmosis were already in use for seawater desalination (although not applied to conventional water).

This is because, on the one hand, the normative requirements were rarely met by conventional processes for contaminated formed water, and, on the other hand, because the public increasingly demanded access to quality water, which motivated designers to deploy those new technologies. Veolia was the first company to install nanofiltration membranes on mine water in 1995 and on surface water in 1998.

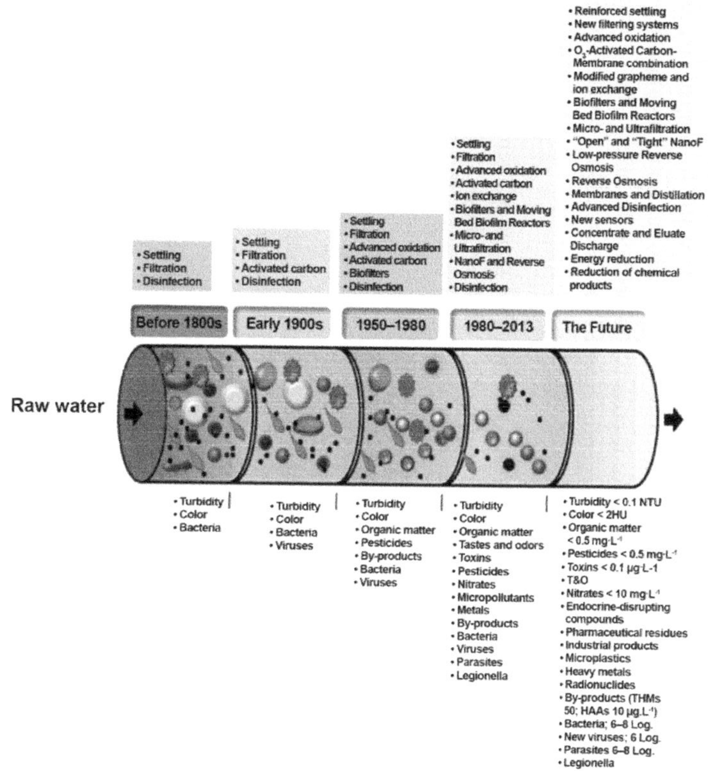


**Figure 25.1.** *The future of water. For a color version of this figure, see www.iste.co.uk/gaid/watertreatment5.zip*

Because of all these advances, designers today have a range of technologies at their disposal to ensure drinking water quality. The technological processes are adapted to the content of raw water and to remove or reduce organic substances, nitrates, heavy metals, bacteriology, etc.

Pharmaceutical residues, endocrine-disrupting compounds and industrial products are added to pesticides, while the range of hazardous substances analyzed in recent years in wastewater continues to expand. This is done in consonance with our evolving consumption habits. The low concentrations of these micropollutants do not make them any less worrying. Chemical analyses produce a partial view of contamination: only the substances sought are measured. Even if exhaustive, they would remain insufficient to establish a causal link with the impact on living organisms because the known toxicity data for each substance are fragmentary, and above all, because they do not make it possible to predict the effect of a cocktail of such substances. Bio-assays are urgently needed to observe the reaction of living organisms to one or to a mixture of these substances.

Newly appearing viruses raise fears as to whether some of them will be found in high concentrations in water, and especially, whether they could be resistant to current disinfection techniques.

## 25.2. Will there be enough water?

Climate change inducing an increase in temperatures modifies meteorological conditions, and therefore evaporation and condensation phenomena. Freshwater supplies can decrease and accentuate water stress in the regions already affected.

Over the past few decades, advances in the treatment of conventional waters such as rivers, lakes and groundwater have made it possible to maximize their use. However, in some rapidly growing coastal regions, the demand for the "creation" of new water resources from seawater or brackish water, as well as the reuse of wastewater, will become increasingly widespread. Desalination technologies benefit from research and innovation investments, making them even more profitable.

What does this mean for anticipating contamination in the future?

The technology will detect contaminants at very low concentrations. But our understanding of the health consequences of such trace contaminants may still be slow to advance.

Waterborne diseases due to bacteriological contamination are still very common today. And bacterial counting methods are always lengthy. Meanwhile, as water continues to be distributed, plant operators are hindered from taking adequate measures soon enough. In that sense, it is the consumers who will trigger the alarm signal and also social networks, which echo the concerns of populations. For all these reasons, it is important for online analytical methods to be developed as a fast and accurate response so that control is enforced in almost real time, making it possible to intervene early enough to avoid large-scale contaminations.

Microbiological sensor technology must evolve rapidly, which is why the real-time microbiological monitoring of water is becoming increasingly necessary, especially in the case of bacteria. For their part, viruses and parasites pose specific challenges to the implementation of sensors due, for example, to the small size of viruses and the weak signals given by parasites before contamination. These types of sensors are part of a more rigorous approach to water quality management.

The future of water is also set in a context demanding the modification of operating conditions, with treatment processes requiring less energy and chemicals, while reaching the desired quality of water.

The aging of water distribution systems in industrialized countries requires adaptation to the new context involving a modification of daily management, with a more rigorous monitoring of performance.

The developments of online controls of the microbiological quality of water are added to those which are necessary in relation to the vulnerability of resources by integrating them into safety plans.

Meeting consumer expectations in terms of drinking water means developing and having a reliable multi-barrier approach, in which monitoring plays a fundamental role. It also enables operators and distributors to ensure permanent security and to respond immediately to accidental and/or intentional incidents.

In the future, water technologies will have to successfully address security challenges, event detection and the control and improvement of operations, meet increasingly stringent regulatory aspects and develop more efficient analytical tools for identifying new molecules, something which will undoubtedly require cutting-edge innovations.

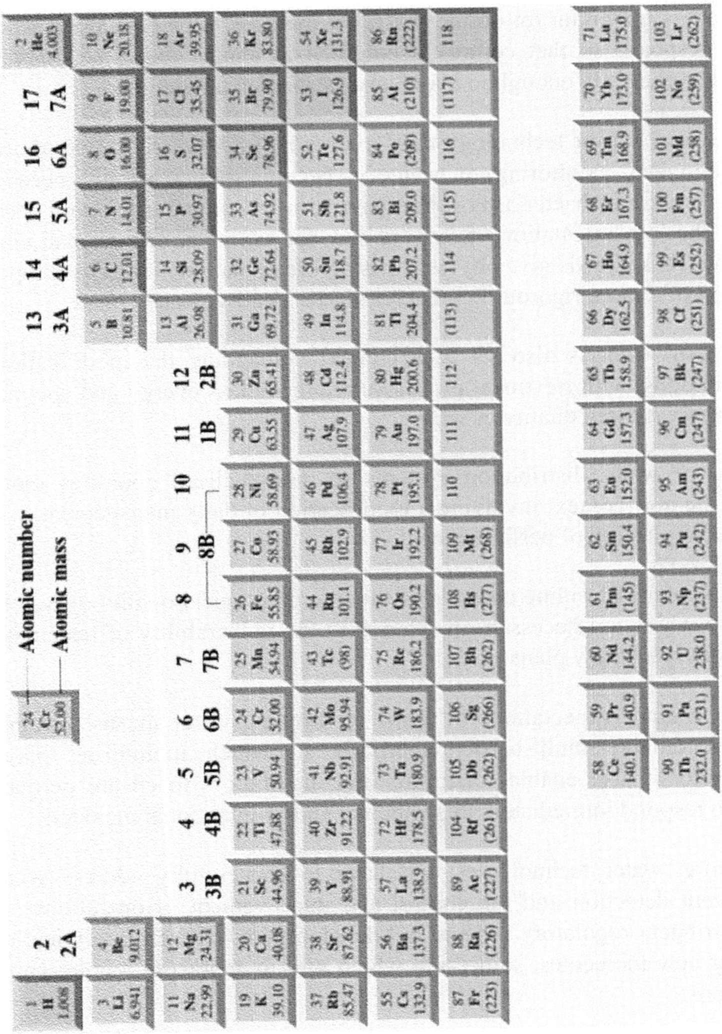


**Figure 25.2.** *Periodic table of elements*

# Index

## A, B, C

aeration, 40, 42–44, 240
algae, 90, 102, 107, 108, 128, 160, 169, 223, 272, 313, 317
ammonia, 135–137, 161, 164, 167, 248
bacteria, 99–101, 104–106, 108, 112, 146, 147, 155, 156, 159, 164, 181, 189, 204, 208, 210, 252, 319, 328, 331
bleach, 121, 122, 133, 138, 139, 143
break-point, 136, 137
cascade, 43, 314, 315
centrifugation, 291, 293, 294
chlorine, 65, 109, 111, 113–116, 118–124, 127–129, 133, 135–141, 143–157, 159–167, 169, 183–187, 189, 198, 201, 208–211, 219, 222–225, 227, 228, 230, 233, 235–239, 242, 248–252, 311, 313, 314, 317, 319
*Cryptosporidium*, 103, 106, 107, 147, 155, 165, 168, 182, 199, 204–206, 208–210

## D, E, F, G

denitrification, 270, 281, 296
disinfection, 99, 121, 150, 168, 187, 313
drying beds, 296, 299, 300
endocrine disruptors, 330
filter
  bags, 295
  band, 294
  plate, 288, 289, 304
Filtraflo®, 77
flocculation, 63, 160, 280, 283, 285, 311, 313–315, 318
*Giardia*, 103, 106, 107, 115, 130, 146, 147, 155, 156, 165, 168, 182, 204, 205, 209, 210

## H, I, L, M

hardness, 14, 29, 38, 51, 61, 63, 67–70, 73, 76, 264
iron, 32, 33, 52, 200–202, 223, 264, 266, 269, 272, 310, 311, 317
lagoon treatment, 302, 304

lime, 27, 29, 39, 46–48, 51–54, 56–65, 85, 86, 88, 89, 91, 125, 149, 261, 262, 265, 280, 288, 289, 291, 292, 314, 315
microbiological risks, 103
micropollutants, 169, 330

## O, P, R

organic matter, 116, 118, 128, 135, 136, 148, 151, 160, 167, 184, 219, 223, 225, 227, 230, 233, 239, 246, 252, 266, 267, 269, 271, 280, 287, 310, 317, 319, 328
ozone, 111, 113, 115, 118–121, 159, 160, 167–172, 174–178, 180–183, 189, 210, 211, 222, 223, 227, 235, 250–254, 317, 319
parasite, 99, 101, 103, 105, 108, 112, 115, 118, 159, 168, 182, 204, 208–210, 328, 331
pesticides, 169, 254, 310, 311, 319, 328, 330
powdered activated carbon, 242, 243, 264, 266, 314, 315
remineralization, 47, 60, 62, 63, 65–67, 69–73, 76–78, 82, 84
resin, 138

## S, T, V

sand, 63, 66, 67, 246, 297, 300, 318
sludge, 261–277, 279–284, 286–289, 291–297, 299–304, 307, 317, 318
soda, 27, 39, 46, 49, 50, 60, 61, 66, 68, 70, 73, 79, 82, 83, 86–90, 92, 138, 141, 152, 153, 175, 314, 315
softening, 38, 40, 46, 261, 262, 264, 266, 268–271, 273, 281, 282, 288, 291, 293, 295, 296
Spidflow®, 266, 313
strainers, 77
stripping, 240
thickening, 262, 266, 273, 274, 276, 277, 280, 281, 284, 285, 287, 291, 294–300, 302–304
TOC, 225, 231, 235–237, 239, 252, 253, 271
viruses, 99, 100, 103–105, 112, 146, 147, 155, 156, 160, 164, 165, 167, 168, 181, 187, 189, 204, 205, 319, 328, 330, 331

# Summary of Volume 1


**Chapter 1. Introduction**

**Chapter 2. Physicochemical and Microbiological Composition of Raw Water**

2.1. Water resources
2.1.1. Physicochemical parameters
2.1.2. Algae (including cyanobacteria and cyanotoxins)
2.1.3. Tastes and odors
2.1.4. Micropollutants
2.2. Microbiology
2.2.1. Bacteria
2.2.2. Viruses
2.2.3. Parasites (Cryptosporidium and Giardia)
2.3. Quality of water intended for human consumption
2.3.1. Microbiological parameters
2.4. References

**Chapter 3. Aeration and Stripping**

3.1. Cascade aeration
3.1.1. Characteristics
3.2. Operating principle of a cascade aerator system
3.2.1. Data
3.2.2. Goals
3.2.3. Results

3.3. Aeration by fine bubble diffusers
3.3.1. Air diffusers
3.3.2. Air insufflation by oxytube
3.4. Stripping
3.4.1. Stripping tower design
3.4.2. Description
3.4.3. $CO_2$ removal
3.4.4. Tetrachlorethylene and trichlorethylene removal
3.5. Synthesis of aeration systems
3.6. References

## Chapter 4. Coagulation–flocculation

4.1. Colloidal matter
4.2. Coagulation
4.2.1. Double-layer compression
4.2.2. Adsorption and interparticle bridging (flocculation)
4.3. Flocculation
4.3.1. Perikinetic flocculation
4.3.2. Orthokinetic flocculation
4.3.3. The influence of agitation
4.3.4. G and t
4.4. Coagulants
4.4.1. Metallic coagulants
4.4.2. Synthetic organic coagulants
4.5. Flocculants
4.5.1. Natural organic and synthetic flocculants
4.5.2. Flocculation adjuvants
4.6. Factors affecting coagulation and flocculation
4.6.1. Influence of water temperature
4.6.2. Influence of pH
4.6.3. Coagulation and flocculation times
4.7. How to choose the best coagulant?
4.7.1. How to choose the optimal dose of coagulant?
4.8. Residual aluminum
4.9. Alkalinity consumption
4.9.1. Aluminum and alkalinity
4.9.2. Iron and alkalinity
4.10. Reduction efficiency of some water constituents
4.10.1. Turbidity and SS
4.10.2. Microorganism removal

4.11. Jar tests
4.11.1. The specific case of jar test for Actiflo®
4.12. References

## Chapter 5. Settling

5.1. The principles of settling
5.2. Horizontal settlers
5.2.1. Principle
5.2.2. Design
5.2.3. Implementation
5.3. Lamella settlers
5.3.1. Theory and principle
5.3.2. Basic design for lamella settlers
5.3.3. Implementation
5.4. Veolia technologies
5.4.1. Lamella settlers: Multiflo® settler
5.4.2. Ballasted floc settlers
5.5. References

## Chapter 6. Flotation

6.1. The scope of DAF
6.2. The main stages of a flotation process
6.2.1. Coagulation
6.2.2. Flocculation
6.2.3. Flotation
6.3. The fundamental mechanisms of flotation
6.3.1. Coagulation
6.3.2. Flocculation
6.3.3. Contact zone
6.3.4. Separation zone
6.4. Design parameters
6.4.1. Coagulation
6.4.2. Flocculation
6.4.3. Flotation zone
6.4.4. Contact time
6.4.5. Temperature
6.4.6. Air saturation tank
6.4.7. Mass balance
6.4.8. Recirculation and injection nozzles

6.5. Operating parameters affecting flotation performance
6.5.1. Choice of coagulant
6.5.2. Rise velocity
6.5.3. Contact time
6.5.4. Bubble volume concentration
6.5.5. Gas solubility
6.5.6. Hydraulic efficiency
6.5.7. Air/water ratio
6.6. Performance and monitoring
6.6.1. Treatment monitoring
6.6.2. Performance of DAF systems in relation to algae removal
6.6.3. Performance against parasites
6.6.4. Performance with the addition of PAC
6.7. Veolia technologies using flotation
6.7.1. Spidflow®: principle
6.7.2. Advantages and limitations of DAF systems
6.7.3. Spidflow® filter
6.7.4. Ozoflot®
6.7.5. Flottazone®
6.7.6. Packaged solutions: Spidflow® Pack
6.8. References

Index

# Summary of Volume 2

## Chapter 7. Removal of Natural Organic Matter

7.1. Natural organic matter: humic substances
7.2. Methods of quantification and assessment of organic substances in water
7.2.1. Total organic carbon
7.2.2. Absorbance of ultraviolet light at 254 nm (UV 254)
7.2.3. Specific UV absorbance
7.2.4. Liquid chromatography
7.2.5. Hydrophobic organic carbon
7.2.6. Fluorescence
7.3. Conditions for the removal of NOM
7.4. NOM removal techniques
7.4.1. Coagulation–flocculation
7.5. Adsorption on activated carbon
7.5.1. Mechanism of NOM removal by activated carbon
7.5.2. Implementation of activated carbon adsorption for NOM removal
7.6. Ozonation
7.6.1. Preozonation
7.6.2. Interozonation
7.7. Biological treatment
7.7.1. Biological treatment with PAC: operating principle of the Opaline® B process (Veolia)
7.7.2. Biological treatment combining ozone and GAC
7.8. Treatment of ion exchange resins
7.8.1. Use of resins for NOM removal: mechanism of NOM removal

7.8.2. Parameters affecting the performance of resins for the removal of NOM
7.8.3. Resin regeneration
7.8.4. The Opalix® process
7.9. NOM removal by high-pressure membranes
7.10. References

## Chapter 8. Filtration

8.1. Rapid filters and very high-rate filters (TGV)
8.2. Multimedia filters
8.3. Direct filtration
8.4. Pressurized filters
8.5. Filtration mechanisms
8.5.1. General principle
8.5.2. Mechanisms involved in filtration
8.6. Implementation parameters
8.6.1. Materials
8.6.2. Material height/$d_{10}$ ratio
8.6.3. ES ratio
8.7. Sizing parameters: filtration rate and material height
8.8. Operating parameters
8.8.1. Pressure loss in a clean filter
8.8.2. Pressure loss during clogging
8.8.3. Estimation of the turbidity of filtered water in single-layer filters
8.8.4. Retention capacity
8.8.5. Filter washing conditions: theory and calculations
8.8.6. Cycle time
8.8.7. Main parameters involved in filter sizing
8.9. Veolia filtration technologies: general information
8.9.1. Characteristics of open gravity filters
8.10. Regulation systems
8.10.1. Hydraulic regulators
8.10.2. Control valves
8.11. Recycling and microbiological risks
8.12. Monitoring the operation and performance of filters
8.12.1. Turbidity
8.12.2. Particle count
8.12.3. Problems with the operation of filters
8.13. References

## Chapter 9. Adsorption on Activated Carbon

9.1. Activation processes of activated carbon
9.1.1. Chemical activation
9.1.2. Physical activation
9.2. Physicochemical properties of activated carbon
9.2.1. Parameters influencing adsorption
9.3. Transport process in activated carbon: mass transfer
9.3.1. Adsorption mechanisms and isotherms
9.4. The different forms of conditioning of activated carbons
9.4.1. Powdered activated carbon
9.4.2. Micrograin activated carbon (µgrain)
9.4.3. Granular activated carbon
9.5. Adsorption reactors on activated carbon: removal process
9.6. PAC reactors: description of PAC reactors
9.6.1. Mass balance
9.6.2. GAC reactors in adsorption mode: GAC filters
9.6.3. Fluidized bed activated carbon reactors
9.6.4. GAC-µgrain-PAC comparison
9.6.5. Hybrid process with a PAC reactor associated with UF membranes
9.7. Veolia technologies: treatment process with PAC reactors
9.7.1. Implementation in a settler without PAC recirculation
9.7.2. Implementation with PAC recirculation
9.7.3. Opaline® C process: PAC-membrane hybrid
9.8. Micrograin activated carbon reactors
9.8.1. Integration of µgrain activated carbon in the treatment process
9.8.2. The Filtraflo® Carb
9.8.3. Opacarb®FL
9.8.4. Opacarb®MG
9.9. Fixed bed reactors – GAC filters
9.9.1. The GAC filter (Veolia)
9.9.2. GAC filters in series
9.9.3. Implementation modes for GAC filters
9.10. Pressurized GAC filters (Opacarb™ filters)
9.11. References

## Index

# Summary of Volume 3


**Chapter 10. Removal of Micropollutants**

10.1. Introduction
10.2. Pesticides
10.3. Pharmaceuticals and industrial waste
10.4. Pesticide removal technologies and emerging MPs
  10.4.1. Adsorption onto activated carbon (AC)
  10.4.2. Ozonation
  10.4.3. Ozone-activated carbon combination
  10.4.4. Advanced chemical oxidation
  10.4.5. Nanofiltration and reverse osmosis membranes
10.5. Frogbox®: an effective monitoring and control tool
10.6. The evolution of micropollutants in drinking water plants
10.7. References

**Chapter 11. Removal of Perfluorinated Compounds**

11.1. Physicochemical properties
11.2. Presence in the water
11.3. Drinking water regulations
11.4. Treatments
  11.4.1. Coagulation–flocculation–settling (or flotation)
  11.4.2. Chemical oxidation
  11.4.3. UV oxidation
  11.4.4. Activated carbon
  11.4.5. High-pressure membranes: nanofiltration and reverse osmosis
11.5. Conclusion
11.6. References

## Chapter 12. Biological Removal of Ammonia

12.1. The principle of biological nitrification
12.2. Design parameters
12.2.1. Dissolved oxygen
12.2.2. Filtration rate
12.2.3. $NH_4^+$ concentration removed as a function of temperature (°C)
12.2.4. Applicable volume load
12.2.5. Contact time
12.2.6. Material height
12.3. Factors limiting oxygen
12.3.1. Mineral carbon
12.3.2. pH
12.3.3. Temperature
12.3.4. Other elements
12.3.5. Biological filter washing
12.4. Implementation
12.4.1. Sand filtration
12.5. Biofilters (Biocarbon® process)
12.6. Water treatment stations
12.6.1. Treatment stations with conventional sand, dual media or GAC filtration
12.6.2. Treatment stations with Biocarbon® filters
12.7. References

## Chapter 13. Nitrate Removal

13.1. Biological treatment
13.1.1. Biochemical reactions
13.1.2. Nitrite formation
13.1.3. The bacteriological aspect
13.1.4. Biofilter description (Biodenit® process)
13.1.5. Water treatment stations including biological denitrification
13.1.6. Factors affecting biological denitrification
13.1.7. Design parameters: applied volumic load
13.1.8. Design parameters: minimum contact time (tc min)
13.1.9. Design parameters: height of Biodagen® material (m)
13.1.10. Design parameters: material volume ($m^3$)
13.1.11. Partial treatment
13.1.12. Sludge production
13.1.13. The reagents
13.1.14. Biological denitrification implementation and exploitation

13.2. Treatment with ion exchange resins
13.2.1. General exchange mechanism (Ecodenit® process)
13.2.2. Ecodenit® process technology
13.2.3. Integration into the water treatment station
13.2.4. Packaged solutions
13.3. Nitrate removal by high pressure membranes
13.4. References

## Chapter 14. Removal of Perchlorates

14.1. General aspects
14.2. Main processes for removing perchlorate ions
14.2.1. Ion exchange resins
14.2.2. Nanofiltration membranes
14.3. Conclusions regarding the removal of perchlorates
14.4. References

## Chapter 15. Water Softening

15.1. Water hardness
15.2. Alkalinity
15.3. Langelier index (LI or LSI)
15.4. Drinking water hardness goals
15.5. General principles of water softening
15.5.1. The main chemical reactions with lime and soda
15.6. Water softening chemical processes
15.6.1. Limitations of the process and empirical considerations
15.7. Veolia water softening technologies
15.7.1. Clariflocculator
15.7.2. Actiflo® softening
15.7.3. Multiflo® softening
15.7.4. Catalytic water softening
15.8. Saphira® process
15.9. Water softening using high-pressure membranes
15.10. Water softening using ion exchange resins
15.10.1. Resin resistance: use precautions
15.10.2. $Ca^{2+}$ and $Mg^{2+}$ removal
15.10.3. Bicarbonate removal
15.10.4. Resin operating capacity
15.10.5. Operating parameters

15.11. Comparison between the four water softening solutions discussed
15.11.1. Advantages and limitations of the different solutions
15.12. References

## Chapter 16. Metal Removal

16.1. Iron and manganese removal: general aspects
16.1.1. Presence of iron and manganese at the resource level
16.1.2. Presence of iron and manganese at the production level
16.1.3. Regulatory aspects of iron and manganese
16.1.4. Iron and manganese treatments
16.2. Arsenic removal
16.2.1. Arsenic chemistry
16.2.2. Arsenic treatments
16.3. Removal of selenium (Se)
16.3.1. The chemistry of selenium
16.3.2. Selenium removal treatments
16.4. Nickel removal
16.4.1. The chemistry of nickel
16.4.2. Nickel removal treatment
16.5. References

## Index

# Summary of Volume 4

## Chapter 17. Microfiltration and Ultrafiltration


17.1. UF and MF: cut-off
17.2. UF and MF: materials
17.2.1. Cellulose acetate
17.2.2. Polypropylene
17.2.3. Polyacrylonitrile
17.2.4. Polyether sulfone/polysulfone
17.2.5. Polyvinylidene fluoride
17.3. UF and MF: membrane types
17.4. UF and MF: implementation of membranes under pressure
17.4.1. Horizontal-vertical configuration
17.4.2. Submerged membranes
17.5. Filtration modes: frontal or tangential
17.5.1. Batch operation: filtration-backwash
17.5.2. Filtration direction
17.6. Sizing parameters: membrane selection
17.7. Sizing parameters: horizontal or vertical configuration
17.8. Sizing parameters: flow
17.8.1. Instantaneous flow and net flow
17.8.2. Transmembrane pressure
17.8.3. Resistance
17.8.4. Permeability
17.8.5. Principle of the calculation of the membrane surface and water losses
17.8.6. Pre-filters
17.9. Operating parameters
17.9.1. Evolution of the permeability
17.9.2. Clogging

17.9.3. Frequency and conditions of hydraulic and chemical backwashing
17.9.4. Frequency and conditions of CIP
17.9.5. Membrane integrity
17.10. MF and UF's place in a treatment process
17.10.1. Turbidity and SS
17.10.2. TOC (and UV254)
17.10.3. Algae
17.10.4. Iron and manganese
17.11. Combination of coagulation and UF membranes
17.12. Combination of PAC and UF
17.13. Performance and guarantees
17.13.1. Turbidity
17.13.2. Supplier warranty on the life of the membranes
17.14. Advantages of MF and UF
17.15. Veolia's experience
17.16. Appendix: sheets
17.17. References

## Chapter 18. Nanofiltration and Reverse Osmosis

18.1. Membranes
18.1.1. Materials
18.1.2. Membrane element configurations
18.2. Principles of operation and separation
18.2.1. Conceptual principle
18.2.2. Molecular weight cut-off
18.3. Treatment process including high-pressure membranes and parameters to be considered
18.3.1. Particulates and SS
18.3.2. Particle count
18.3.3. Conductivity
18.3.4. The SDI or MFI: clogging indices
18.3.5. The SDI
18.3.6. The MFI
18.3.7. Salts and metals
18.3.8. Biological clogging
18.3.9. Undesirable substances
18.3.10. Limit values of compounds at the inlet of high-pressure membranes
18.4. Sizing parameters
18.4.1. Temperature

18.4.2. Implementation configuration
18.4.3. Calculation of the osmotic pressure
18.4.4. Mass flow diagram
18.4.5. Salt passage
18.4.6. Concentration factor
18.4.7. Hydraulic pressure loss
18.4.8. Pressure tubes and number of modules per tube
18.5. Chemical conditioning of pre-treated water
18.5.1. Calculation of saturation indices and antiscalant dosage
18.5.2. Choice and implementation of the antiscalant
18.5.3. pH adjustment at the membrane inlet
18.5.4. Choice and application of the acid
18.5.5. Influence of sulfates
18.6. Design and implementation
18.6.1. Pre-treatment
18.6.2. Treatment processes
18.6.3. Membrane station
18.6.4. Post-treatment
18.6.5. Cleaning units in place
18.7. Functional and operating parameters
18.7.1. Basic principles
18.7.2. Permeability (Lp)
18.7.3. Longitudinal pressure drop (ΔPfc)
18.7.4. Hydraulic resistance
18.7.5. Energy
18.7.6. SDI
18.7.7. Chemical cleaning
18.7.8. The fate of concentrates and used washing solutions
18.7.9. Methods for assessing the impact of concentrate discharges in the natural environment
18.8. High-pressure membrane performance
18.8.1. Organic matter
18.8.2. Pesticides, drug residues, endocrine disruptors and industrial residues
18.8.3. Various toxic and undesirable substances
18.8.4. Salts
18.8.5. Micro-organisms
18.8.6. Overall performance
18.9. Lifetime warranties
18.10. Parameters affecting the performance of NF membranes
18.10.1. Taking clogging into account
18.11. Monitoring and control parameters: standardization of raw data
18.12. Veolia's experience: examples of treatment processes

18.12.1. Surface water No. 1
18.12.2. Surface water No. 2
18.12.3. Groundwater No. 1
18.12.4. Groundwater No. 2
18.13. References

## Chapter 19. Desalination by Reverse Osmosis

19.1. Characterization of the water to be treated
19.1.1. Physical characteristics
19.1.2. Chemical composition: ionic content
19.1.3. Chemical composition: organic substances
19.2. Fields of application
19.3. Operating principle of RO
19.4. The membranes used in desalination
19.5. Sizing parameters
19.5.1. Flow
19.5.2. Concentration polarization
19.5.3. Conversion rate
19.5.4. Passage rate and rejection rate in salts
19.5.5. Influence of the temperature
19.5.6. Determining the number of modules and pressure tubes
19.6. Implementation
19.6.1. Membranes
19.6.2. Pressure tube
19.6.3. Pass
19.7. Pre-treatment
19.7.1. Pre-treatment selection
19.7.2. Pre-treatment systems
19.8. Pre-chlorination
19.8.1. Pre-chlorination and development of micro- and macro-organisms
19.8.2. Implementation of chlorination
19.8.3. pH adjustment
19.8.4. Direct filtration
19.8.5. Chemical conditions of implementation
19.8.6. Flotation
19.8.7. Settling
19.8.8. Membranes (UF and MF)
19.8.9. Conclusions on pre-treatment with UF membranes
19.9. Energy consumption
19.9.1. Energy consumption without recovery

19.9.2. Energy consumption with recovery
19.9.3. Hydraulic exchanger systems
19.10. Operating parameters
19.10.1. Relationship between conductivity and salt concentration
19.10.2. Controlling RO membrane clogging
19.11. Performance of RO membranes used in desalination
19.11.1. Boron removal
19.12. Post-treatment
19.12.1. Indicators characterizing the aggressiveness or corrosiveness of the water
19.12.2. Application to desalinated water
19.12.3. Treatments
19.13. Monitoring and control parameters
19.13.1. Standardization of raw data
19.13.2. Bromates
19.14. Veolia's new processes applied to seawater desalination
19.14.1. Flotation with the Spidflow® process
19.14.2. Spidflow® filter process applied to seawater desalination
19.14.3. BiopROtector
19.14.4. Barrel (SIDEM Veolia)
19.14.5. Hiprode
19.15. Packaged solutions in desalination
19.16. Veolia's experience (HP membranes)
19.17. References

## Index

Printed and bound by CPI Group (UK) Ltd, Croydon, CR0 4YY
13/09/2026
14961041-0005